Environmental Impact of Polymers

Series Editor
Jack Legrand

Environmental Impact of Polymers

Edited by

Thierry Hamaide
Rémi Deterre
Jean-François Feller

iSTE

WILEY

First published 2014 in Great Britain and the United States by ISTE Ltd and John Wiley & Sons, Inc.

ISTE Ltd
27-37 St George's Road
London SW19 4EU
UK

www.iste.co.uk

John Wiley & Sons, Inc.
111 River Street
Hoboken, NJ 07030
USA

www.wiley.com

Library of Congress Control Number: 2014941853

British Library Cataloguing-in-Publication Data
A CIP record for this book is available from the British Library
ISBN 978-1-84821-621-1

MIX
Paper from responsible sources
FSC® C013604
www.fsc.org

Printed and bound in Great Britain by CPI Group (UK) Ltd., Croydon, Surrey CR0 4YY

Contents

Preface

Better known by its acronym GFP, the French group for the study and applications of polymers (www.gfp.asso.fr) is a non-profit organization established in November 1970. Its mission, recognized as a public utility in 1990, is to promote the development of polymers, not only in universities and research centers but also in industry.

The GFP Commission focuses on promoting all aspects of the teaching of "polymer science", both in France and abroad. A significant part of its activity is devoted to the periodic organization of educational internships for teachers willing to update their knowledge in all the areas specific to polymers. Following each of these courses a textbook is published, which is available to all members of the GFP. The list of available titles can be found on the Website of the GFP (www.gfp.asso.fr).

The Education Commission therefore organized several meetings between teachers, researchers and industrialists on the theme of *Polymers and the environment: the impact of polymers on society and the eco-design of plastic materials. What solutions can be proposed for the societal impacts of polymeric materials*?

The main objective of these meetings was to go beyond the traditional analysis of the lifecycle of polymers and to provide for everyone willing to acquire the basic concepts of the environmental, and therefore societal, impacts of polymers and the means to reduce their effects, in order to update their knowledge and to be able to integrate it optimally into their lecturing. They wanted to share the solutions already in existence and implemented by industry and the prospects for the short and medium term. This book collates several contributions originating during these periods. It provides solutions already in existence, as well as predictable or desirable developments. It concludes with reflections upon the role of education and training in order to ensure and guide the future.

Thierry HAMAIDE

President of the Education Commission of the GFP

June 2014

Introduction: The Societal Impact of Polymers and Plastic Materials: Solutions and Perspectives

I.1. Polymers, an ever young science in step with the economy and employment

Whatever their origin, natural or synthetic, polymers involve chemistry and/or process engineering at one time or another in their lifecycle. Chemistry and process engineering are fundamental sciences for sustainable development; they are fundamental disciplines for understanding the world around us, at all scales, and for mastering its transformations. They are able to strongly interact with other disciplines – physics, biology, mathematics. Moreover, they are in a huge number of fields of application in our daily life: energy, materials, information and the living world.

If organic chemistry is essential to describe the majority of molecules around us, it is analytical chemistry that helps us identify them and catalysis is at the heart of transformation of materials and of energy savings; inorganic chemistry is strongly involved in the issues of production and energy storage. Process engineering – which includes different levels of physical chemistry – is essential for the feasibility, economics and environmental performance of production.

Finally, the interaction of humans with chemistry encourages a better understanding of our reading of the world, our opinions, beliefs, historical creations, representations and evolution.

Introduction written by Bernard SILLION, Thierry HAMAIDE, Gérard PIGNAULT and Jean-Charles MOUGENEL.

This crossroads of science, economics and human values constitutes precisely the zone of overlap which is described by the term sustainable development. The evolution of how people view polymers, more commonly called plastics, is exemplary in many respects [BER 95] and illustrates this concept.

Polymers are used in more than 90% of materials, and their cohesion and properties depend closely on the chemical structure and the organization of chains which can be governed by the methods of production. Thus, polymer science consists essentially of three scientific disciplines, which are chemistry, physics and mechanics. Chemistry is involved in the transformation of raw materials and also in the production processes, which becomes possible due to chemical engineering [FON 08, FON 10, MER 96, ODI 04, ETI 12]. Physics leads to the development of analytical tools that make available the characterization of these macromolecules of different sizes, unveiling their organization. Finally, because of mechanics, properties of application of polymeric materials can be explored [OUD 94, FON 08, FON 10].

Polymer science is a relative young science, and the concept of macromolecules for understanding the properties of polymeric materials was introduced by H. Staudinger in 1919 and developed in the 1920s, in particular on cellulosic materials. The industry had certainly previously produced vulcanized rubber, a process invented by Goodyear in 1839, celluloid in 1865 with the Hayatt brothers, bakelite in 1910 developed by Baekeland, but the true nature of the chemical species was at that time not yet revealed. From 1930, a better understanding of the macromolecular structure led to the fast development of different chemical families: low-density polyethylene by radical polymerization synthesized in 1933, and the works of Carothers on polycondensation led to polyamides in 1938.

The period following the Second World War saw the emergence, with an accelerated speed, of new polymerization methods: in 1953–1954, polymerization catalysis by coordination was developed by K. Ziegler and G. Natta (Nobel Prize, 1963), which led to for high-density polyethylene (PEHD) and polypropylene (PP). Anionic polymerization and the concept of living polymerization proposed by M. Szwarc in 1956 led to the design of blocks copolymers and the first macromolecular architectures. We then saw the emergence of catalysis by metallocene in 1980 by W. Kaminski. Radical polymerization controlled by M. Sawamoto and K. Matyjaszewski in 1994 combined the benefits of radical and ionic polymerization without the drawbacks of the former.

Meanwhile, analytical physical chemistry made progress at the same speed. Significant evolution in chromatographic methods and the study of interactions between matter and radiation helped in analysis with higher precision of the polymer microstructure, and to revisit the intimate mechanisms of polymerization [TAN 00].

As in all areas, fundamental discoveries have often been generated by industrial developments which tackle global problems of the planet such as energy and water resources as well as those experienced at the level of individuals in everyday life such as health and hygiene, nutrition, comfort, communication, recreation and so on, and therefore we can claim that society has entered the "age of polymers", which is explained in the first section of introduction.

At the same time, this entry into the age of polymers goes hand in hand with an overall negative societal impact [LEM 10]: polymers in general suffer from double penalties, namely the image of plastic materials in addition to those of chemistry. For society, the chemical aspect is the one that generates questions or concerns, because it highlights the impacts on the environment of waste which does not easily degrade biologically or leads by combustion to hazardous materials such as dioxins. We are also concerned about health risks caused by the release of toxic materials during use.

All of these concerns should be taken into account, but the requested analysis of risks should also be carried out in terms of risk-benefit assessments. It is thus important for society to recognize the developments in science and in industry in the field of polymers. It is not generally known that plastics consume only 4% of global oil production and that their widespread use makes it possible to save a much larger quantity. In addition to this, at the end of their life they produce less waste, because they are lighter (less than 1% of the total weight of waste in Europe). The study by Claude Duval (see Chapter 2) allows us to address the intimate link between plastic waste and the environment. Thus, despite appearances, plastics undoubtedly contribute to sustainable development, and are beneficial for the future, as Michel Loubry attempts to demonstrate (see Chapter 4).

The industrial world in close collaboration with academic research strives to develop new methods of developing plastic materials. We are therefore moving to an eco-design approach which consists in integrating the environment into the fabrication phase of products, requiring us to take into account the whole steps of the lifecycle of products [ADE]. In other words, the lifecycle analysis must integrate to the maximum extent possible the use of new sources of raw materials (and wisely) and of new methods of production, and no longer be limited to issues concerning the end of life related to recycling and valorization of waste.

Eco-design is not really a new profession or an established university discipline but a new approach, acquired through specialization by individuals already owning the skills needed in the different core disciplines, especially in chemistry related to polymers. The training needs are clearly identified. As a result, this approach must also now enter into the world of education, through a course suited for all ages. The essential role of the teacher is discussed at the end of this chapter.

However, to go further in this reflection, we should write Education with a capital E, because while the vast majority of people have little, or no, impact on the eco-design, the whole society has a major role in the gathering and recycling of plastic waste. Are the "plastic continents" attributable to industrial plastics or to the consumers of these plastics? In this regard, the study by François Galgani (see Chapter 3) allows us to judge the effect of polymers and litter on the environment and the societal assessment. Another example from everyday life concerns carrier bags. Are bags which are displayed as biodegradable (but are they really?) designed to deal with the incivility of some of our citizens or to follow a fashion?

I.2. The polymer industry: its role in the economy and the workforce

The production of plastics, which was 1.5 million tons in 1950, rose to 230 million tons in 2009, going through a record production of 260 million tons in 2007. This increase is much larger than what is observed for other materials. Europe is well positioned in the world with a production of 55 million tons in 2009.

With regard to employment, taking into account the whole sector, that is to say production, construction of machines dedicated to processing and the processing industry, the sector employed 1.6 million people, which places this industry at the top of those who contribute to the economy of Europe [PLA 10]. For example, the chemical industry in 2011 generated only 1.19 million jobs in the European Union [CEF].

French plasturgy is a growing business which embraces all industrial sectors. Today it ranks fourth in the world and second in Europe with 4,500 thousand tons of products, representing a turnover of 29 billion euros, with a quarter for export. It employs more than 140,000 employees in 3,850 companies, of which more than 90% being small and medium businesses [LAPa].

However, the crisis has affected more, in terms of production, the polymers and specialty chemicals than the basic chemistry, as shown in Figure I.1 where growths in different domains are compared [CEF]. It reveals the clear reduction in the growth in 2012.

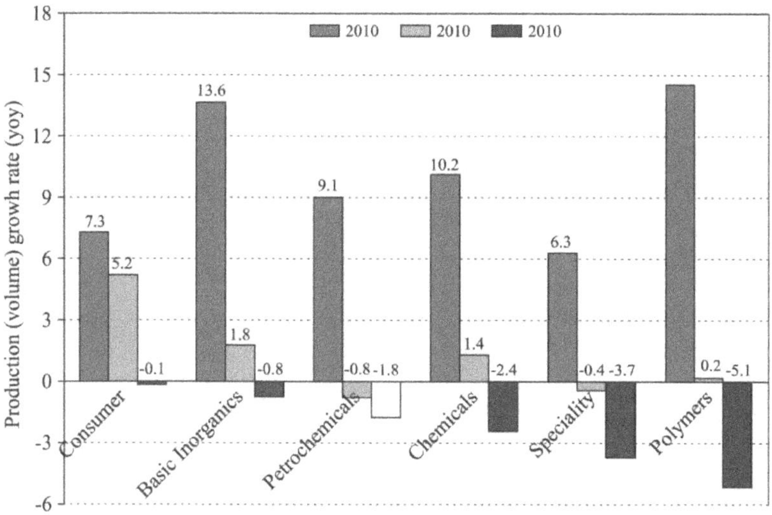

Figure I.1. *Growth reduction in 2012*

I.3. Polymeric materials to assist society in addressing the big problems facing the planet: water and energy

I.3.1. *Polymers for water*

The world's population is currently 7.2 billion. The curent issue of food supply, already happening today, will be particularly important in 2050 for a world population around 9.6 billion. 2013 was the international year of cooperation in the domain of water for UNESCO [UNE]. Water is abundant on Earth, but 97.2% (sea water) is unsuitable for human consumption and the rest (fresh water) is very poorly distributed in the world since 2.5% is in the form of polar ice. The water available for use, 0.3% (river, groundwater, lakes), is itself distributed very unevenly across the world: 800 million people are deprived of drinking water; 2.4 million have no sanitation and 450 million experience periods of shortage (according to UNESCO [AQU]). The need for water is rising twice as fast as the population, which leads to a mobilization to save and recycle consumed water, but also requires the development of processes for producing fresh water and drinking water. The use of polymers has become essential for the production of drinking water and for the management of water in agriculture.

For the production of drinking water from saltwater, techniques of membrane treatment have been developed. Different methods of membrane separation help in separating water from objects in suspension or solution according to their sizes

[MAU 10], ranging from conventional filtration with dimensions greater than 10 μm to nanofiltration with dimensions of the order of the nm.

Osmosis is the phenomenon that occurs when two solutions of different salinities are separated by a membrane permeable to water. For example, the desalination plant of Ashkelon in Israel produces 320,000 m^3 of drinking water per day, or 108 million m^3 per year (the consumption of a city of 1.4 million inhabitants). After a pre-treatment (bilayer filtration and microfiltration), the water is gradually desalinated through 32 stages of reverse osmosis. The concentration of dissolved salt coming out of the plant is 30 $mg.L^{-1}$, 1,000 times less than in the water pumped from the sea [LIN].

Several families of polymers may be used for this application, including cellulose acetates, aromatic polyamides and polysulfones, but progress remains to be made regarding the nature of membranes and their implementation to control the morphologies of the interface.

For water management in agriculture, cross-linked polyacrylamides are used for water retention and for soil moisture [HOL 05]. In contrast, polymers and copolymers of ethylene glycol are used to promote infiltration of water. The recycling of wastewater uses for the step of clarification (flocculation and sludge conditioning) polymers derived from polyamines or acrylamide.

I.3.2. *Polymers for energy management*

The energy consumption of the world in 2009 was 8.4×10^9 toe (ton of oil equivalent; 1 toe = 11,628 kWh), but for this final consumption of energy, it was necessary to actually produce 12.3×10^9 toe, the difference of 3.9×10^9 toe being used for transformation, transport, etc. The needs for energy throughout the world are not the same in all parts of the world, as shown in Figure I.2 from International Energy Agency (IEA) documents [BRO 11].

For 40 years, energy consumption has increased by more than 40%, and the development of emerging countries accelerates this trend, which raises two main problems: the scarcity of raw materials: oil, coal and gas; and the production of carbon dioxide, which contributes to climate change. The production of primary energy is at the beginning of this century based on fossil fuels, as shown in Figure I.3, extracted from IEA statistics for 2008.

We see from this figure that more than 80% of primary energy is fossil fuels. In 2010, the amount of CO_2 released into the atmosphere was more than 30 gigatons, which leads to a concentration in the atmosphere of the order of 385 ppm; these releases are due in part to the transformation of fossil energy.

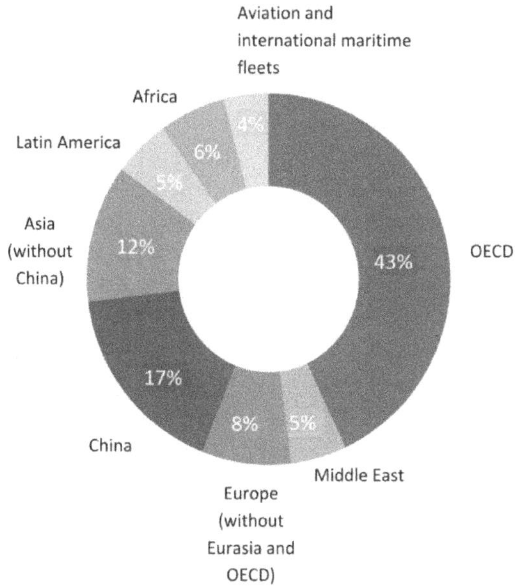

Figure I.2. *Distribution of the total consumption of final energy (8,353 Mtoe) in 2009 [BRO 11]*

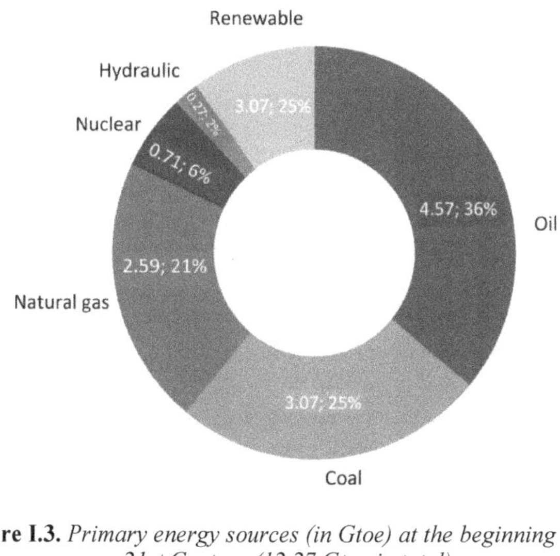

Figure I.3. *Primary energy sources (in Gtoe) at the beginning of the 21st Century (12.27 Gtoe in total)*

The thinking behind the evolution of primary energy consumption in the world is the subject of several hypotheses that take into account not only the evolution of populations and the evolution of GDP in different countries, but also the establishment of policies to fight against global warming, that is to say, the control of emissions of carbon dioxide [BRO 11, BOU 09]. In 2050, fossil fuels will still hold an important role in primary energy because resources are still significant: oil between 180 and 200 gigatons, gas 150 Tm^3 ($T = 10^{12}$) or about 150 Gtoe and coal between 700 and 900 Gtoe. To this must be added the reserves of unconventional gas (shale and coal gas with reserves of around 380 Tm^3) which could ensure production for between 120 and 150 years [TOTAL].

Among the scenarios used to predict the supply needs in 2050, a hypothesis based on the control of global warming involves a reduction in rich countries so as not to exceed a consumption of 20 Gtoe [BOU 09, WIE 05], or 2.4 times that in 2009. However, it is likely that this consumption will be exceeded given the difficulties in signing and implementing agreements against global warming.

I.3.2.1. *The role of polymers in the production of energy*

If we examine the role that polymers play and can play in the field of energy, we can distinguish improvements which focus on the production of fossil fuels and others which allow the saving of energy.

It is known that the efficiency of oil recovery from a field is only about 30% and it was found that the injection of water made more viscous by the dissolution of polymers, including biopolymers obtained by fermentation, helps to increase the production of the reservoir. The injection of biodegradable biopolymers for fracturing would contribute toward making the exploitation of shale gas less risky for the environment.

Beyond saving fossil fuels, polymers are key materials for the development of new energies such as composites for wind turbine blades by the resin transfer molding process. The development of composites is also a determining factor for the transport sector, as shown below.

Another renewable energy, which can be described as permanent and which is rapidly developing, is solar energy. The power available above the Earth's atmosphere is 340 $W.m^{-2}$, but only 170 $W.m^{-2}$ can be recovered at ground level, the rest being reflected or absorbed and used during the water cycle. The transformation of solar light energy into electrical energy occurs in inorganic semiconductor materials based on silicon (made of silicon doped with phosphorus, which provides negative charges and boron, which creates a deficiency of electrons within the silicon). Two types of polymers are involved in protecting the doped silicon: a film of ethylene vinyl acetate on the front of the plate and a film of polyvinyl fluoride

(Tedlar ®) at the back [WIE 05]. Research seeks to develop totally organic photovoltaic cells with the help of semiconducting polymers. The power available from photovoltaic cells in 2011 was 63.3 GW, which is still very low.

I.3.2.2. *The role of polymers in the storage of energy*

The development of more efficient batteries is crucial for the development of hybrid cars or cars running with electric motors. In 1978, the team of Michel Armand [ARM 87] studied the complex between polyether glycol and lithium to produce solid electrolytes. The batteries of polymer electrolytes derived from poly(ethylene oxide) offer many advantages, including non-volatility, and the possibility of using Li metal, but they require an operating temperature of 80°C. A lot of work to improve the conductivity of the polymer was undertaken and led to the use of fluorinated copolymers doped with electrolyte [COL]. The Bolloré Group and Renault in 2011 announced the launch of an electric car based on lithium polymer technology. In aviation, the problems with the 787 batteries now seem to have been resolved.

I.3.2.3. *The role of polymers in the saving of energy*

Housing and transportation are the most important areas in which to achieve energy savings. With regard to housing, 40% of primary energy is used for heating and cooling, and the question of building insulation is paramount. Polymers provide a surface thermal resistance (heat resistance R per unit area in $m^2.KW^{-1}$) greater than that of mineral insulators. Expanded polystyrene would probably be used more often, given its low price, if it did not have a flammable nature.

In transport, polymers have definitely brought the greatest energy savings, i.e. the lightening of motor vehicles and aircraft.

In the 1990s and 2000s, the weight of automobiles increased by about 25 kg per year in order to improve comfort by increasing the size and also to increase safety. It is known that reducing weight by 100 kg results in the reduction of consumption by 0.6 liters/100 km and, more generally, decreasing the weight by 10% leads to a decrease in consumption of the vehicle of 5–6%. Any decrease in consumption correlatively reduces the CO_2 emissions. In 2012, the European Parliament deliberated that the average new car should not emit more than 130 $g.km^{-1}$ in 2015 and no more than 95 $g.km^{-1}$ in 2020 [ADE]. The strategy for lowering the weight of vehicles relies on the decrease in the thickness of steel plate (decreasing the thickness by 1/10th brings a weight benefit of 10%), and on the use of polymer materials; the weight of the vehicle can therefore be decreased by 10–15%. Some examples are reported:

– wings of the C4 are a polyamide/poly(phenyl oxide) alloy, which provides a gain of 1.5 to 2 kg and a resistance to small impacts, because this mixture has a shape memory. In the same spirit, we can cite the low-support bumpers which prevent injury to pedestrians. Tanks (fuel, washer fluid) use PE, the dashboard is made of ABS, the floor mat is of PVC, the bumpers of PP, which is more easily recyclable than the ABS resins. The two studies on the recycling of PP by Frédéric Viot (see Chapter 11) and Valérie Massardier (see Chapter 12) concretely illustrate these aspects;

– with a view to reducing CO_2 emissions from transport, the electric car (*vide supra*) offers the most efficient solution, but there is a factor of 15 between the energy contained in 1 kg of battery and 1 liter of petrol (180 $Wh.kg^{-1}$ and 3000 $Wh.kg^{-1}$) [TAR 10] and the issues of the autonomy of the vehicle and the recharge time have not yet been resolved;

– the fuel cell is based on the principle of the combination of oxygen and hydrogen with the formation of water [FOU 02]. The cell consists of two compartments separated by a membrane of Nafion, a fluoropolymer produced by Dupont de Nemours. The hydrogen oxidized into a proton in the anodic compartment crosses the membrane to reduce the oxygen at the cathode. This membrane is effective for hydrogen but not suitable for methanol which has greater ease of storage in a conventional tank of an automobile because it is a liquid;

– in the field of air transport, polymers were introduced into aircraft construction in order to reduce weight and to provide more seats and/or consume less kerosene. The most recent examples are the *dreamliner*, the Boeing 787, of which 50% of the mass is due to composites; the replacement of aluminum leads to a weight reduction of 20% and a saving in kerosene of around 20% during operation. In Europe, Airbus built the A350 with 52% composites, thereby reducing the weight by 10–15 t. For these applications, scientists are developing new high-performance composite materials based on carbon fibers and polymers.

I.4. Polymers in daily life

It would be too long and too tedious to make an exhaustive list of all the applications for which polymers play a fundamental role in everyday life. The most important is packaging which uses more than 40% of the production of plastics, we shall simply mention three domains less well known to the public: food, health, clothing, sports and leisure.

I.4.1. *Polymers and food*

Plastic materials in contact with food, and the duration of contact, are constantly increasing. The more lively debates on the real safety of food in contact with materials, popularized by magazines more or less able to master the concepts used to protect the consumer (low toxicity, low migration of the substance, low exposure of the consumers, etc.), do not promote consumer confidence in plastic materials used for packaging. Successive regulations on polycarbonate materials, on residual bisphenol A, illustrate this idea perfectly. This controversy and the accompanying media coverage are analyzed in the study by Laura Maxim (see Chapter 1).

There is less emphasis on the fact that polymers have experienced a strong growth in the field of food packaging, as the barrier effect to oxygen indeed provides good conservation of oxidizable products and avoids contamination of meat and vegetables, with more transparency of the packaging allowing examination of the product. In their study (see Chapter 13), Olivier Vitrac and Audrey Goujon emphasize the universal character of the diffusion of the constituents of materials. This risk, as with many risks involving chronic chemical contaminants, requires specific attention and a proper consideration going beyond regulatory requirements alone, helped by a reliable prediction using simulations and modeling.

I.4.2. *Polymers and health: diagnosis, treatment and surgery, vectorization*

Polymers have become essential in all areas of health, from the solid materials used, for example, for prostheses, catheters and blood bags to the nanosystems developed for the vectorization of active substances and for diagnostics [GAU 03]. Among all these devices, some are single use (syringes, infusion sets, etc.), while others are introduced into the body for a period which is *a priori* indefinite (articular prostheses, vascular substitutes, artificial crystalline lenses, stitches thread, etc.).

I.4.2.1. *Surgery*

For hip surgeries, the protection of the acetabulum is provided by a deposit on the metal of the prosthesis of a layer of polyethylene with a very high cross-linked mass using γ radiation. Inguinal hernias are treated by the introduction of a polypropylene membrane. Cataracts are treated by the extraction of the natural crystalline lens, and replacing it with a substitute in polymethylmethacrylate.

In the cardiovascular area, PET has a dominant position with heart valves. Expanded polytetrafluoroethylene, known for its chemical inertia and its high-thermal stability, possesses other advantages compared with PET, especially its biostability and the possibility of sterilization by heat. Finally, substitutes based on synthetic textiles now open up an alternative therapy for the replacement of injured arterial segments. Plastic surgery utilizes silicones but with the risks that have come to light during recent cases.

I.4.2.2. *Vectorization and the controlled release of active ingredients*

Vectorization consists of encapsulating substances of interest (solid or liquid) within a carrier material and transporting them to their area of action, where they will then be released. Vectorization techniques principally use materials based on natural or synthetic polymers. In parallel with conjugated polymers – the active ingredient – the galenic form of the polymer is involved in the activity of the active ingredient. We can distinguish the self-assembled polymers (for example, micelles) and the nanoparticles such as nanoshells, wherein the active ingredient is scattered in a polymer matrix or in nanocapsules consisting of a heart and a polymer membrane [SHA 12, VIL 12]. The concept of "polymer therapy" has emerged from this synergy.

The choice of which type of particle and polymer to use depends on the desired effect. The support material helps in changing the layout to control the immediate or delayed release of the active ingredients. For example, it has been shown that for certain leukemias, the drug treatment (gemcitabine) alone was less effective than when the drug was administered by nanoparticles [DEL 03].

In this particular field, it is noted that polymers are not used as structural materials, strictly speaking, but as a tool for formulation (polymer formulation as opposed to the formulation of polymers commonly used). They are found here ranging from the micrometric scale as a component of nanoparticles to the micelle, or even the macromolecule in the case of nonionic polymer surfactants for the stabilization of particles [HAM 11, TAD 01]. A self-association leads to particle sizes from 50 to 80 nm in water in the presence of poly(ethylene glycol). Nanoparticles can also be obtained with associative polymers which spontaneously self-assemble [GRE 11].

I.4.2.3. *The diagnosis*

The principle is to use physicochemical interactions between a complex molecule (hormones, antigen antibodies, etc.) contained in a biological fluid and a specific reagent bound to a polymer. The target molecule binds to the site of the polymer and is then recognized by the same reagent fixed on a specific enzyme which causes a color reaction, facilitating the measurement [DEL 03]. This technique is used for pregnancy tests, for example.

I.4.3. *Polymers, clothing and sport*

The garment industry has in less than 80 years been totally transformed by the arrival of polymers. The requirements which are becoming increasingly important in innovative new textiles with high added value are clearly expressed not only by consumers but also by European producers which must meet their expectations and face the competition, which is becoming increasingly significant from countries with cheap labor.

A response to these needs, giving the companies a wide range of technological solutions, consists of the multi-functionality of these soft materials, that is to say, the integration of multiple functions. Innovation in this area is wide open, switching toward technical and functional textiles and adaptive textiles (smart textiles). It is possible to view a vision of this development by consulting a special issue of *Actualite Chimique 2012* [NEM 12].

The desired effects are in the fields of waterproofing, dyes, traceability and bacteriostatic applications, through molecular or nanostructured entities, personal care such as the release of cosmetic active ingredients or drugs for example, grafting sensors for people under medical surveillance, UV protection.

Technical textiles are manufactured primarily for their technical performance and functional properties rather than for their aesthetic or decorative characteristics. Only a small proportion of technical textiles (about 2–3% by volume) use high-performance fibers such as carbon fibers, aramid fibers (Kevlar, Nomex) or ultra-high-molecular-weight polyethylene. The vast majority of these textiles are made on the basis of polymers and classical materials such as polyester, polypropylene, polyamide (nylon), viscose, cotton and so on. Probably the most popular developments are waterproof breathable textiles that let vapor travel from the body to the outside, but are impermeable to water droplets due to the application of a Teflon layer ® culminating in Gore -Tex ®, and also the application of a layer of fleece which traps air within the polyester creating a thermal barrier. Some fleeces come from recycled PET, certified by labels (Polartec ®, for example). It takes about 27 plastic bottles to create a sweater. Sports equipment has benefited from advances in materials for aeronautics. Skis, surfboards, snowshoes, pedal boats, etc. all originate from synthetic polymeric materials.

I.5. What does the future hold?

I.5.1. *Substitution of hazardous substances, development of new polymers and "new" processes*

Society has largely accepted the penetration of polymers into daily life, but strong criticism began to emerge for some applications, for example as a result of the excessive use of packaging bags obtained from non-biodegradable materials causing sustainable pollution by their release into the environment, or regarding bisphenol, a raw material for polycarbonate, which is an endocrine disruptor, which will be banned in France for all applications involving contact with food. More generally, it has been shown that certain polymers present problems because they can release toxic substances during use [YAN 11].

This issue of the impact of chemicals on human beings and the environment is the subject of a European Regulation known as REACH [REA 06] (registration, evaluation and authorization of chemicals), of which the first version dates back to 2006. Since then it has been amended several times and has been accompanied by a large number of additional documents including a guide for polymers which was modified for the last time in April 2012.

Under certain conditions, a polymer substance can be considered as a complex substance itself composed of several polymeric substances (in the strict chemical sense) and accompanied by monomer residues, additives and various impurities. Polymers are currently not subject to most of the obligations because they do not exhibit toxicity to human beings owing to their dimensions which prevent them from crossing cell walls. However, some hazardous substances, including monomers and additives, are either already subject to authorization or some limitation of use or are included on a list of potential products for such constraints.

Expertise acquired by the ARN on the consequences of the application of REACH on industries producing, using or recycling polymers emphasizes the need for knowledge on the physicochemical evolution of polymers in terms of their use, on the need for setting priorities regarding the substitution reagents and formulation additives based on the tonnage, and the impact on the economy and on the characterization techniques to be developed for the recycling industry [SIL 12].

For all these reasons, a major research effort is required to substitute from hazardous substances which enter in their chemical compositions and which we may find existence over the lifetime of the polymer, as shown in Figure I.4 [SIL 12], to substances which are less or totally harmless. This effort should also extend to consider the processes of production that will prove to be cleaner and take into account the use and the end of life of the polymers.

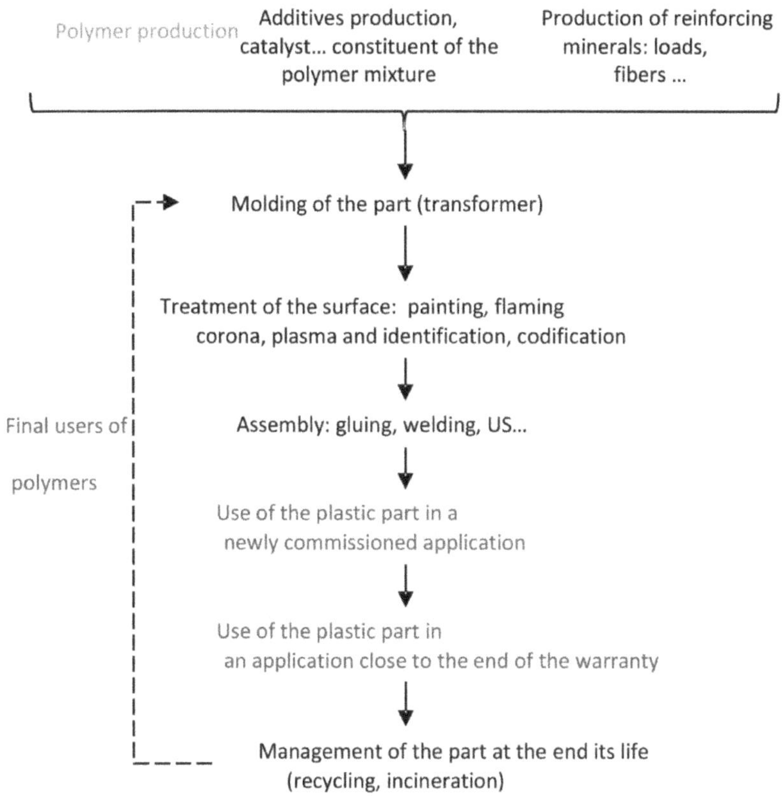

Figure I.4. *Major steps in the value chain of plastic parts*

I.5.1.1. *New polymers, physical forms and sources of raw materials*

An example of substitution by a bio-based product is given by isosorbide, a molecule bio-sourced from starch. Isosorbide (Figure I.5) is considered to be a possible alternative, as well as fatty esters of isosorbide, for the replacement of phthalates as plasticizers of PVC [ROQ].

The use of biodegradable polymers provides another promising path [RUT 04]. The concept of biodegradability is often used incorrectly (sometimes even to justify action in sustainable development) and should be precisely defined. We will read about this subject in the study by Guy César (see Chapter 14).

In his study, Luc Avérous (see Chapter 9) reports on the formulation and the implementation of multiphase biodegradable materials based on plasticized starch.

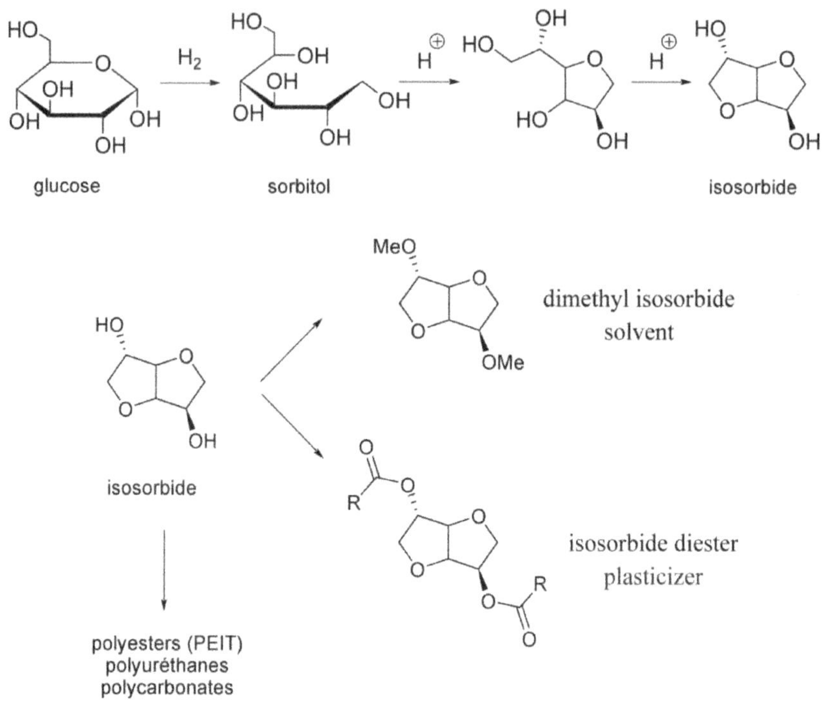

Figure I.5. *Synthesis and use of isosorbide*

Alongside the new molecules, nanomaterials now allow us to eradicate a number of technological barriers, and offer solutions to many problems, especially in the field of health. These promising developments should not obscure the risks associated with these nanosciences and call for new regulation, as highlighted by Eric Juet (see Chapter 15).

A question often raised is the use of biomass to replace fossil materials for chemistry in general and therefore also for the production of polymers. Although the consumption of fossil fuels for polymers is less than 10% of the total consumption which is mainly for energy, should the world of polymers submit to the trendy "green movement"? Does sustainable development imply bio-based products? Following the same train of thought, is a "bio-product" "green"? The article by Olivier Talon (see Chapter 6) provides some relevant thoughts on this topic.

The urgent need for the planet is to develop modes of energy to satisfy industrial and domestic needs, and it is not certain that this can be done by converting the 172×10^9 tons of lignocellulosic materials produced each year, on the one hand because the soil will be more and more consumed to feed the world (more than 9 billion in 2050) and on the other hand because the conversion of the biological material will be energy intensive. The production of large families of polymers will continue to be based on fossil materials for much longer period [SIL 11].

Biomass is otherwise of great interest in other areas because it provides molecules whose functionalities are useful for fine chemicals and pharmaceuticals, and some polymers obtained by fermentation or from monomers have already found specific applications and it is in this direction that research should be focused.

I.5.1.2. *The processes*

Besides their chemical activity, polymerization processes have also contributed for a while (without claiming so) to sustainable development. Moreover, the synthesis of polymers in water offers a previously known solution to these new concerns.

Water does not dissolve hydrophobic monomers, which therefore require organic solvents. Polymerizations in a dispersed medium, which are already old processes, constitute a particularly interesting alternative because they allow us to control the temperature, since the heat of reaction is easily exchanged with the continuous aqueous phase, and also allow us to greatly limit the overall viscosity of the reaction medium. Finally, and this is not the least of the advantages, they allow us to recover at the end of the reaction polymers in a "ready-to-use" form. For example, the suspension polymerization of styrene leads to polystyrene beads which are directly usable for ion exchange resins and for expanded polystyrene. In the paint industry, the emulsion polymerization of vinyl acetate or acrylic allows us to replace organic binding agents with binders in aqueous dispersion. These paints dry by evaporation of the water and limit the volatile organic solvents.

I.5.1.3. *End of life*

This concerns the plastics themselves and the objects containing plastic. Plastic materials must satisfy environmental concerns: the distribution of non-biodegradable plastic bags has been reduced since January 2010. Vehicles contain more and more plastic materials. The rate of reuse or energetic valorization concerning used vehicles is expected to reach 95% in 2015.

These innovations require an increase in the number of formulations to improve product quality (compounds), new shapes, and an increasing use of composite materials, nanomaterials and bioplastics. It is also necessary to rely on process innovations. The extruder is now considered not only as a tool for the manufacture of the polymer allowing, for example, the introduction of various loads, but as a chemical reactor in which we can operate polymerizations or chemical modifications of polymers (hence the concept, now accepted, of reactive extrusion).

I.6. Eco-design: toward an eco-innovation

The concept of sustainable development is already old, as developed by the Brundtland Commission of the United Nations in 1987, in order that economic development is not accompanied by a deterioration in the quality of life. The recommendations of the committee focused on the necessity to limit global warming by controlling emissions of carbon dioxide, to seek alternatives to fossil fuels and to reduce the risks for human health from industrial activities and new products.

There is therefore now a real awareness of a need for sustainable development around various societal and economic issues. Imposed by regulations and required by customers, analysis of the lifecycle and "green" chemistry for a new environment have become essential and require new tools of innovation, as demonstrated by Sylvain Caillol (see Chapter 5).

The previous sections have given a number of examples demonstrating the involvement of polymers at various stages of their production and their use in the context of sustainable development. It is still necessary to incorporate these elements in a coherent and effective way and to be careful not to omit anything that is important. This is the goal of eco-design methodology which is necessary regardless of the specific project. Eco-design (ISO 14062) is not only interested in analyzing the lifecycle of materials, but also in the whole environmental criteria. It consists of integrating the environment into the design stage of the products, by reducing the impact through a global and multi-criteria approach to the environment. This integration is based on taking into account all the stages of the lifecycle of the products (e.g. how to identify the toxicity of a component, how to determine the rate of recyclability or valorization of a plastic material, how to know the carbon footprint of this product?), which therefore defines the criteria of choice for materials [ADE]. Stéphane Le Pochat and Arnaud Roquesalane (see Chapter 7) clearly demonstrate the necessity of the environmental characterization of materials among the selection criteria. The example of bottled water proposed by Rémi Deterre (see Chapter 8) illustrates this. For a broader view, the interested reader is referred to the different strategies of developing eco-plastic described in the study by Jean-François Feller (see Chapter 10).

All companies that can act directly or indirectly on the design or the improvement of products are involved in eco-design. Conversely, eco-design can develop only through the involvement of companies, which has many considerations in terms of economic gains, image and market differentiation. At the same time, we need to aim:

– to better manage the risks and costs associated with the lifecycle of products;

– to anticipate the emerging expectations of the purchasers or consumers, in favor of a better consideration for the environment;

– to make the environment a new factor of stimulation and creativity in the processes of designing and manufacturing products.

There are many benefits for both the company and the consumer (with more sustainable and cost-effective products, less harmful to the environment and to health).

I.7. The need for a multidisciplinary education for eco-design. What is the role of the course and of the teacher?

I.7.1. *The needs and the objectives*

The consideration of sustainable development affects many skills and the ability to exercise these skills, hence the necessity for an adapted education, based on a few important concepts that go beyond the domain of polymers and involve all sectors of chemistry and materials.

Even if the training needs are clearly identified, it should be noted that both sustainable development and eco-design are not academic disciplines, nor a truly new occupation, but an additional expertise acquired through specialization by people who have already acquired the knowledge and skills needed for different core businesses. A relevant initial degree is therefore not the most essential experience when looking for eco-design skills. The necessary prerequisites make teaching this subject cross-curricular in nature.

The creation of courses oriented toward sustainable development or sustainable chemistry in science may challenge the lecturer in chemistry insofar as the concepts of "sustainability" and "development" rely more on societal postulates than on real scientific truths. The notion of "green chemistry: has at least the advantage of showing that chemists are aware of the sometimes irresponsible use which has been made of their science, and that they have the will and the means to address such use. Thus, chemistry presents a positive aspect often overlooked in the media, which may justify a renewal of interest by young people. This is demonstrated in many areas of current research that can be recounted:

– research into the development of new catalysts, into new synthetic paths which save carbon atoms, by energetically optimized reaction pathways, by economizing on the use of solvents or even using an aqueous medium, has been a concern for the chemist for many years. More recently, new specialties have been developed around biomaterials, biofuels, recycling and the fight against pollution;

– in the field of process engineering, the search for savings in energy and in materials or for a reduction in pollutant emissions has long been a major constraint imposed by strong economic imperatives.

However, for industrial enterprises, regardless of the basis of a truth, if it is socially accepted, the company has no other choice but to comply or to face a media frenzy that could jeopardize its business.

A symposium for education in chemistry for sustainable industrial development was organized in September 2011, jointly by the Division of Industrial Chemistry and the Education Division of the *Société Chimique de France* (French Chemical Society). It was found that for industry, the priority of the training is primarily to provide competent chemists. They will be capable of integrating the concept of sustainability into the practice of their profession. The industrial world has reservations about the introduction of the various courses with options about "sustainability" for which we cannot really define the specific content. Industry asks for courses which have less specialized training in chemistry, but with a stronger multidisciplinary nature and which integrate training in toxicology.

If we ignore the scientific relevance of the concept, teaching must be based on solid scientific foundations. The teaching of chemistry should be broad and comprehensive, in order to provide students with the language which will enable them to broach the key application areas of the future. In addition, such a teaching oriented toward sustainable development requires common references and terminology able to be used without any confusion by everyone dealing with these preoccupations. This can be found in the IUPAC Terminology for Biorelated Polymers and Applications [VER 12]. It is thus important to avoid early specialization, show the universality of the methods and promote the ability to learn by ourselves. Chemistry – a fundamental science – has indeed developed a considerable *corpus* that only an organization of observations, a comprehension of the mechanisms, allows us to understand. This combination of a broad education but at the same time a huge *corpus* which is yet largely to be discovered is a challenge for educational choice. It is also in this context that Serge Walter (see Chapter 16) insists on the fact that education for sustainable development is primarily a warning, an incitement to scientific caution rather than transmission of established and duly verified knowledge.

Two aspects are equally important:

– the role of practice: it is also an aspect of sustainable development to seek a permanent anchor in reality, in the factual, and not just in an abstract understanding.

This helps to understand the role of complexity, impurities, "helping hands", and competitive and unstable transient phenomena which are components of a complex reality;

– the role of highly developed research which seeks to simultaneously simplify mechanisms by unifying them, and by making the developments more bountiful by exploring new paths. This also explains the high proportion of engineers extending their studies with a PhD.

I.7.2. *Status of achievements*

Therefore, it is difficult to define the specificities of a "sustainable chemistry" option as this concern is an integral part of all training, as is that of "security". However, several courses already incorporate it clearly within their training, probably guided by the objective of strengthening their attractiveness among students by enjoying a media-friendly profile, but integrating the multidisciplinary approach sought by industry.

All chemistry schools within the Gay Lussac Federation and especially CPE Lyon share this vision of training.

Universities offer courses at all levels dedicated to sustainable development and eco-design. Course subjects available nationally where training is dedicated to materials and/or the environment are "plasturgy and composite materials" and "industrial production". There are currently more than 800 degree programs ranging from BEP to Master's claiming the term "environment" with keywords such as rural, urban and industrial, drinking water, air, pollution, waste management sorting and energy. Most of these courses offer general training, but few actually look at the plasturgy and plastic materials. There are a few courses which devote more than 20 hours to eco-design in the broadest sense.

Similarly, there are a large number of courses dedicated to polymers, plastics and composites. Some include the concept of recycling within their description, but the specializations or options actually dealing with eco-design appear only very rarely.

To meet the needs clearly expressed by French plasturgy, training tools have been developed for companies and training centers, with the aim of informing and educating young people in training and employees on the issues of sustainable development in plasturgy. The Website www.plasturgie-durable.fr is dedicated to sustainable development within the training of plastics, and for teachers to feed their training modules on the theme of "sustainable development", whether or not they are specialized in plasturgy [LAPb].

In conclusion, the modern world demands more energy, but less expensive, less polluting and inexhaustible, light materials which are also reusable or recyclable, better health and an appropriate treatment when required, and a greater respect for all things and human beings through a better understanding of their operations and interactions.

Our ability to modify the planet needs an ethic of this power, and therefore requires extreme skill, which must pass through an intersection, or a fusion, of science, economics and human values.

I.8. Bibliography

[ARM 87] ARMAND M.B., "Current state of PEO-based electrolytes", in MCCALLUM J.T., VINCENT C.A. (ed.), *Polymer Electrolyte Reviews*, Elsevier, Paris, pp. 1–22, 1987.

[BER 95] BERTOLINI G., *Homo Plasticus. Les plastiques, défi écologique*, Le Sang de la terre, Paris, 1995.

[BOU 09] BOUNEAU S., *et al.*, "Construction d'un monde énergétique en 2050", *Ann. Phys. Fr.*, vol. 34, no. 1, 2009.

[BRO 11] BROWN J.A., *et al.*, "Energetic limits to economic growth", *Bioscience*, vol. 61, no. 1, pp. 19–26, 2011.

[DEL 03] DELAIR T., ELAISSARI A., PERRIN A., *et al.*, "Les polymères de synthèse, supports du diagnostic médical", *L'Actualité Chimique*, vol. 11–12, pp. 79–83, 2003.

[ETI 12] ETIENNE S., DAVID L., *Introduction à la physique des polymères*, Dunod, Paris, 2012.

[FON 08] FONTANILLE F., GNANOU Y., *Organic and Physical Chemistry of polymers*, John Wiley & Sons, New York, 2008.

[FON 10] FONTANILLE F., GNANOU Y., *Chimie et physico chimie des polymères*, Dunod, Paris, 2010.

[FOU 02] FOULETIER J., "Piles à combustibles", *L'Actualité Chimique*, vol. 3, pp. 138–143, 2002.

[GAU 03] GAUDUEL Y., "De la molécule au lit du malade...", *L'Actualité Chimique*, vol. 269–270, pp. 5–7, 2003.

[GRE 11] GREF R., COUVREUR P., "Nouveaux matériaux pour la vectorisation des médicaments", *L'Actualité Chimique*, vol. 353–354, pp. 88–91, 2011.

[HAM 11] HAMAIDE T., VAUTHIER C., LEPAIS M., *et al.*, "Synthetic polymer surfactants for nanotechnologies in the biomedical field", in POPA M., OTTENBRITE R.M., UGLEA C.V. (ed.), *Medical Applications of Polymers*, vol. 2, pp. 1–51, 2011.

[HOL 05] HOLLIMAN P.J., CLARK J.A., WILLIAMSON J.C., *et al.*, *Science of the Total Environment*, vol. 336, pp. 13–24, 2005.

[LEM 10] LE MONDE MAGAZINE, Plastique, l'ennemi intime, 19 September 2010.

[MAU 10] MAUREL A., "Les techniques séparatives à membranes considérations théoriques", *Les Techniques de l'Ingénieur*, vol. J 2, p. 790, 2010.

[MER 96] MERCIER J.P., MARECHAL E., *Traité des matériaux, Chimie des polymères. Synthèses, réaction, dégradations*, Presses Polytechniques et Universitaires Romandes, Lausanne, vol. 13, 1996.

[MIN 10] MINISTÈRE DE L'ÉCOLOGIE, DE L'ÉNERGIE, DU DÉVELOPPEMENT DURABLE ET DE LA MER, "Plan de mobilisation des filières et des territoires", *Conférence nationale sur les métiers de la croissance verte*, http://www.developpement-durable.gouv.fr/IMG/pdf/Dossier_de_presse-2.pdf, 28 January 2010.

[ODI 04] ODIAN G., *Principles of Polymerization*, 4th ed., John Wiley & Sons, New York, 2004.

[OUD 94] OUDET C., *Polymères, structure et propriétés introduction*, Masson, Paris, 1994.

[PLA 10] PLASTICS EUROPE, *Plastiques 2010: faits et chiffres*, http://www.plasticseurop.org, also see: www.plastcseurope.fr/industrie-du-plastique.asppx, 2010.

[REA 06] REACH, Registration Evaluation Authorization of Chemical, Règlement CE no. 1 907/2006, 18 December 2006.

[RUT 04] RUTOT D., DUBOIS P., "Les (bio)polymères biodégradables : l'enjeu de demain?" *Chimie nouvelle*, vol. 86, pp. 66–75, 2004.

[SHA 12] SHAIK M.R., KORSAPATI M., PANATI D., "Polymers in controlled drug delivery systems", *International Journal of Pharma Science*, vol. 2, no. 4, pp. 112–116, 2012.

[SIL 11] SILLION B., "REACH : un outil pour améliorer le dialogue entre chimie et société", *L'Actualité Chimique*, vol. 355, pp. 37–39, 2011.

[SIL 12] SILLION B., NORMAND A., GUILLEBON A., "Conséquences de l'application du règlement REACH pour les polymers, Compte rendu d'une expertise demandée par l'ANR", *L'Actualité Chimique*, vol. 359, pp. 4–7, 2012.

[TAD 01] TADROS F.T., "Polymeric surfactants", *Handbook of Applied Surface and Colloid Chemistry*, J. Wiley & Sons, pp 373–384, 2001.

[TAN 00] TANAKA T., *Experimental Methods In Polymer Science*, Academic Press, San Diego, CA, 2000.

[TAR 10] TARASCON J.M., Leçon inaugurale au Collège de France, 9 December 2010.

[UNE 03] UNESCO, L'eau pour l'homme, l'eau pour la vie, unesdoc.unesco.org/images/0012/001295/129556f.pdf, 2003.

[VER 12] VERT M., DOI Y., HELLWICH K.-H., *et al.*, IUPAC Terminology for Biorelated Polymers and Applications, *Pure Appl. Chem.*, vol. 84, no.2, pp. 377–410, 2012.

[VIL 12] VILARA G., TULLA-PUCHEA J., ALBERICIO F., "Polymers and drug delivery systems", *Current Drug Delivery*, vol. 9, pp. 1–28, 2012.

[WIE 05] WIESENFELD B., *L'énergie en 2050, nouveaux défis et faux espoirs*, EDP Sciences, Les Ulis, 2005.

[YAN 11] YANG C.Z., YANIGER S.I., JORDAN V.C., *et al.*, "Most plastic products release estrogenic chemicals: a potential health problem that can be solved", *Environ. Health Perspect*, vol. 119, p. 989, 2011.

I.8.1 *Websites*

[ADE] http://www2.Ademe.fr.

[AQU] http://www.aquawal.be/xml/liste-IDC-100-.html.

[CEF] http://www.cefic.org/facts-and-figures.

[COL] http://www.college-de-france.fr/media/jean-marie-tarascon.

[LAPa] http://www.laplasturgie.fr/communiques/rapport_developpement_durable_2011.pd.

[LAPb] http://www.laplasturgie.fr.

[LIN] http://www.linternaute.com, l'eau potable, nouvelles ressources : dessaler l'eau de mer.

[ROQ] http://www.roquette.fr.

[TOT] http://www.total.com/fr/dossiers/gaz-de-schiste/.

1

Some Notes on Two Controversies around Plastic Materials and their Media Coverage

1.1. Introduction

This chapter addresses two controversies associated with plastics, namely the fate of plastic waste in the environment and the effects of substances contained in plastics and suspected of having endocrine-disrupting effects.

Section 1.2 is designed to be a non-exhaustive introduction to the treatment by social science, and more generally by the interdisciplinary science of risk, to the question of the effects of plastics on health and the environment. Section 1.3 provides an empirical analysis of the media coverage of these two controversies, with a *corpus* chosen as an example (French newspapers and two television documentaries). On this basis, we draw conclusions about the role, actual and potential, of researchers in chemistry in relation to the analyzed controversies.

1.2. Socio-political aspects of the two controversies in the scientific literature

The controversy surrounding the regulation of the risks of endocrine disruptors has crystallized around conflicting arguments regarding the role of low-dose toxicity of chemical substances in general. Scientific confrontation between the classical vision in toxicology of postulating the monotonous character of the relationship between dose and effect, and the emergence of research showing the existence of non-monotonic [VAN 09] relationships, raises questions about the adequacy of the

Chapter written by Laura MAXIM.

regulatory protocols of risk assessment. Used for many years, these protocols are built upon the basis of the "traditional" vision of risk.

One of the best known controversies of endocrine disruptors concerns bisphenol-A (BPA). For researchers who conducted toxicological research on BPA, the scientific debate touches upon several aspects of risk. Generally for endocrine disrupters, the points in contradiction concern the existence of non-monotonic dose–response relationships, the relevance of proof of the physiological nature which seems to confirm the effects of low doses, and the importance of critical windows of exposure. Vandenberg *et al.* [VAN 09] analyzed in detail the scientific literature on BPA, including whether:

– the mechanisms of action of the substance are understood for low doses;

– the population is exposed to levels high enough for an effect to occur;

– the oral route of exposure is the only relevant route;

– BPA is conjugated by the digestive system;

– animal studies can be extrapolated to humans;

– the effects observed in animals exposed to BPA are consistent from one study to another;

– BPA induces cancer.

The way we consider each of these arguments may have a decisive influence on how to assess the risk of BPA. Depending upon the specialists they welcomed into their midst, committees of various experts have made different assessments of the relevance of the knowledge for each of these elements. The result is a situation that may seem paradoxical: various expert committees have resulted in completely opposite conclusions from the same starting data: if for some "there is no risk from BPA" for others "there is a general risk from BPA for the entire population" [BER 10].

The literature in social science is relatively poor in studies of the controversy surrounding endocrine disruptors and is even poorer concerning the controversy about the future of plastic waste. A search of the literature of the CNRS in social science with the keywords "plastic"/"plastics" in the title and "controversy" in the body of the article results in a single article dealing with a controversy about a major fire in a plastics recycling factory in Canada. A search with the word "plastic" in the title and "media" + "environment" in the body of the article leads to no responses.

The first systematic analysis of public attitudes to the controversy surrounding BPA was recently published by Brewer and Ley [BRE 11]. The authors studied four categories of public attitudes, namely familiarity with BPA, concerns toward its toxicity, support for bans and behaviors to reduce the personal exposure to plastics containing this substance. The study showed that these attitudes are directly correlated to two factors:

– the use of media (i.e. newspapers and the Internet);

– the level of trust in the key players in the controversy (i.e. scientists, the food safety and environmental agency primarily concerned with controlling the risk of chemical substances and industry).

The use of traditional media has an influence which is directly proportional to the level of familiarity with the subject and to the protective individual behavior. More than newspapers, the use of the Internet has been found to be a major factor influencing concerns, pro-ban attitudes and protective behaviors. TV is not correlated to any of the attitudes studied.

Yet the media is not an essential factor for favorable attitudes toward banning the substance. The less we have confidence in the US Food and Drug Administration (FDA) and industry, the more we are in favor of banning BPA and the more we are concerned with toxicity. Similarly, the higher the confidence in science, the more attitudes are in favor of banning BPA and demonstrating concern for its toxicity.

Somewhat paradoxically, confidence in the FDA does not play an important role in protective behaviors (whereas the role of such an agency is precisely to induce such behavior when a potential hazard exists or, in contrast, to assert their lack of necessity when the risk is not real).

The use of the media (traditional and the Internet) and the trust (or not) in the players having a central role to play represent, therefore, "shortcuts" (*heuristics*) for the public that help non-specialists to form an opinion during scientific controversies and to take decisions in relation to their individual consumption practices.

Among the English-language writers, Sheldon Krimsky was among the first to study the history of the controversy surrounding endocrine disruptors in the United States in his book *Hormonal Chaos* [KRI 00]. Another recent sociological and historical analysis of the risk assessment of BPA had a significant impact in the United States. This is the work of Sarah Vogel of Columbia University, "The Politics of Plastics: The Making and Unmaking of Bisphenol A 'Safety'" [VOG 09]. Her description of the controversy emphasizes the disparity between the standards of risk assessment, still in force after about 20 years in the United States, and the scientific advances on the effects of doses. A second striking element of this analysis is the polarization of the industry studies and the public research on the toxicity of BPA – the first showing no effect, while the vast majority of the second highlights various changes in the prostate of the fetus and in the development of the mammary gland, genetic, immune and metabolic perturbations, and changes at neurological and behavioral levels. This "funding effect" cited by Vogel [VOG 09] is particularly well demonstrated by a study of the literature conducted by Vom Saal and Hughes [SAA 05], which showed that 90% of the 115 studies financed by government sources reported effects of one form or another owing to BPA exposure of the (regulatory) reference dose or below this dose, whereas none of the 11 studies funded by industry showed an effect.

In the field of social sciences in France, a recent research report describes the emergence of "low doses" as a cornerstone of a paradigm change in terms of risks to health and the environment. Analyzing several controversies and on the basis of a series of interviews with epistemologists, toxicologists and biologists, Chateauraynaud *et al.* [CHA 11] describe the history of the question (sociological and political as well as scientific) of low doses in three waves (the 1960s with the radiological risk and the inability to determine a threshold below which an exposure to radioelements would be safe; the major health crises of the 1990s and the rise of the precautionary principle; and finally the controversy surrounding endocrine disruptors with the questioning of dose–monotonous effect relationships). The study concludes with a "reconfiguration of the relationship between fundamental research, risk assessment and public management of vulnerabilities".

Its funding from the PNRPE program of the Department of Ecology, research has been done by Jean-Paul Gaudillère (INSERM) and Nathalie Jas (INRA), into a comparison of the trajectories of the controversy surrounding endocrine disruptors in France and in the United States. The project began at the time when endocrine disruptors were not yet becoming a media controversy and were not yet visible in France. The central hypothesis of the project was the difference between France (some controversy) and the United States (strong controversy for several years) and was not so much about the presence or absence of certain research skills but rather concerned the practical expertise and public debate and whether or not it facilitated the convergence between ecology, medicine and agriculture.[1]

Finally, the communication to the general public of scientific uncertainty about the effects of endocrine disruptors on male fertility has been studied in a recent research project and published in part in Maxim *et al.* [MAX 12].

1 Project "knowledge, expertise, regulation: how the 'endocrine disruptors' became objects of government? A comparison between France and the United States".

The author is not aware of media coverage in France of any analysis of the controversies surrounding plastics.

1.3. Plastics in the French media: a small sample

Two main topics recur in recent media treatment of the environmental and health effects of plastics, especially the environmental effects of plastic waste (particularly to the oceans) and the effects on public health of toxic substances contained within plastics.

For an analysis of the written press, the results presented in section 1.2 are from a search using the LexisNexis database, which is one of the largest collections of French and international print media, with is updated as soon as the newspapers are published and provides almost exhaustive access to national and regional sources in France.

For televisual news, the results are from a non-systematic analysis of two documentaries, namely “Addicted to plastic” and “Plastics: a toxic warning”.

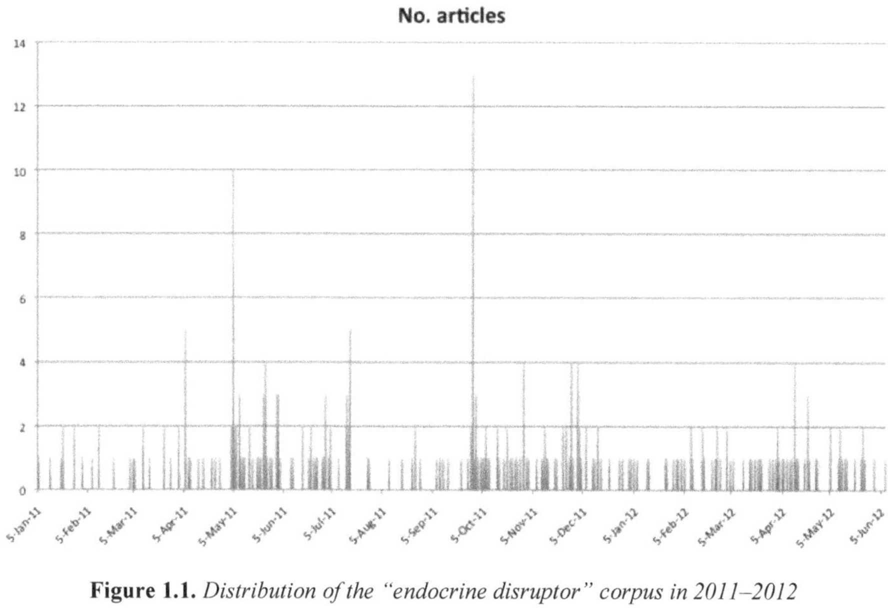

Figure 1.1. *Distribution of the “endocrine disruptor” corpus in 2011–2012*

1.3.1. *The written press: endocrine disruption*

The corpus used for analyzing the media coverage of endocrine disruptors has been obtained by using the keywords "disruptor" and "endocrine" for the period from 1st January 2011 to 10th June 2012. A number of articles, 278, were identified in newspapers, magazines and internet media. Mail, legal news and the specialized press were excluded. This controversy appeared with relatively constant frequency throughout the corpus of the year 2011 (Figure 1.1).

The four peaks which can be observed (Figure 1.1) are associated with specific events:

– 5th April 2011 corresponds to the publication of a study showing that distylbene has transgenerational effects on the grandchildren of women treated with it;

– 4th May, when the National Assembly passed a bill banning the use of endocrine disrupters such as phthalates and parabens;

– 15th July, at the occasion of the Senate publication on endocrine disrupters;

– 28th September 2011, when ANSES proposed the labeling and substitution of BPA, on the occasion of the publication of two reports by the agency on health effects and uses.

National and regional press releases are represented relatively evenly during the media peaks (Figure 1.2).

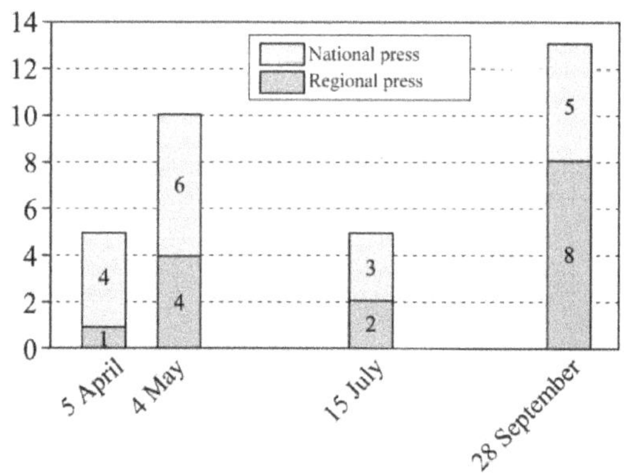

Figure 1.2. *Distribution of peaks of articles between different types of media*

None of the articles gives an important role to substitution, to the innovation of substances "benign by design", or to chemical research; the debate is focused on the assessment and regulatory control of hazards in consumer products.

1.3.2. *Plastics in general*

All the national and regional French press, public and professional, was included in the constitution of the *corpus* (news excluded). The research was conducted for the month of March 2011. It resulted in a number of articles, 355, of which the vast majority were in the regional press (83%) and a minority were in the national press (17%). Similarly, most of the articles were published by the mainstream press, with a few published by other media.

Among the identified categories, most articles refer to collective cleaning of natural areas through the collection of waste of any kind, including plastics (24%). In this category of articles, the theme of "waste" (pollution, recycling, sorting) and the theme of "plastic" almost coincide. Almost all articles in this majority category were published in regional newspapers.

A similar category which is well represented (17%) is that containing articles citing educational activities around the recycling of waste, including plastics, intended for children or young people in general, but also for consumers, farmers or traders.

The category of "business life" which discusses the productive situation (for example, the economic situation, the market position and the cessation of activity) or the social situation (for example, strikes) is also present, with 17% of articles.

Articles in the national press are found mostly in the business press (*Les Echos, La Tribune, L'Agefi quotidien*) and address the economic situation of enterprises producing or working with plastics (for example, the increase in raw material prices and their impact on the price of plastic, innovations in the field of plastics), with only a few articles in the mainstream press:

– Japanese toys without phthalates (*Le Monde*);

– the use of bio-based plastic for water bottles (Brita bottles) or drinks bottles (Coca-Cola, PepsiCo, Danone, Evian) (AFP, *La Tribune*);

– artistic works around recycling (*La Croix*);

– the movie *Notre poison quotidien* and the presence of toxic substances in consumer products (*Aujourd'hui en France, Le Monde, Libération, AFP*);

– recycling (*La Croix*).

With regard to research in chemistry and green chemistry, we can observe (Figure 1.3) that:

– the contribution of plastics to sustainable development (including insulation, which aims to reduce energy consumption) is discussed in 3% of articles;

– the research into green chemistry appears very little, with only 2% of the articles referring to it;

– companies which make green chemistry the heart of their activities are not well represented either, with only 5% of the articles.

Voluntary corporate actions, identified as limiting the environmental effects of plastics, are mentioned in 4% of the articles, but the message of an environmental nature is difficult to separate from the message with commercial advertising purposes.

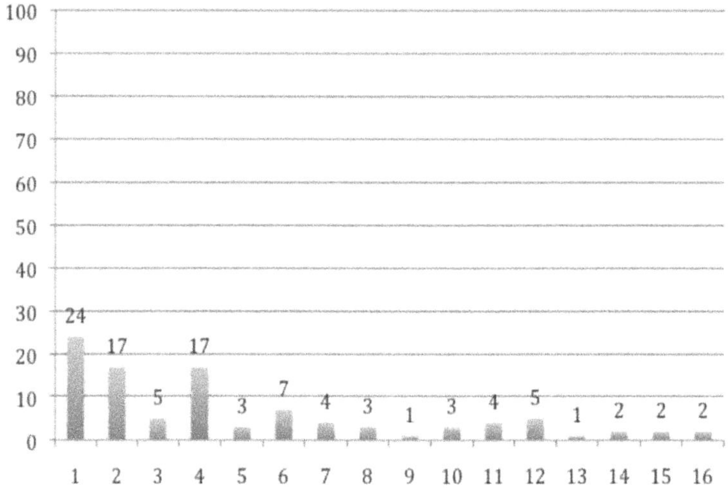

The categories identified are: 1. waste collection including plastics; 2. business life; 3. green chemistry companies; 4. educational activities around recycling waste including plastics; 5. educational activities around chemistry where the issue of plastics is addressed; 6. artistic actions enhancing plastics; 7. substances considered to be toxic contained within plastics; 8. contribution of plastics to sustainable development; 9. chemical accidents involving plastics or occupational toxicity in the production of plastics; 10. call for waste collection; 11. voluntary corporate actions identified as limiting the environmental effects of plastics; 12. humanitarian or social services (for example, rehabilitation) associated with recycling plastics; 13. impact of plastic waste on the environment; 14. incivilities associated with plastic waste; 15. research into green chemistry; and 16. description and analysis of the state of sorting and recycling of plastics in France.

Figure 1.3. *Articles published on the "plastic" theme in the national press by category (horizontal axis) and percentage of total (vertical axis)*

It is interesting to note that the articles dealing mainly with substances considered to be toxic contained within plastics (category 7 in Figure 1.3) like phthalate or BPA represent only 4% of the total number of articles, while the impact of plastics upon the environment or health is the main subject for 1% of the articles.

Systematic analysis of articles on the topic of plastics shows that their toxicity and their impact on the environment (which we might call "guilt-laden" articles) is represented only very secondarily as the main subject, and are widely surpassed by incentives or positive messages calling for waste management or "green" innovation.

1.3.3. *An analysis of two television documentaries*

Two documentaries briefly analyzed in sections 1.3.3.1 and 1.3.3.2 are "Addicted to plastic" and "Plastics: a toxic warning".

1.3.3.1. *Endocrine disruptors*

The media coverage is characterized in both television documentaries in particular by:

– increased attention being given to the discrepancy between the results obtained by scientists (science) and the regulators of risk assessment (politics), as well as scientific advances which explain this discrepancy (for example, the non-monotonic relationship between dose–effect);

– the opposite treatment by large chemical industries and small industries, with special cases of companies of a rather small or medium size occupying very specific market sectors such as food containers for babies and toys for children. The attitude of big business, producing significant quantities of incriminating substances, with regard to the results from toxicology is described in the two documentaries as resistance somewhat motivated by financial interest, the latter being characterized as dominant when compared with concerns to public health. Lobbying activities seem to play an important role;

– the scientists who are the most prominent are toxicologists; researchers in chemistry appear briefly. Their role to find intrinsically hazardous substances is not highlighted, even if the name of John Warner is mentioned briefly in connection with green chemistry;

– at the opposite of the controversy on plastic waste, with toxic substances the responsibility of the consumer is hardly present; he is described entirely as a victim of risk who remains invisible, especially since there is no labeling of chemicals (BPA, phthalates, etc.). The responsibility is entirely associated with industry and authorities of expertise (health agencies).

There is a repetitive logic in this media treatment, with moderate criticism accompanied by positive messages about alternative solutions:

– presentation of negative effects on health and the environment;

– identification of chemical risks which contribute to them;

– questioning the origins of the identified problem and the existence of alternatives;

– questioning "if alternatives exist, why do things not change"?

– identification of barriers to the level of consumption, to production and to the economic policies and risks;

– positive examples of the implementation of alternative solutions, at the level of individual companies or at the scientific level.

The overall message is rather optimistic–educational–incentive rather than pessimistic–catastrophic–"hopeless".

1.3.3.2. *The fate of waste*

The mediatic treatment of the subject has some specific features:

– a change of perspective where the liability is no longer with the end user (mostly consumers) or where the regulator is supposed to control the "secondary effects" of the markets (i.e. pollution). The liability lies more specifically with the stage of production, hence the manufacturer is responsible for environmental and health effects;

– the researchers in chemistry become people who find solutions (bio-plastics, waste treatment) and play the "good role" in history, with particular reference to green chemistry, to the point that some controversies such as those concerning the choice between the use of farmland to produce either resources for plant chemistry or food for feeding purposes almost do not appear.

1.4. Conclusion

One goal of the workshop which permitted the production of this text was to promote a better understanding of the role of chemists in relation to controversies related to the effects of plastics on health and the environment. In light of the social science research discussed and our media analysis, a difference can be observed in the scientific and media processing of researchers in chemistry (with a neutral or positive role) and that of industrial chemists and researchers working for "big industry", whose image is negative.

In general, it appears that research chemists are either absent or present in a positive way, as a potential source of solutions to existing problems.

1.5. Appendix: equations of research to identify the "plastic" *corpus*

Plastic AND (eco-design OR ecodesign OR environment OR wildlife OR ocean OR phthalate OR recycling OR ecopackaging OR eco-packaging OR packaging OR ecology OR waste OR animals OR animal OR disruptor OR bisphenol OR pollution OR health OR food OR polymer OR resins) OR bioplastic OR bio-plastic OR polyethylene OR polypropylene OR polystyrene OR PVC OR polyvinylchloride OR polyester OR PET OR polyamide OR nylon OR polycarbonate OR polyurethane OR polyurethane OR polyacrylates OR vinyl OR vinylic paint OR (vinylic S/4 paint) OR acrylic paints OR epoxy OR epoxides OR synthetic textile OR (synthetic S/4 textile).

1.6. Bibliography

[BER 10] BERONIUS A., RUDEN C., HAKANSSON H., *et al.*, "Risk to all or none? A comparative analysis of controversies in the health risk assessment of Bisphenol A", *Reproductive Toxicology*, vol. 29, no. 2, pp. 132–146, 2010.

[BRE 11] BREWER P.R., LEY B.L., "Multiple exposures: scientific controversy, the media, and public responses to bisphenol A", *Science Communication*, vol. 33, no. 1, pp. 76–97, 2011.

[CHA 11] CHATEAURAYNAUD F., DEBAZ J., FINTZ M., La dose fait-elle toujours le poison? Une analyse sociologique des mondes de la recherche et de l'expertise à l'épreuve des faibles doses, Product under the HANDLES/EHESS convention report (GSPR), 2011.

[KRI 00] KRIMSKY S., Hormonal chaos. *The Scientific and Social Origins of the Environmental Endocrine Hypothesis*, The John Hopkins University Press, Baltimore, MD, 2000.

[MAX 12] MAXIM L., MANSIER P., GRABAR N., "Public reception of scientific uncertainty in the endocrine disrupter controversy: the case of male fertility", *Journal of Risk Research*, vol. 16, no. 6, pp. 677–695, 2012.

[SAA 05] VOM SAAL F.S., HUGHES C., "An extensive new literature concerning low-dose effects of bisphenol A shows the need for a new risk assessment", *Environmental Health Perspectives*, vol. 113, no. 8, pp. 926–93, 2005.

[VAN 09] VANDENBERG L.N., MAFFINI M.V., SONNENSCHEIN C., *et al.*, "Bisphenol-A and the great divide: a review of controversies in the field of endocrine disruption", *Endocrine Reviews*, vol. 30, no. 1, pp. 75–95, 2009.

[VOG 09] VOGEL S., "The politics of plastics: the making and unmaking of bisphenol A 'Safety'", *American Journal of Public Health*, vol. 99, no. S3, pp. S559–S566, 2009.

2

Plastic Waste and the Environment

2.1. Introduction: waste and the environment

No longer can industry now lose interest in the end of life of the objects it places on the market place. This is particularly true of the plastics industry which is frequently attacked for its real or imagined implications for environmental issues.

However, focusing on the problems of waste is probably not entirely justified[1].

Indeed, the total waste, to take a significant example, contributes only 8% to the creation of greenhouse gas emissions (Table 2.1). Road transport on its own contributes 33%. It may be noted that there are important contributions from housing, industry and agriculture. As for plastic waste, which represents 11% of waste, its contribution to the greenhouse effect does not exceed 0.73%. This figure is also less surprising when we remember that plastics consume about 4% of crude oil for their synthesis. This number also places into context the interest in replacing oil with biomass as a source of raw materials for the synthesis of polymers.

2.2. The end of life of plastic parts

2.2.1. *Reduction at source*

It is good to remember that the best waste is that which we avoided producing and indeed it seems logical to try to reduce waste as much as possible. However,

---

Chapter written by Claude DUVAL.

1 The cost of the treatment of plastic waste depends on the recycling rate of the material. It goes from € 0.32 kg for a recycling rate of 10% to € 0.80 kg at a recycling rate of 50% (according to Neil Mayne and Herbert Fisch, APME, Brussels, February 2002).

when we go from great principles to practice, it becomes clear that the routes to reduction are not very many [GUI 08b]. Indeed, the nature and the quantity of waste are a reflection of our civilization and only a call into question of our lifestyle[2] could considerably change the production of waste.

Sector	Greenhouse effect (%)
Road transport	35
Housing	14
Industry	12
Farming	7
Energy	7
Service sector	7
Industrial processes	5
WASTE	8

Table 2.1. *The part played by waste in the greenhouse gas effect, Chauveau L. (2000), Environnement Magazine*

However, efforts have been made in this direction by industry. Moreover, these actions to reduce waste are, in general, sources of savings.

In this sense, we have seen in several areas, and especially in the areas of packaging and the automotive industry, an effort to reduce the mass [FRI 98] of objects leading, while providing the same service, to an automatic reduction of the mass of the waste, but also saving energy, especially during transport.

The plastic industry, like other sectors, also seeks to minimize waste production. For this, rationalization systems have been developed (hot channels removing carrots and channels supplying materials for injection, software preventing the wrong parts being made especially at startup, etc.) or systems for immediate recovery and immediate recycling after grinding in front of the press.

We may also mention the use of industrial packaging (most often in plastic) which is reusable [GIL 08, LIG 09, LAV 97] a number of times before being repaired and finally recycled. This application required the development of standards

2 For example, the very large number of households with one person, the reduced availability of women to cook an elaborate cuisine, the increased concern for hygiene, the sale of ready meals leads to a multiplication of packaging of prepared food portion, which leads to an increase in the mass of packaging waste.

which optimize the loading of transport and the transport of empty containers (Galia standards, foldable packaging and/or stacking).

In an entirely different field, we can welcome the creation of "recycling centers[3]". Objects brought by individuals that can still be used, either as is, or after minor repairs, are offered for sale at attractive prices and therefore contribute to the social economy. The 17 oldest recycling centers, in 2008, succeeded in reusing 12% of the delivery to their landfills. It is therefore an action which is not at all negligible.

However, all methods which seek the reduction of waste at the source that we have just mentioned are relatively limited. The aim should be their generalization. Nevertheless, the actions of reduction at the source will not avoid the necessary care of waste.

2.2.2. *Hierarchy of choice of valorization*

Plastic waste can be recycled in different ways and it is legitimate to try to establish a hierarchy between the various processes. We propose to use the criterion of entropy creation. Natural systems evolve spontaneously toward an increase in entropy. Living things decrease their entropy locally to ensure their survival. Human beings may therefore have as their objective to limit, to the best of their ability, the increase in entropy (Figure 2.1).

Living organisms that gave rise to oil created a strong increase in entropy by dying. The combustion of this oil, leading to CO_2 and H_2O, increases the entropy to a maximum (in terrestrial conditions). In contrast, the distillation of petroleum leading to naphtha, the chemical conversions producing a pure substance, the monomer, are a reduction of entropy. Polymerization and the creation of objects are a specific arrangement of units of monomers and also lower the entropy. These transformations, which are not spontaneous, require investments of all kinds and consumed energy. The obtained objects which "accumulate" these investments are therefore "valuables" and should be treated as such.

Reuse, at the expense of limited efforts, leads to a stability of entropy and appears to be the best solution.

Physical recycling, consuming a little more energy but restoring the level of entropy, seems to constitute a good solution.

Landfill, which will be discussed, has an unclear entropic status.

3 Resource recycling network; www.ressourcerie.fr.

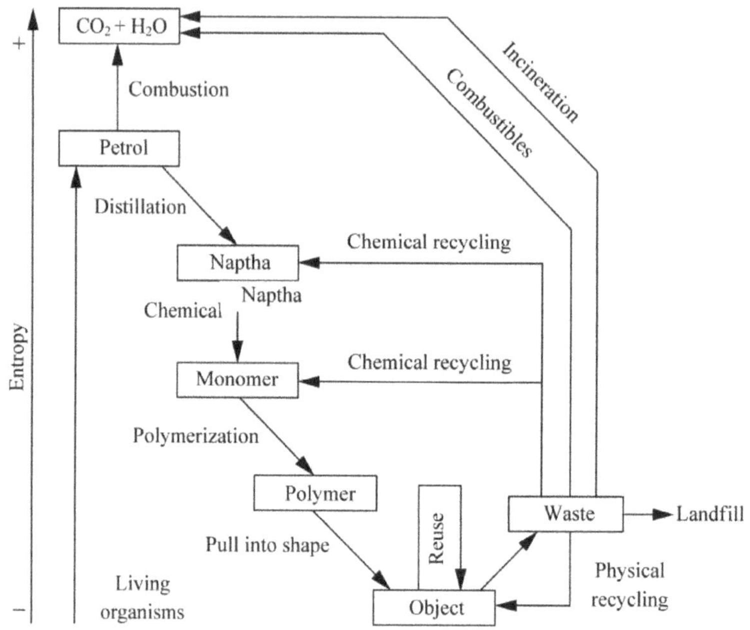

Figure 2.1. *Classification of processes in the valorization of plastic objects as a function of entropy*

Chemical recycling that gives, at best, monomers and, at worst, a mixture of hydrocarbons comparable to naphtha is penalized from the point of view of entropy because it climbs up the ladder seen previously.

Fuel manufacture and waste incineration are solutions which increase entropy the most.

In summary, processes of valorization are not the same from a thermodynamical point of view. The hierarchy that has just been sketched is also variously adopted by different laws.

However, the entropy criterion despite its significant interest cannot be the only criterion used. We must consider health impacts which may prohibit the physical recycling of certain waste because of infection risks or radioactivity. We may want to avoid excessive transportation and therefore process waste on site in an existing factory. Some processes of valorization require a large quantity of waste and a consistent quality that cannot always be maintained. Economic aspects play an important role and may limit the use of a good solution if appropriate incentives are not implemented.

2.2.3. *Inventory*

With regard to the objectives that we can achieve in light of the above, what about the valorization of plastic waste in practice? It is found that few plastics are physically recycled compared with other materials used, which are similar to the case of packaging plastics (Figure 2.2).

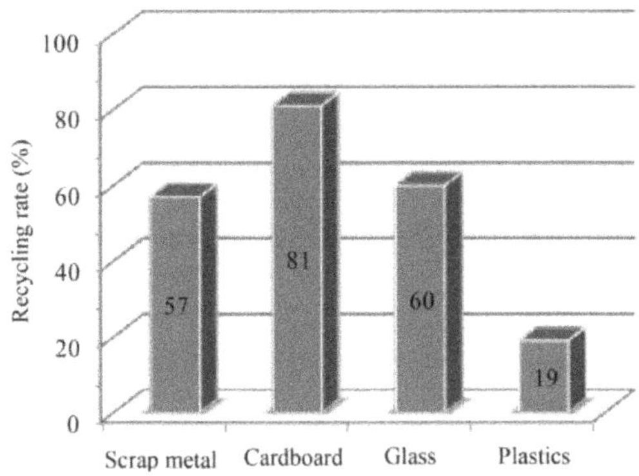

Figure 2.2. *Rate of physical recycling of packaging (data from 2005 – ADEME)*

These already low numbers are even lower if we do not limit ourselves to packaging waste. Then, it is only 7% of plastic waste which is physically recycled. We can, with difficulty, console ourselves by finding that the growth of the recycling of physical packaging was 360% between 1997 and 2004, while the volume of packaging added to the market increased by only 30% (Figure 2.3).

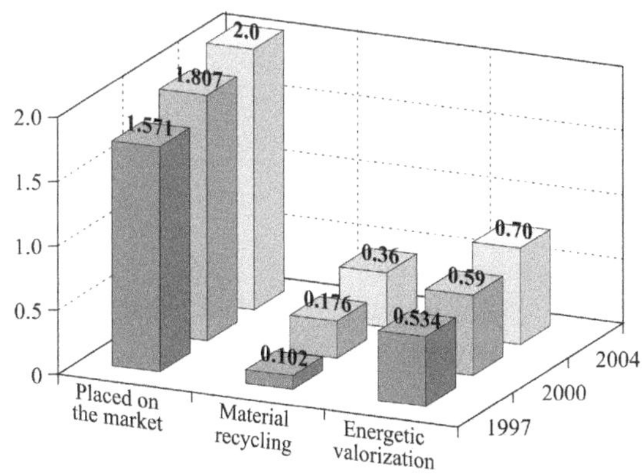

Figure 2.3. *Evolution of how much of plastic packaging waste was placed on the market, how much material was recycled and how much was energetically valorized between 1997 and 2004*

2.2.4. *Specific difficulties with the physical recycling of plastics*

The specific difficulties with the physical recycling of plastics permit us to explain a large part of the observed delay. They are of two kinds.

On the one hand, there are a large number of plastics. There are 20 monomers commonly used with which, by varying the lengths of chain and their distribution as well as by adding various additives to the created polymers, we reach several thousand grades. What is a great asset in adapting the nature of the plastic to its intended use becomes a handicap at the end of the life of the objects.

This wide variety of plastics creates problems, because for thermodynamic reasons inherent to the structure, they are almost always incompatible [BEL 05][4]. Mixtures which are made accidentally are heterogeneous and the bondings between the different domains weak, which results at the macroscopic level in mechanical properties which are much lower than those of the plastics which contribute to the mixtures.

This situation leads us in general to try to separate, to the best of our ability, the diverse polymers, which implies sorting and costly laborious cleanings.

On the other hand, the polymers can degrade themselves during the process of shaping and when in use [CHE 81, FRA 05, LEM 86, LEM 96, VER 84]. Actions of heat, mechanical treatments and UV are particularly significant. Moreover, the adjuvants are able to migrate or to be consumed over time.

For all these reasons, plastics obtained from objects at the end of their life may have reduced properties. We can often compensate for these deficiencies [THE 91], but it requires actions and additional investments which may affect the effective cost of recycling the material.

2.2.5. *The recycling chain*

Despite all the obstacles mentioned above, at least some plastics are actually recycled. In France, with regard to garbage, Eco-Emballage has decided to recycle bottles and plastic vials. The relevant materials are PET and HDPE.

---

4 *Nouveaux polymères, nouvelles combinaisons, nouvelles propriétés*, SFIP-12/1985-Bookstore Lavoisier, www.reduiremesdechets.com/pdf/plastique(9).pdf. *Transition ductile-fragile des mélanges incompatibles à faibles concentrations à base de propylène recyclé*, Sabin-Chiarilli V., Pabiot J., www.sciencedirect.com, http://irap-pari.nrc-cnrc.gc.ca/success/plassep_f.html. *Recyclage et valorisation des déchets plastiques*, http://cubitus.senat.fr/rap/098-415/098-41524.html, http://www.recywall.be/francais/technologies/design/apresvie.htm.

The recycling of material is performed on industrial waste, which is generally well segregated, or on household garbage, which first supposes a selective collection conducted door to door or by voluntary contribution or from a recycling center. From this collection, a first sorting, often manual, results in the making of balls of sorted material which meet the criteria of well-established purity (PTM)[5]. These balls are then resorted and processed in recuperator-regenerator factories. The sorting, grinding, washing and regeneration are complex, and we may wonder whether it would be possible to do without any of them. Therefore, the idea of using plastics in a mixture, possibly with other materials such as paper, regularly returns to fashion. We can compensate for the low mechanical properties of such mixtures by increasing the thickness of the pieces, but it should be limited to specific applications for the replacement of wood or concrete. Studies on the analysis of the life cycle (ALC) have shown that environmental assessments are rather poor.

A recent ALC (2008) on the "bricks" used for packaging milk, soups or fruit juices made headlines [GUI 08a]. These bricks are made in a composite multilayer material comprising cardboard, sometimes waxed, an aluminum film providing the role of barrier to the oxygen, and a polyethylene film to ensure food contact. These intimately linked layers are difficult to separate and their recycling seems impossible. Despite this, and after taking into account the various environmental impacts, the ALC showed that the bricks were the best packaging solution for such contents.

However, a decade ago, a manufacturer of chipboard panels demonstrated the possibility of using these "bricks" to produce panels respecting the rules of the Scientific and Technical Centre for Building for fire resistance and with a rather pleasant appearance. Hence, good recycling solutions without splitting up the components are sometimes possible!

However, in general, we should instead ensure that we use the most efficient method of sorting possible to allow a noblefuture use for the recycled materials.

Manual sorting can be efficiently followed by automated ranking utilizing identification by infrared radiation or visible radiation (to separate colors). Identification software controls robots (robots or air flow) which perform the selective removal of objects according to their nature.

Then, the operations of grinding, washing, flotation, in order to ensure separation between plastics and pollutants (or, in some installations, between polymers of different densities) and drying are usually succeeded by a "regeneration" step by

---

5 Valorplast, minimum technical requirements for plastic bottles, extracted from agreements on Eco-packaging 2005–2010 and Adelphe.

passing the materials through an extruder which provides the filtering (using removable grids with frequent replacement) of the last remaining impurities [March 98] in conjunction with the manufacture of rings which, crushed after cooling, give rise to reusable granules used for various production purposes.

The passing in the extruder can also be an opportunity to add diverse adjuvants which allow the restoration of the desired properties of the materials.

The use of recycled material, alone or mixed with virgin material to make new pieces, constitutes the real recycling stage, the previous stages being the preparation.

2.2.6. *Physical recycling in solution*

A less used method than the one described in the previous section is to handle polymer waste by dissolving it selectively in a suitable solvent. The best-known process which uses this technology is the Vinyloop® [SCH 01] for PVC when the latter is intimately linked to other materials (textile-coated tarpaulins, electrical insulation, flooring, etc.). Grinding and pre-sorting is carried out prior to the dissolution of PVC and most of its additives in methyl ethyl ketone. The undissolved foreign materials are eliminated by centrifugation. An additive which causes the precipitation of the PVC is then added and the solvent is regenerated. The use of a solvent is rather disadvantageous and requires a recovery rate of about 99.9%. The residual water is polluted (Figure 2.4).

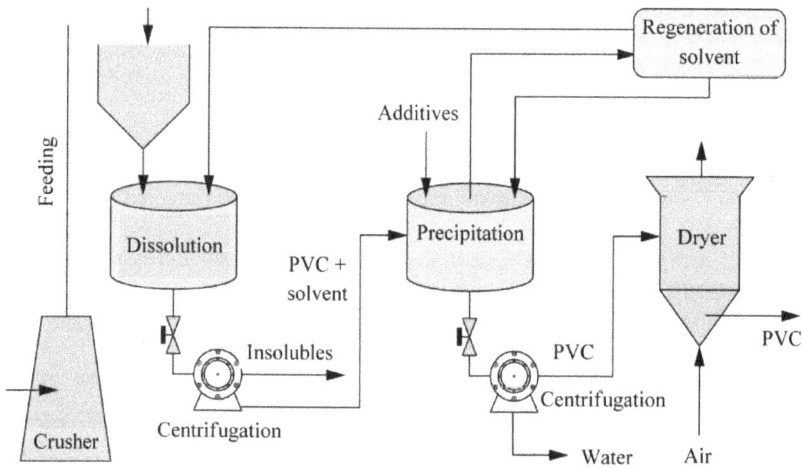

Figure 2.4. *Diagram of the vinyloop® process*

2.2.7. *The use of recycled materials*

The uses of recycled material can be divided into two broad categories.

Almost all applications accept a recycled rate of between 10 and 25%. It should still be verified that the rules are being observed, particularly in the case of food packaging, and that the parts match the specifications well.

Higher rates of 50–100% are reserved for specific applications, among which we can mention: garbage bags, bins, boxes, chips, agricultural films, floor tiles and floor security products (anti-slip) and the fibers used for textile padding.

2.2.8. *Chemical recycling*

This can be achieved in different ways, by degradative extrusion [MIC 02], by solvolysis, or by pyrolysis. We thereby obtain appropriate gases which can be burned or used for new syntheses, or used as oligomers and monomers.

A form of pyrolysis utilized industrially concerns PMMA (Figure 2.5). PMMA waste, which is generally clean and which does not leave the industrial circuits, is collected and thrown into a bath of molten metal. PMMA decomposes itself by restoring 97.2% of monomers, which can be used for the polymerization of new plates of PMMA. The method works less well if the waste is too dirty.

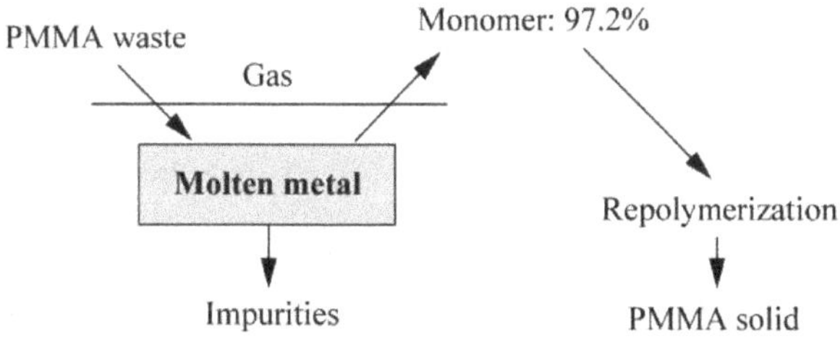

Figure 2.5. *Pyrolysis of PMMA*

However, other polymers, such as polyolefins for example, do not have the same monomer yield (about 50%), and pyrolysis gives rise to too many compounds, which are difficult to separate.

Solvolysis [CAU 02, SER 02, TER 02] is applied to the polymers obtained by polycondensation. Polycondensation consists of reactions which are reversible. By applying well-chosen conditions (and ones contrary to those of the polymerization) starting monomers or related compounds can be obtained. Numerous tests on several polymers have been conducted (Figure 2.6).

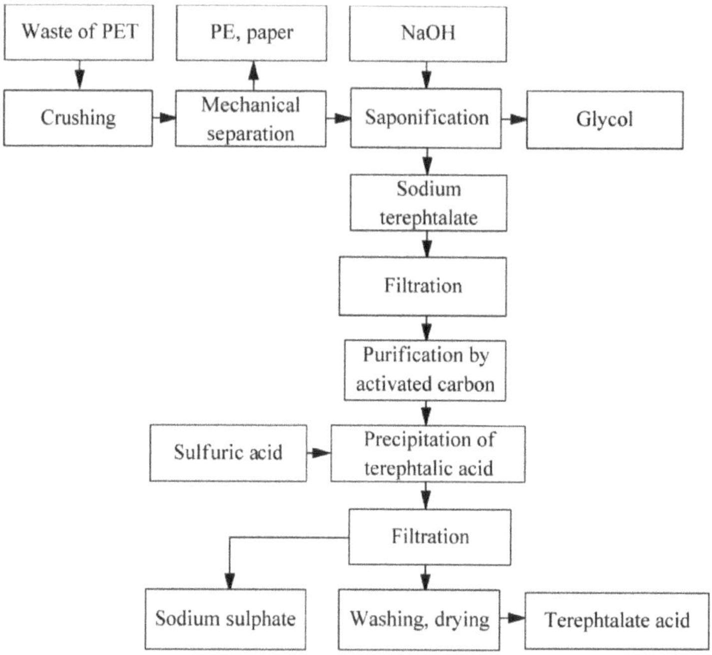

Figure 2.6. *Example of solvolysis*

Technical aspects seem close to being solved, but the economic viability is questionable.

Except in special cases, chemical recycling involves the input of clean waste which has been sorted and well characterized. To make the process profitable, we need to work with large quantities, which are difficult to handle. As a result, chemical recycling currently appears to be rather a marginal sector.

2.2.9. *Energetic valorization*

Plastics are combustible. Waste made up of plastic parts may be used by certain industries seeking inexpensive combustibles.

Cement factories burned such waste, but plastics are rivaled in this application by other waste such as animal meal (which is better subsidized and has constant properties). Moreover, despite the denials of cement manufacturers, we cannot exclude the release of toxic substances [SAR 05].

Various boilers can use plastic waste as fuel. However, the requirements on the regularity of the particle size and on the calorific value limit their applications.

In blast furnaces, a proportion of coal can be replaced in its role by plastic waste as a reducing agent. In this application, it is desirable that the percentage of PVC is low. In this case as well, the release of toxic substances can be feared, but perhaps no more than in usual operation!

Plastic waste and especially garbage are generally incinerated within dedicated facilities (waste incineration plants). These installations include a furnace inside which the incineration occurs and a boiler which helps in the production of steam. This steam allows us to rotate an alternator, thereby producing electricity. The steam is then, eventually, used to heat nearby buildings. The performance of the facility rarely exceeds 25% when we only recover electricity, and 60% when the steam is recovered for heating. These yields must be applied to the chemical energy of the burned material. In the case of plastics, this share is only a third of the energy stored within objects. The overall yield of the valorization of plastics is therefore around 20%.

Moreover, if the incineration process reduces the volume of waste by 75%, it gives birth to residues: steel which is recovered, clinker whose harmlessness must be checked before use in public works and waste which must be prevented from being incinerated (REFIOMS) and instead must be stored in landfill for hazardous waste. A flue treatment is required. It must reduce the levels of dust, HCl, SO_2, volatile metals and dioxins. The safety checks required for the installation (necessary twice a year) appear to be slight, given the potential risks.

The example of Sydom of Jura (in 2003, but the numbers have changed little since) shows that plastic waste accounted for 11% of garbage waste. Twenty percent is recycled by physical processes and 60% is incinerated. The rest went to landfills. In 2010, 18% of waste was recycled and 40% energetically valorized in France.

2.2.10. *Landfilling*

Landfill appears to be a non-valorization of waste. Yet in France it accounts for almost half of the waste.

Mixing plastics with other waste is undoubtedly bad for the entropy balance. However, buried and protected from light, plastics remain intact within landfills. "Destocking" operations of landfills were performed repeatedly (landfill mining)

[LAC 08], but the companies who performed these operations essentially simply recovered aluminum.

We can conceive of the landfilling of plastics only. This dedicated storage would permit, probably in a few decades, depending on the rarity and the price of oil, the exploitation of plastic waste which cannot be currently treated in an economically viable way.

2.3. Conclusion

The impact of waste on the environment is not negligible, but it does not, by a long way, hold the first place. The share of plastics in waste is about 11%. The part of plastics in oil consumption is 4%. Waste treatments are not the same from a thermodynamic point of view. Waste prevention is a satisfactory concept but has limited application. Plastics are especially landfilled in mixture with other garbage or incinerated. They are not well recycled (7%) despite a rapid and positive evolution in the rate of recycling in the years 1997–2007. Recycling therefore remains an important activity to provide in the coming years.

2.4. Bibliography

[BEL 05] BELLE S., DAVIS B., JAWAID A., *et al.*, Rapport final sur la conception de produits recyclables, www.nrcan-rncan.gc.ca/mms-smm/busi-indu/rad/pdf/2005-42(cf)cc-fra.pdf, 2005.

[CAU 02] CAUDRON J.C., Etat mondial du recyclage chimique des MP, Filières et recyclage, ADEME, 2002.

[CHE 81] CHEVALIER J.L., BALME A., SALLEE H., "Durabilité des matériaux plastiques transparents utilisés en couverture de capteurs solaires plans", *Colloque Cl*, supplément au no. 1, tome 42, p. Cl-401, *Journal de physique*, January 1981.

[FRA 05] FRAÏSSE F., Etude du recyclage de mélanges PET/PC en vue de définir un procédé respectueux de l'environnement pour l'utilisation de profilés dans le bâtiment, Thesis, Blaise Pascal University, Clermont-Ferrand, 15 September 2005.

[FRI 98] FRITSCH A., "Emballage et environnement se marient dans la norme", *Industries et techniques*, no. 799, December 1998.

[GUI 08a] GUICHARDAZ O., "Tetra Pak, une ACV, une polémique et l'inaction", *Environnement et technique*, no. 277, June 2008.

[GUI 08b] GUICHARDAZ O., "Prévention des déchets: les limites d'un concept", *Environnement et technique*, no. 281, November 2008.

[LAC 08] LAC C., "Faisabilité du déstockage d'anciennes décharges", INDDIGO, *Déchets et écologie industrielle*, TRIVALO, 2008.

[LAV 97] LAVABRE S., "Emballage industriel: le plastique en "pool" position", *Caoutchouc et plastiques*, no. 759, May 1997.

[LEM 86] LEMAIRE J., ARNAUD R., GARDETTE J.L., *et al.*, "Zuverlässigkeit der methode der photo-schnellalterung bei polymeren. (Reliability of the accelerated photoageing method)", *Kunststoffe, German Plastics (Int Ed.)*, vol. 76, pp. 149–153, 1986.

[LEM 96] LEMAIRE J., "Predicting polymer durability", *Chemtech,* 42–47, October 1996.

[LES 91] LESEUR B., PABIOT J., PIETRASANTA Y., *et al.*, Ecologie, économie des polymères, Ecole des Mines de Douai, 17–18 October 1991.

[LIG 09] LIGHT A.L., Green-pallets nest in Linpac product range, PRW.com du 6/02/2009.

[MAR 98] MARRE S., Le passage en extrudeuse permet une amélioration de 10% des propriétés mécaniques de contrainte au seuil et d'allongement à la rupture, CNAM plastiques-Paris, 1998.

[MIC 02] MICHEL A., Extrusion réactive: les avantages du recyclage mécanique et chimique réunis, Filières et recyclage, ADEME, 2002.

[SAR 05] SARAZIN M., Les dangers de la combustion de déchets et de l'adjonction de mâchefers d'incinération d'ordures ménagères, dans les cimenteries, APPEL, March 2005.

[SCH 01] SCHUT J.H., New ways to salvage plastic waste, http://www.ptonline.com, Vinyloop, www.solvay.com, 2001.

[SER 02] SERVOL M., Glycolyse du PET: de nouveaux produits pour l'isolation, Filières et recyclage, ADEME, 2002.

[TER 02] TERSAC G., Les composites thermodurcissables enfin recyclés? Filières et recyclage, ADEME, 2002.

[VER 84] VERDU J., *Vieillissement des plastiques*, AFNOR, 1984.

3

Polymers and Marine Litter

3.1. Introduction

Litter quantities in the ocean exceed way beyond a billion tonnes: an endless inventory, consisting of all types of materials with shapes of all kinds, e.g. plastics, glass, metal or ceramics. It is generally considered that the floating litter, stranded or submerged, is solid and visible to the naked eye. The recent demonstration of the existence of microparticles at sea, mainly microplastics, sometimes not visible, required a more appropriate characterization. Of less than 5 mm in size, these microplastic items (Figure 3.1) are mainly originated from the abrasion and degradation of larger debris, triggered by light and oxygen. The scientific community and the international authorities recently characterized litter in continental marine aquatic environments as any material or fabricated objects discarded or abandoned in aquatic environments directly or indirectly, intentionally or unintentionally [CHE 09, GAL 10]. This definition excludes unprocessed elements of natural origin (vegetation, algae, various organic debris, etc.). It also does not include semi-solids, such as paraffins and chemical products, which are sometimes stranded or floating (tar balls, for example).

The evolution of waste production is directly linked with that of the population, industry, consumer products, lifestyles and eating habits. Our models of socio-economic development generate an increasing production of consumer goods of all kinds facilitated by the petrochemical capacity to offer a more diverse range of materials. What is interesting about polymers, their strength and lightness, is unfortunately a problem for the environment. The current economic policy, based on a lifecycle for products, which is becoming shorter and shorter, leads to the development of objects with a single use and which are disposable. One sees paradoxically, there is a mingling of efforts to reduce waste at source, and, at the

---

Chapter written by François GALGANI.

same time, an indirect incentive to increase the production of waste by encouraging consumption.

Figure 3.1. *Diversity of microparticles at sea. Sample collected in the North West Mediterranean. (Courtesy of Credit F Galgani, J.H. Hecq, 2010)*

At sea, the nature of the waste varies greatly depending on whether we consider the beaches, the surface or the seabed, and according to the activities. As a result, managers must consider the specific types of waste characteristics of an area, fishing or tourism, for example and the specific types of activity, industrial, agricultural, urban or harbor. Thus, for a global definition, the diversity of activities, places and future plans requires the consideration of local particularities.

For polymers, the impact is mainly known at the individual level and not at the population species or ecosystem level. The distribution of plastics at sea, however, remains worldwide, and the potential risk for alterations in communities by transporting invasive species is a reality which can modifiy the balance of ecosystems. The diversity and the complexity of the problem require a thorough knowledge of the nature of litter at sea, transport mechanisms, future possible impacts, and effects on the environment.

3.2. The cycle of litter at sea

3.2.1. *Methods*

The quantification of marine litter is complicated by the size of the oceans and surfaces or volumes to be sampled, seasonal variations in intake and the presence of currents. Observation, trawling, direct counting and filtration, overflight, video imagery, aerial photography, the use of submersibles and scuba diving are the main possible methods and approaches to quantify litter at sea, each corresponding to a site, a region or to a particular biotope.

On beaches, the main advantage is, in general, easy access for sampling. This is the reason why the largest number of scientific studies concern this compartment of the marine environment. Sampling concerns surfaces or the coastline, with samples of up to 5 km. In monitoring programs, the categories can be very detailed, up to more than 120 types of litter in the case of the Protocol of the Convention of the states of the North East Atlantic [OSP 09].

At sea, the implementation of the counting method is related to weather factors, and requires substantial resources. Direct counting on the surface is suitable for small areas, and in some work, on the scale of a bay, litter was even counted individually to better understand its transport. The use of video imagery is new and facilitates, with embedded systems, the capture of millions of images, but requires images to be sorted before analysis. Current computer technology has enabled the development of "ferry boxes" for the collection of data from large ships on regular routes or when cruising. The recent development of autonomous vehicles, real underwater drones, is expected to soon provide tools to meet the needs of automation.

For the seafloor, the equipment required is heavier. Trawling, TV video operations and submersibles for greater depths are complementary to deep scuba diving in shallow waters. These protocols, which are sometimes very sophisticated, currently use imaging recognition or laser *in situ* measurements in order to quantify the litter on the seabed but also that buried or hidden by vegetation. The development of sounders, sonar and other acoustic systems, as well as the implementation of satellites for the detection of very large objects, is limited to the evaluation of large items without being able to distinguish whether it consists of polymers or not. Numerical modeling completes methodological approaches, identifying the probable areas of accumulation and the future possible impact at sea, especially the elements due to cross-border transport or dispersion of microplastics.

For these microplastics, most of the approaches are based on sampling with a plankton net for microparticle sizes of a few hundred microns. Smaller samples were observed up to 1.5 μ, but required large volumes of filtration, possibly on board, and a formal characterization of the nature of the polymers. Biological sampling is possible because the polymer particles are ingested by some species such as birds, including fulmars (Fulmarus glacialis), and filter feeders such as mussels or some scavengers or planktonic crustaceans [BRO 08, COL 13]. These methods are subject to the same monitoring programs [FRA 11].

The characterization of polymers often requires the separation of the other components, including organic components. It is based on the use of density gradients or organic solvents. Quantification using imaging is adapted for large objects, but there is a problem in the case of microparticles. In this case, the use of infrared spectroscopy (in transmission or reflection mode) which compares the spectra of samples to the spectra of a reference is the most appropriate method to characterize polypropylenes, polyethylenes and polyesters. Raman spectroscopy is sometimes used, as well as calorimetry, for the characterization of the fumes after combustion.

3.2.2. *Nature and quantity of litter reaching the sea*

On land, the fate of waste is incineration, recycling, landfill or being abandoned in the wild. The contents of our garbage bins are made up of around 30% of putrescible waste, of paper and cardboard (20%), and about 10% of glass, 10% of plastic, and 10% of textiles, and the rest consists of metals and unclassified composites. The share of packaging has declined slightly in recent years owing to recycling programs.

In all, with waste coming from communities and businesses, 850 million tonnes of waste are generated each year in France. Incineration, the solution preferred by some countries, allows the recovery of energy, but nevertheless generates a production of CO_2 greater than that of power plants. Risks relating to public health also exist, even if the operation of the plant has improved in recent years. The alternative to recycling is interesting because of the reduction in the volume of waste and the conservation of natural resources. In this sense, it contributes to the reduction of inputs but some materials such as elastomers, which are very commonly used, are however still not recyclable and are a real environmental problem.

Landfills, especially marine sites or those near the coast, sometimes pose serious environmental problems because they can give rise to deferred impacts. In the open air, storms can dump thousands of plastic bags close to the sites, sometimes into the sea. These contributions, however, remain localized, in contrast to those due to the abandonment of waste which is mainly related to human behavior. The latter constitute a significant part but one which is not quantified in the statistics. Experts speak of around 10% of waste that would eventually finish up in the sea. Discharges from ships and rivers constitute a significant part.

Emerging in the 1950s and revolutionary at the time, polymers represent the largest proportion of marine litter which can constitute up to 100% in the case of floating debris. Their diversity and their properties make the understanding of the evolution of this litter in the sea more complex [THO 09]. The typological analysis of waste produced in recent years in European regions shows the widespread importance of plastics [GAL 11b]. In the North Sea, followed by the OSPAR program, and in the Mediterranean, under the International Clean up Campaign (ICC) project, synthetic materials represent 75 and 83% of litter, respectively. Wooden objects represent 6.0 and 8.3%, respectively. In the North Sea, the other litter is mainly glass (8.3%) and paper or paperboard (5.4%), while in the Mediterranean, fabrics (7.8%, sometimes synthetic) form the core of this other litter. Other categories in both areas represent less than 3% of litter.

Degradation products from macrolitter have different properties at sea, different effects and different futures from the original products. Eventually, they form microparticles, usually made of plastic, which constitute a very important aspect of the problem of marine litter. They comprise an assembly of objects of varying size, shape, color, density and chemical composition [HID 12]. With densities ranging from 0.8 to 1.4 g cm^{-3}, the plastics in the sea and on the surface are mainly polyethylene (density 0.92–0.97 g cm^{-3}), polypropylenes (0.85–0.94 g cm^{-3}) and polystyrenes (from less than 0.05 in the case of foamed polystyrene to 1.00 g cm^{-3}) [LES 11]. The denser plastics such as poly (vinyl chloride) (PVC) and polycarbonates, if they do not aggregate with the organic material, tend to sink.

The biological compartment plays an important role in the cycle of plastics as the latter can be ingested passively or actively, some organisms mistaking them for their prey. More than 660 species, from plankton to whales, are known to swallow litter, especially microplastics, further complicating the problem because of their possible interactions at the trophic webs. The cycle of litter of large size is somewhat different. On the sub-surface or by sinking to the bottom, they are eaten by some large organisms. The analyis of data in France, performed during the initial evaluation of the process used for the strategy framework of the marine environment, showed that turtles are particularly sensitive, while ingestion rates by large cetaceans are limited with less than 1% of organisms being affected.

3.2.3. *Sources*

Identifying the sources of most polymers is difficult because the point of entry is often diffuse, their fate being determined by the currents and their tracking being complex. Plastics are transported directly or indirectly to the sea by rivers and streams, by treated water, or during extreme events such as floods. The land-based litter includes widespread sources such as runoff, but also those related to economic activities such as tourism, industry and urban and port activities. According to the French Ministry of the Environment, beach users produce on average one liter of waste per person per day. Some are abandoned and quickly find themselves in the sand or in the sea.

Oceanic sources include ships, boating and fishing, offshore installations, and, locally the aquaculture facilities. Port activity also generates large amounts of waste of all kinds. They come from losses during cargo handling at the docks, as well as the maintenance of boats, but also the abandonment of household garbage. Fishing and shellfish farming generate characteristic waste (synthetic ropes, nylon line, buoys, nets, polystyrene, cans, etc.), which often end up on beaches.

Accidental intake, whether or not linked to natural phenomena, can be a significant source of litter (http://www.cedre.fr/fr/rejet/macro/provenance.php [DEC 00]. The tsunami in Japan in 2011 is the most representative example and the most recent. Around this event, approximately 25 million tonnes of litter arrived at the sea, and it is estimated that 2.5 million tonnes of litter, mainly plastic, drifted out on the sea surface. These events are, in reality, frequent (the earthquake in Haiti in 2010, Hurricane Katrina, with 75 million cubic meters of waste, etc.) and can be linked to the negligence of human beings, especially in the case of the collapse of coastal landfill sites.

Some litter can be linked to a single source which permits being able to consider prevention measures and waste collection prior to its arrival at sea. The additional quantities of waste in the municipalities located near the beaches of the Mediterranean are estimated to be 10–15%, and the total volume of waste increases by 75% in summer. At a more local scale, plastic pellets lost in production and during transport or distribution become litter before even having been used by the plastics industry.

The review of litter ingested by marine organisms has shown the importance of waste from household consumption. In the case of the North Sea, 90% of debris comes from the plastic of packaging, bags or bottles, associated with practices of consumption. Industrial waste represents only 10% of inputs and consists mainly of granules. Data analysis on the seabed also shows the importance of different hydrodynamic, geomorphological and human factors on the distribution and contribution of litter. Rivers, according to their flow, modulate these contributions by carrying off most of the litter, sometimes hundreds of kilometers. For small coastal rivers, flows are too low to allow this transport, but promote accumulation on the adjacent shores.

Recent work has characterized the microparticles and identified their sources other than those from the degradation of materials. For example, textile fibers, sometimes coming directly from the water of washing machines [BRO 11], are clearly identified as a source constituting a significant contribution to the sea. UNESCO also assesses that purification sludge stations are also an important source of microplastics.

3.2.4. *Fate and distribution*

Various studies show that transport by the sea sometimes occurs over long distances, and observations have noted its presence as far as mid-distant ocean areas, in Antarctic waters and those beyond the Arctic Circle, on the surface and on the seabed [BAR 05]. This oceanic litter is of any kind, but concerns principally plastics. Currents widely modulate its distribution. The global dispersion of these floating plastics on all the world's oceans and its large quantity, beyond hundreds of billions of objects in each basin, raises the question of its possible impact on the balance of the ecosystems. Imagine the influx of dozens of species attached to floating plastic and acclimatizing to an area by disrupting interactions between species, especially the organization of trophic structures. This risk is real, and the influx of 54 new species enumerated in Canada in 2012 on a single piece of floating plastic dock torn off by the tsunami a few months earlier in Japan is the best example. Furthermore, what does one think of nanoparticles whose presence at sea is more likely? The degradation time of plastics and its transportation over thousands of kilometers suggest the possibility of surprising discoveries in this field

Further offshore and deeper, the presence of zones of accumulation of litter in the trenches has been demonstrated, but the phenomenon is still poorly described (Figure 3.2).

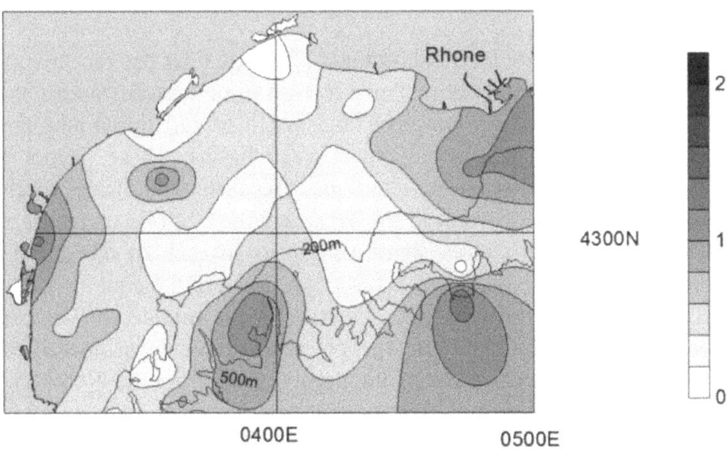

Figure 3.2. *The distribution of plastics on the continental shelf of the Gulf of Lion. Average value (1994–2009) by density (number per hectare). Areas with high concentrations are often located beyond the contours of 200 or 500 meters of depth, where the litter is no longer under the influence of currents and can deposit. (Courtesy of F. Galgani/Ifremer)*

Even if they do not represent spectacular landfills, the accumulation of several tonnes of litter has been demonstrated in some underwater coastal areas, sometimes tens of miles offshore. These are mainly heavy litter, metal, glass, dense plastic or packaging made of PVC and polycarbonate. High densities were found in very deep trenches and very distant from the coasts, in the Hausgarten area off Greenland, 79° N [GAL 10]. Close to the coast and in deep water, the phenomena of "cascades" were discovered, carrying in particular tens of thousands of tonnes of sediments because of the bedload of the major rivers, and carrying litter, including the heaviest, toward the bottom. The deep currents are poorly known in ocean basins and transport predictions are still debatable; it is still one of the challenges of current research.

Examples of cross-border transport are also very numerous. From North America to Europe, *via* the Gulf Stream, from the Channel toward the North Sea and from the North Sea to the Kattlegat at the entrance of the Baltic Sea, beaches accumulate plastics. In the Mediterranean, plastic floats from the coast of France to the south to Spain, Morocco, Algeria, Tunisia and Italy or from the coast of Egypt to the Aegean Sea through Israel, Lebanon, Syria and Turkey. The problem is international and can only be resolved within international institutions.

3.2.5. *Oceanic convergence zone*

Reported since the 1990s, the floating mass of plastic litter has recently made the news: "something" that extends from Japan to Hawaii or stretches across the entire North Pacific, the great garbage patch or large patch of garbage, a new continent. "Continents" of plastic also exist in other ocean basins. These phenomena are related to the dynamics of geostrophic currents. Because of the trade winds at the Equator, the accumulation of water in the West and the Coriolis force, the induced flow to the north or south causes significant shifts in each of the ocean basins. Returning to the East, these bodies of water come down to the equator and form an oceanic whirpool which functions like a vortex, in the middle of which floating objects are accumulated. Five convergence zones have been identified, each corresponding to the five core areas of the ocean basins from the North and South Atlantic, from the North and South Pacific and from the Indian Ocean, only in the southern hemisphere.

The collective psyche is very sensitive to this information but it is greatly exaggerated, and the actual amounts in the gyres represent a few thousand tonnes, or tens of thousands by counting macroplastics and microplastics. In the Atlantic convergence zone, the best known from a scientific point of view, the figures show only 1,100 tons of microplastics [LAW 09]. This region represents an area equal to that of Europe, centered on the Sargasso Sea, of concentrated floating algae, like debris, in this zone and having given their name to this "ocean sea". By counting microplastics and macroplastics, densities can sometimes be significant, of the order of several kilograms per square kilometer, but most often associated with large-sized objects, and not with microplastics whose weight is negligible. Their number remains considerable and if in some coastal areas it is true that the densities increase for many years; this is not the case for the microplastics in these oceanic convergence zones. In 30 years, the quantities have changed very little in the North Atlantic gyre. It is still too much and the consequences of the dispersion of these plastics and their subservient species or whatever is fixed above remain a concern. Nonetheless, the fact remains that the problems at the coast can be of more consequence, because the quantities of plastics are sometimes more significant than in the convergence zones [COL 11].

Recent years have seen the development of prediction tools for the transport layers of plastic litter. These tools provide the facility to follow the trajectories of clusters that may be washed ashore. Eventually, the development of an atlas of risk sensitivity will facilitate the improvment of the management of accidental risks. However, simulations performed globally over several years help in the identification of the most sensitive zones at the global scale, such as the Mediterranean, the Atlantic and Pacific convergence areas, and the Bay of Bengal. These areas on their own should receive within 30 years between 35 and 60% of all litter at sea [LEB 11].

3.3. The degradation of litter at sea

The degradation time of polymers in water is subject to external factors such as:

1) brightness which affects the photolysis of certain materials;

2) the presence of oxygen which promotes aerobic bacterial degradation;

3) the possibility of abrasion, especially on the beaches.

At least a few years for thin plastics, durations are highly variable and can reach a persistence estimated to several hundred years. Work in the laboratory or experiments on the surface or in the open air have been carried out and the true values may be somewhat different in the sea. Studies notably show a significant increase in the degradation time at depth, where oxygen concentrations are low and where the light is absent. The problem of the degradation of litter into microparticles and the known environmental consequences may be related to a process which in return accelerates the degradation due to the increase in surfaces. These fragments have a much larger surface area relative to the volume and therefore a greater potential for the transport and release of the contaminant.

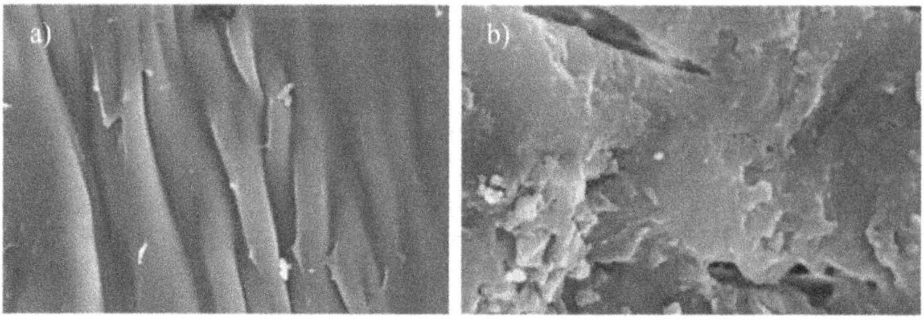

Figure 3.3. *Electronic microscopy (×5,000 magnification) of a fresh surface a) and degraded b) industrial pellets of polyethylene on the beaches. (Courtesy of [FOT 12])*

A phone card or ski pass can take a thousand years to degrade, beyond the fishing nylon thread which will remain for only 600 years. Expanded polystyrene is also likely to persist for hundreds of years but the record for the length of degradation time belongs to glass and all its forms which are likely to persist for thousands of years to the point where it is considered to be an inert material. *A priori*, we would think that the degradation of litter at sea causes the disappearance of its impact. The problem is however much more complex because fragmentation makes effects possible on smaller marine organisms, likely to ingest them. The inclusion of degradable polymers in order to promote the degradation of packaging has emerged as a solution, a few years ago. In reality, only the organic constituents, most often polysaccharides such as starch, degrade rapidly and promote the formation of small size fragments and tatters which affect more marine organisms and those of the smaller size, down to plankton where they were identified in the digestive tract. Under these conditions and to be rewarding for the environment, degradation must involve all materials and not only a part of them.

It seems difficult not to imagine that the degradation of microplastics, coming from marine litter itself, does not lead to the existence of invisible particles made of polymers. However, are they likely to persist in the oceans? In the air and water, nanoparticles are often aggregated and can be consumed by filters. They have a size comparable to microorganisms and to algae of Pico plankton, the most important compartment of the plankton, which plays a key role in primary production. Interactions with these species are therefore crucial. The main known nanoparticles which are oxides or carbon nanotubes are toxic for algae, zooplankton, bivalves, fish embryos and adult fish. The extrapolation of particles coming from the degradation of polymers at sea is difficult, even if these particles can adsorb on their surface contaminants which can be released in amounts much larger than the microplastics. Their critical size makes them also prey which might be phagocytosed by certain cells and they can act as vectors of contamination.

3.4. The effect of marine litter on the environment

Over the last 15 years, the impacts observed at sea have increased by about 40% and the number of species affected went from 247 to 663. Half of them relate to catches in fishing gear or the ingestion of litter and are principally localized in the developed regions. However, effects have been described in areas considered as virgin like the Arctic or Antarctic. These effects are related to the nature of the litter and their shape. Fishing gear and synthetic materials such as nets, creels and abandoned traps are the litter that impact the wildlife most because they are designed for this purpose. Fishing lines, ropes, packaging and plastic bags are also an important cause of death or injury due principally to the risk of ingestion. For large marine animals, such as whales, pinnipeds and especially turtles, the ingestion of objects can lead to obstructions or digestive occlusions, the latter leading inevitably to death. With the effects of catches by fishing gear, they are the most severe impacts on species that breathe and which may drown when entangled in nets. Direct mortality can affect entire populations, affecting thousands of individuals, as in the case of seals caught incidentally in active or abandoned nets along the coast of Alaska.

Figure 3.4. *Leatherback turtle strangled by a fishing net. (Photo courtesy of RTMAE/CESTM)*

The proportion of polymers in these objects is important. Their use has been widespread in recent years for the manufacture of nets, ropes and lines. In France, 36% of stranded individuals of the species *Caretta caretta* contain litter in their stomachs of which 80% is plastic mistaken for jellyfish. In the estuary of the Rio de la Plata, a very urban area, 37% of catfish contain litter principally constituted of plastics. In the case of the fulmars of the North Sea, [FRA 11] evaluates the average weight of ingested microplastics to be 0.38 g per individual, with 92% of individuals being affected and with an average of 37 pieces per individual. This value is not significant in itself, but if extrapolated to a 70 kg human, this would represent around 600 grams per person. Undoubtedly, such an amount in humans would have led to stringent measures and regulations for many years. In the case of birds, intergenerational ingestion is even possible because of the role played by some adults in feeding their juveniles.

The Food and Agriculture Organization (FAO) and the organization of the United Nations for the Environment [UNE 09] mention the increased impacts due to increased fishing activities and the generalization of resistant synthetic materials. In general, their presence is not deliberate but attributed to losses due to bad weather or tangles on the seabed. In some fisheries, lost creels represent tens of thousands of traps and can fish for many years, including the capture of crabs. For example, the Norwegian authorities have collected about 500 abandoned nets per year since 1983, which has led to mandatory reporting of losses since 1989.

Habitat modification is an example of the alteration of the seabed. Impacts in this direction have been demonstrated, especially for the ecosystems of the coral reef. By the action of abandoned nets or the mere presence of debris from fishing, 70% of the biotopes within a region may be affected. Conversely, litter, especially intact packaging and nets or their fragments, is a potential habitat for many species. It promotes the colonization of new areas, sometimes separated by several thousand kilometers [BAR 02]. This result is particularly important for plant or animal species whose mode of reproduction, asexual or not, is heavily promoted by the presence of early stage plankton moving on potential supports. Measuring the extent of the phenomenon remains difficult. Among the hundreds of species present on the supports originated from litter, only five examples of truly invasive species have been described. Some terrestrial species have even been transported and now we have enough evidence to say that the colonization of the pelagic environment by benthic species is widely favored by litter.

Reused miscellaneous debris is a potential source of support and shelter, especially for fragile species such as spin lobsters. Similarly, the organisms living in the deep, like cuttlefish or squid, frequently lay their eggs on old submerged debris. Microparticles, from the decomposition of litter at sea, can be ingested by plankton or small fish. Species living on the surface are the most affected, but other species such as sardines feeding on plankton ingest it. Ingestion rates currently remain anecdotal. At the cellular level, the phagocytosis of particles by intestinal cells or the toxicological impact of chemical constituents associated with plastics have been demonstrated in the laboratory.

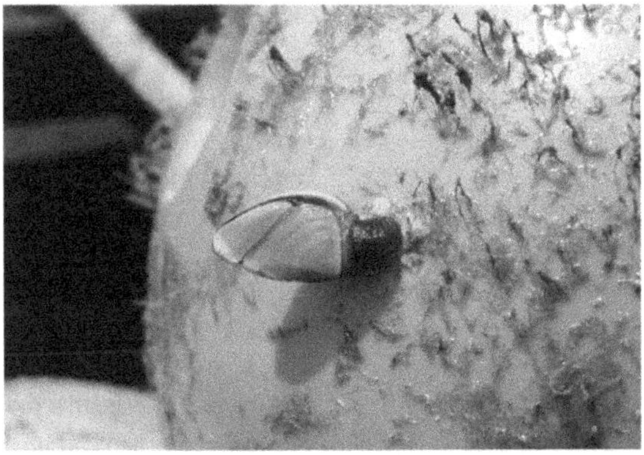

Figure 3.5. *Transport of cirripedes (pedunculate crustaceans) and algae on plastic buoys lost at sea. The presence of millions of billions of plastic fragments at sea multiplies the number of possible vectors for the spread of anchored species. (Courtesy of Credit F. Galgani [GAL 00])*

From a chemical point of view, and without taking into account accidental massive inflows, the most significant risks are associated with polymers and their properties. Between the specific constituents of these polymers (polyethylene, polypropylene, polycarbonates, polyvinyls, etc.) and the plasticizing agents which are principally phthalates, monomers which are sometimes associated like the Bisphenol A and chemical compounds adsorbed to the plastic, it can be said that many molecules pose a potential risk to marine organisms, even to human beings. These molecules can be dissolved following the deterioration of materials or simply detached from their support in the case of attached chemical contaminants, especially for persistent organic contaminants. This contaminant source is however not significant compared with traditional inputs from the sea, runoff, rivers or the atmosphere. Seawater contains molecules as diverse as insecticides, pesticides, hydrocarbons and other emerging contaminants. Some of these pollutants such as poly chloro biphenyls (PCBs), poly bromo diphenyl ethers, and the per-fluoro octanoic acids have solubility constants which favor their adsorption on polymers, sometimes exceeding 10,000 times, in particular on polyethylene. For example, concentrations greater than 10 μg/g of plastic and 0.7 μg/g of plastic have been found on the polymers harvested on beaches for oil and insecticides of the Dichlorodiphenyltrichloroethane (DDT) family. Furthermore, the degradation of plastics promotes the attachment of metals, particularly in the crevices of the altered surfaces. The toxicity of these compounds is known and described in the scientific literature. In the case of certain species of seabirds, the release of adsorbed chemicals on the litter results in an increase in their concentration in the blood. This problem does not yet concern whole populations of organisms but it has been proved by scientists, especially for the PCBs.

Whether it is grounded or floating, litter can affect human health. Broken glass, syringes and medical waste may be a cause of injury or even contamination. In the UK, up to 4% of injuries are related to litter on beaches. Choking swimmers and divers in fish nets have also been described, although they are uncommon, and most often concern invisible nets which are nylon monofilaments. The risk is also important when accidental and massive inflows are considered. Thousands of plastic bags containing pesticides stranded on the Channel beaches after the "Sherbro" accident in 1993 are significant for the incurred risks which are sometimes unrelated to the nature of the packagings but rather to the contents.

At the oceanographic scale, the most sensitive issue is the one concerning the transport of pathogenic species, especially microorganisms. The multiplication of floating supports of all sizes has recently increased the risk of dispersion. Various plastics can fix vibrios, pathogenic for farmed fish, but the risks to human health are still being assessed.

Studies on the economic impact of litter at sea are quite rare. The lack of information on this topic is due to a very recent concern with the issue of environmental degradation. Since the 2000s, national or international institutions began to lay the groundwork for an emerging public policy toward this new highly complex environmental issue. European regulation reinforces this trend with the Marine Strategy Framework Directive (MSFD) in 2008. Costs and losses are related to the maintenance of beaches and bodies of water (collection, transport, treatment), to underwater interventions in the bottom of ports, to awareness programs, to cleaning and maintenance of nets and to ship repairs. These costs represent each year millions of euros for some municipalities, especially when trying to prevent any impact on tourism. Macrolitter indeed disfigures seascapes and this pollution affects the aesthetics of landscapes and affects seaside tourism. The most bulky litter poses risks to navigation in case of collision or winding into a propeller as well as obstructions in cooling circuits. Port authorities estimate the number of incidents or accidents of navigation at several hundreds in Europe each year, representing costs of about 150–500 Euros for simple cleaning to thousands of euros for more complex operations (towing or cranes). Ninty-five percent of the fleet in some parts of Europe is affected by this problem [MOU 10].

3.5. Socio-economic aspects

3.5.1. *Legal aspects (laws, conventions and directives)*

The waste stream that reaches the marine environment represents only a small percentage of the quantity of waste produced annually. However, it is the continued increase in waste production and its persistence related to its slow degradation which leads to an increasing accumulation in the marine environment up to the point where it constitutes a real worldwide concern [GAL 11b].

Although it is difficult to account precisely for the extent of the negative impacts of the macrolitter, because we cannot estimate well the economic value of a beach and the costs of its environmental degradation, there is principally an aesthetic nuisance associated with a potential real deficit of image and therefore an impact on the number of tourists (Figure 3.6).

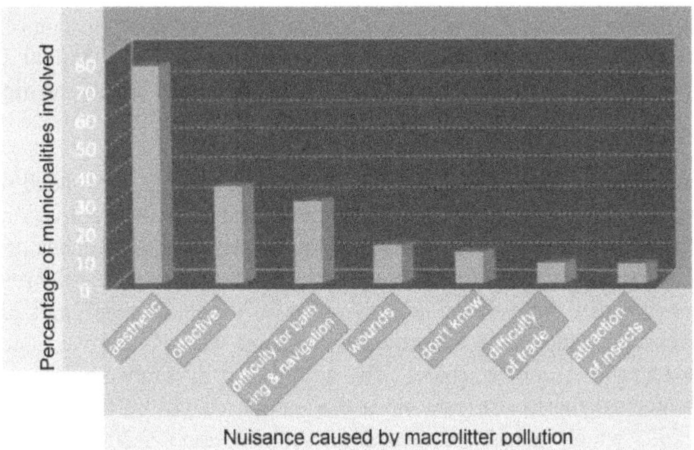

Figure 3.6. *Results of a survey of communities and municipalities on the*

The responsibility for the presence of litter is both individual and collective, and the corresponding regulation is part of the waste sector policy with a strong association with the water policy and nature conservation.

Deliberate or accidental infringement in these areas or of these protected species may be the object of sanctions (Article L. 218-73 of the Environmental Code). This regulation thus concerns the potential impact on marine ecosystems due to the accumulation of litter at sea but these provisions remain general to all forms of pollution and do not specifically apply to solid waste.

Under French law, the law of 15 July 1975, completed in 1992, is integrated into the Environmental Code Articles L 541 and following, and concerns mainly waste from land-based activities. It punishes the dumping of waste. In the Code of seaports, the article R 353-4 provides that non-compliance with locations used for waste disposal is also liable to a fine. Municipalities also have the obligation to organize and implement the collection and treatment of waste (article L 2224-13 of the Code of local authorities), the obligation to clean up the beaches and coastal areas frequented by the public (decree of 7 May 1974), and the mayors have a special police power over the strip of land 300 meters from the shore.

The article L 216-6 paragraph 3 (Water Law of 3 January 1992) makes it an offense to "throw or dump waste in large quantities in territorial marine waters, beaches or the shores of the sea". These provisions do not apply to dumping at sea from ships and penalties are two years imprisonment and a € 75,000 fine. The application of these provisions remains difficult because of the difficulties in assessing the too vague concept of "significant quantities".

International conventions in territorial waters and beyond are related mainly to releases from ships (pleasure, fishing, commerce). The MARPOL convention and its Annex V relating to litter at sea prohibited, for vessels over 200 tonnes or carrying more than 10 passengers, the rejection at sea of plastics and limited the rejection of other types of waste from ships. It also refers to "special areas", notably in the Mediterranean, in which only the rejection of food waste is allowed and only at more than 12 miles from the coasts. The application by France of the MARPOL convention is particularly stringent since the Act of March 9, 2004 and the Act of August 1, 2008 which include one year imprisonment and a € 200,000 fine, for any captain guilty of infringements of the provisions of rules 3, 4 and 5 of Annex V, relating among other things to bans on the discharges of waste.

Various international conventions (GA/RES/60/30 of the United Nations, Barcelona Convention, OSPAR Convention) supplement these legal provisions and publish guidelines for the management of litter at sea and also publish on the state of affairs (http://www.unep.org/regionalseas/marinelitter/ [TEN 09].

The Grenelle de l'environnement summit in France places macrolitter in an emerging theme. Launched in 2007, it strengthens certain provisions of the Water Framework Directive and the Marine Strategy supports. It sets out as a goal the implementation of a program of measures to achieve good ecological status for 2/3 of surface of water masses. The bill states that all measures will be implemented to strengthen the fight against illegal practices, reduction at source, and prevent marine pollution, including macrolitter and floating litter. Sixty-five recommendations developed by the operational "litter" committee were related to macrolitter. The Grenelle de la mer summit announced on the 27th February 2009, completes the commitments of the Grenelle de l'environnement and covers a wider field on the theme of the sea and the contribution to the development of sustainable activities at sea. The Grenelle de la mer summit should eventually lead to a framework law of the sea and a coherent and coordinated organization of the coastline.

The MSFD, established in June 2008, sets out for the first time, a comprehensive policy for the protection of the marine environment, focusing on ecosystems and habitats. It also introduces the basic concept of a good ecological status (GES) for the marine environment. Its objective approach, with the commitment of Member States to achieve or maintain GES 2020 is based on the measurement of a set of 11 descriptors. Its geographical extension involves the three million square kilometers of marine area under the jurisdiction of the European states and its approach at regional level will be part of an integrated management shared between the member states. Of the 11 criteria or descriptors of GES, descriptor No. 10 and four associated indicators relate to macrolitter (the properties and quantities of marine litter do not cause harm to the coastal and marine environment). The development and implementation of a monitoring program in 2014 will be followed by a program of measures in 2016 to achieve GES by 2020.

3.5.2. *Initiatives*

The importance of the subject and its relevance has recently led political institutions to take action in support of environmental protection.

The program of the United Nations for the Environment has initiated several operations of collection worldwide and set guidelines for the management of coastal litter for regional seas. The best recent example of a comprehensive solution is the Honolulu approval signed in 2011, which defines a set of clear objectives for the actions of governments and all stakeholders in the environment.

This commitment of 163 recommendations will facilitate the sharing of technical, legal and economic solutions to reduce litter pollution. It is a first step toward developing a worldwide platform for the prevention, reduction and management of marine debris called the "strategy of Honolulu". Europe is also very active in the fight against pollution caused by macrolitter through the setting up of guidelines and the support of research work. The European Commission also supports non-governmental organizations (*Surfrider foundation*, *Watch the Litter*, *Seas at Risk*, *Clean up the Ocean*, etc.).

At the national level, support is given to certain operations sponsoring awareness or collections by public institutions and decentralized services of the State. French research organizations work on the topic of litter and contribute to enhancing knowledge and participate in the implementation of quality monitoring [GAL 11a].

Local authorities also hold a key power to reduce marine pollution by litter. General and regional councils organize awareness or cleaning operations and fund certain organizations. Municipalities typically manage the problem of cleaning on their own or as part of municipal associations. Based on tourism issues, some municipalities do not just clean the beaches, they carry out aerial surveillance in the season in order to recover surface floating litter, and set up safety nets to protect their beaches.

Many associations for the protection of the environment or sports associations coordinate cleanup actions. Downstream, the improvement and the coordination of the collection, including manual or automated collection of litter on beaches or that aggregating in ports, by setting up floating booms involving managers of *Les voies Navigables de France* and *Electricité de France* as well as arranging the involvement of fishermen in the form of initiatives paying for the collection. Public awareness in order to encourage people to change their practices, streamlining the treatment to reduce losses or fluxes of waste, strengthening the regulation and administrative or criminal framework, the establishment of specific treatment processes especially for fishing, intervention on sensitive sites such as some beaches in summer, and the adaptation of treatment plants allowing the disposal of waste before discharging wastewater are important aspects of public policy.

From an environmental point of view, the management and taxation measures adopted since the 2000s in favor of limiting the use of bags provided when leaving the checkout, enabled the reduction of the amount of packaging bags in supermarkets by 90% in 10 years in France. The impact on the marine environment is difficult to assess, but lower quantities of litter were observed over the last 15 years in the only existing experimental monitoring network, situated on the seabed of the Gulf of Lion.

The plastics industry has only relatively recently taken into account the problem of plastic litter at sea, and includes the manufacturer of plastics, processors, recyclers and the waste industries. It is the result of a more global approach toward the protection of the environment and started in the 1980s (http://www.marinelittersolutions.com). Initiatives concerning prevention and waste reduction, recycling policy and the valorization of plastics in all applications (packaging, civil engineering, automotive, electronics, etc.) and the development of eco-design are now part of the industrial "process". Forty-seven players in the industry are committed within the framework of the *5th International Conference on Marine Litter* in 2011 through a "joint declaration for solutions to the problem of marine litter".

Industry, supported by the European Commission, has also committed with the help of the fishing community, the community which is the most affected by marine litter, to recover the floating debris and return it to land in order to sort and value it.

The current operation is an example of institutional and industrial support to the fisheries under a common strategy and this type of initiative is expected to grow.

3.5.3. *Understanding and educating*

The future requires reasonable practices, rationalization of production, the use of less polluting materials, better management and reasoned actions which are carried out according to rules of conduct and good practice. The problem of waste in general and litter at sea is unique in that it concerns everyone. Many associations for environmental protection or sports associations coordinate cleanup operations. Some participate in working groups and national think-tanks and organize educational activities. Some have recently started campaigns at sea for sampling litter by eco-volunteers. If it seems difficult today to imagine going back to a situation where there was no litter in the sea and on the beaches, it seems that everyone realizes the magnitude of the problem, on an aesthetic, environmental, economic and human health level. The development of new intelligent materials, new solutions for collecting waste, efficient collection and appropriate behaviors appear increasingly necessary. Marine litter often consists of objects of mass consumption, packaging and bottles, and its presence is most often related to individual behavior. It is important to limit the excesses, and elected officials, research and industry should consider sustainable solutions.

Densities on the bottom vary greatly but the figures are of limited interest when we do not know the sources and methods of transport. And what will they become? How fast will they degrade? Estimated often in the laboratory, degradation is poorly understood. It was announced that there will be an increased stability for polymers in future materials, without really knowing what might be the consequences on the future of polymers in the environment. When will there be an indicator for the aesthetic impact? How should one measure the social and economic impact of environmental degradation, as it begins to carry out for the accidental intake of oil?

When considering the overall impact, we realize how much consumers are heavily involved in the inputs, and it is an area where we do not really know how to proceed, this is the role of education and behaviors.

3.6. Conclusion

When considering polymers, the nature of the effects suggests an impact mainly at the individual level, sometimes at the population level, but less at the ecosystem level. The potential risk for alterations in communities by the transportation of invasive species is a reality. This is probably the chief risk posed by the presence of litter at sea.

We can easily imagine the future impact of research on novel polymers in today's society. The likely development that polymers can achieve in the coming years depends on the problem of improving knowledge. The challenges include nanoscale architecture of structures and the prediction of properties, with so to speak, "tailored" polymers that will meet the commitment to "sustainable development" which is required by today's society. Natural biodegradable polymers such as starch, cellulose, polylactic acid for packaging, as well as biomaterials or polymers from biological resources rather than petroleum resources should be promoted. Plant raw materials have particularly attractive properties such as biodegradability, biocompatibility, selective permeability or modifiable physical and mechanical properties, and this new interest of the plastic sector fits better within a perspective necessary for the respect for the environment. Recycling is also a source of savings. At sea, there is hope for original materials, having all the properties of lightness and strength necessary for consumers but rapidly degrading, without consequences for the natural environment. The improvement in practices in order to limit inputs, of materials whose degradation limits the number of microfragments, improved treatment of wastewater plant processes to limit losses, effective prevention of waste, streamlining of processes and collection, are all issues to explore. Anchoring forever in the minds of consumers behaviors concerned with the environment remains the main route but the most difficult one.

3.7. Acknowledgment

The author thanks all contributors, direct or indirect, and F. Claro for reviewing the document.

3.8. Bibliography

[BAR 02] BARNES D.K.A., "Invasions by marine life on plastic debris", *Nature*, vol. 416, pp. 808–809, 2002.

[BAR 09] BARNES D.K.A., GALGANI F., THOMPSON R.C., *et al.*, "Accumulation and fragmentation of plastic debris in global environments", *Philosophical Transactions of the Royal Society B*, vol. 364, no. 1526, pp. 1985–1998, 2009.

[BRO 08] BROWNE M.A., DISSANAYAKE A., GALLOWAY T.S., *et al.*, "Ingested microscopic plastic translocates to the circulatory system of the mussel, Mytilus edulis (L.)", *Environmental Science & Technology*, vol. 42, pp. 5026–5031, 2008.

[BRO 11] BROWNE M.A., CRUMP P., NIVEN S.J., *et al.*, "Accumulation of microplastic on shorelines worldwide: sources and sinks", *Environmental Science & Technology*, vol. 45, pp. 9175–9179, 2011.

[CED 00] CEDRE 2000, Etude des stratégies de réponse au problème des macrodéchets rejetés sur le littoral, Final Report, General Secretariat of the Sea, http://www.cedre.fr, 2000.

[CHE 09] CHESHIRE A.C., ADLER, E., BARBIÈRE, J., *et al.*, UNEP/IOC guidelines on survey and monitoring of marine litter, UNEP regional seas reports and studies, no. 186, IOC 41, 2009.

[COL 12] COLLIGNON A., HECQ J.H., GALGANI F., *et al.*, "Neustonic microplastic and zooplankton in the North Western Mediterranean Sea", *Mar. Pollut. Bull.*, vol. 64, pp. 861–864, http://dx.doi.org/10.1016/j.marpolbul.2012.01.011, 2012.

[COL 13] COLE M., LINDEQUE P., FILEMAN E., *et al.*, "Microplastics ingestion by zooplancton", *Environmental Science & Technolgy*, vol. 47, pp. 6646–6655, 2013.

[FOT 12] FOTOPOULOU K.N., KARAPANAGIOTI H.K., "Surface properties of beached plastic pellets", *Marine Environmental Research*, vol. 81, pp. 70–77, 2012.

[FRA 11] VAN FRANEKER J.A., BLAIZE C., DANIELSEN J., *et al.*, "Monitoring plastic ingestion by the northern fulmar Fulmarus glacialis in the North Sea.", *Environmental Pollution*, vol. 159, pp. 2609–2615, 2011.

[GAL 00] GALGANI F., LEAUTE J.P., MOGUEDET P., *et al.*, "Litter on the sea floor along European coasts", *Marine Pollution Bulletin*, vol. 40, no. 6, pp. 516–527, 2000.

[GAL 10] GALGANI F., FLEET D., VAN FRANEKER J., *et al.*, Marine Strategy Framework Directive Task Group 10 report on marine litter, JRC Scientific and technical report, ICES/JRC/IFREMER joint report, N. Zampoukas, 2010.

[GAL 11a] GALGANI F., HENRY M., ORSONI V., *et al.*, MACRODECHETS en Méditerranée française : état des connaissances, analyses des données de la surveillance et recommandations, Report IFREMER, RST.ODE/LER-PAC/11, 2011.

[GAL 11b] GALGANI F., HANKE H.G., WERNER S., *et al.*, Marine litter technical recommendations for the implementation of MSFD requirements, report of the MSFD GES Technical Subgroup on Marine Litter, EUR 25009 EN-2011, 2011.

[HID 12] HIDALGO-RUZ V., GUTOW L., THOMPSON R.C., *et al.*, "Microplastics in the marine environment: a review of the methods used for identification and quantification", *Environmental Science & Technology*, vol. 46, pp. 3060–3075, 2012.

[LAW 10] LAW K.L., MORET-FERGUSON S., MAXIMENKO N.A., *et al.*, "Plastic accumulation in the North Atlantic subtropical gyre", *Science*, vol. 329, pp. 1185–1188, 2010.

[LEB 12] LEBRETON L., GREER S.D., BORRERO J.C., "Numerical modeling of floating debris in the world's oceans", *Marine Pollution Bulletin*, vol. 64, no. 3, pp. 653–666, 2012.

[LES 11] LESLIE H.A., VAN DER MEULEN M.D., KLEISSEN F.M., *et al.*, Microplastic litter in the Dutch marine environment, Deltares report, final report, 1203772-000, 2011.

[MAC 09] MACFADYEN G., HUNTINGTON T., CAPPELL R., Abandoned, lost or otherwise discarded fishing gear, regional seas reports and studies no. 185, fisheries and aquaculture technical paper no. 523, United Nations Environment Programme (UNEP), Rome, UNEP/FAO, 2009.

[MOU 10] MOUAT J., LLOZANO R., BATESON H., Economic impacts of marine litter, Kimo report, http://www.kimo.org, Kimo, 2010.

[MUR 11] MURRAY F., COWIE P.R., "Plastic contamination in the decapod crustacean Nephrops norvegicus (Linnaeus, 1758)", *Marine Pollution Bulletin*, vol. 62, pp. 1207–1217, 2011.

[OSP 09] OSPAR Marine litter in the North-East Atlantic region: assessment and priorities for response, Report of the OSPAR Commission (conventions in London and Paris, www.ospar.org), London, 2009.

[TEN 09] TEN BRINK P., LUTCHMAN I., BASSI S., *et al.*, Guidelines on the use of market-based instruments to address the problem of marine litter, United Nations Environment Programme (UNEP), 2009.

[THO 09] THOMPSON R.C., SWAN S.H., MOORE C.J., *et al.*, "Our plastic age", *Philosophical Transactions Royal Society B*, vol. 364, pp. 1973–1976, 2009.

[UNE 11] UNEP/NOAA, The Honolulu strategy: a global framework for prevention and management of marine debris, http://marinedebris.noaa.gov/projects/pdfs/HonoluluStrategy.pdf, 2011.

4

Between Prejudice and Realities: How Plastics Are Essential for the Future

4.1. From a gloomy picture to a solution for the future

4.1.1. *An antiplastic crisis with often paradoxical consequences*

Surveys conducted at a European level show that the perception of plastics darkens year after year, but not at the same speed across all countries. It is in France that the image of plastics is the most negative: between 1977 and 2011, it fell from 67 to 35%. In Germany, although it chips away a little, it remains positive at 77%; in Italy at 55%; in the UK at 52%; in Spain, it went below 50%.

Among the many cases that illustrate this phenomenon of distrust, the most striking was in France, the anathema against plastic bags. Between 2005 and 2008, they crystallized all the indignation relating to pollution, so much so that supermarkets decided to remove the so-called "single-use" bags they distributed free at the checkout. From 15 billion in 2006, production fell to 750,000 today. But this spectacular result had several paradoxical consequences.

Since the bags called "single-use" had a second life as trash bags, the production – and sales – of garbage bags has increased by 45%. And as consumers were helpless in front of the checkout, supermarkets have established the sale of shopping bags, "reusable" and with a "lifetime warranty". Results: consumers pay between fifty cents and one euro for something that was previously issued free of charge; as for supermarkets, they save € 150 million annually (the single-use bag, distributed free, cost them 1 cent) while making profits on the bags now being marketed. In

---

Chapter written by Michel LOUBRY.

total, the production of plastics for reusable bags and garbage bags has increased significantly.

Such a scenario proves two things: that taking into account environmental expectations is a trigger for profits and also that there is a certain absurdity in the demonization of plastics orchestrated by our media. No, plastics do not disappear so easily. Contact lenses, wind turbine blades, capsules, airbags, ski boots, blenders, telephones, freezer bags, lifeboats, adhesive tape, hearing aids, paintings, strollers, airplane wings, food film, safety helmets, credit cards, ballpoint pens, incubators for premature babies, PlayStation, electric cables, pans, tights, etc.: which materials could replace plastics without the object becoming too heavy, too expensive, less practical, less reliable or impossible?

However, if the global production of plastics has been rising steadily, almost 10% annually, societal expectations lead to policy decisions that may influence this stunning growth. Under the combined effect of prejudice and legitimate questions, they begin to advocate for alternative materials. This is already the case in the construction sector, for example, where some contractors, not necessarily enlightened, impose cast iron pipes instead of PVC for sewerage. That the cost of cast iron is higher, that the corrosion makes it less reliable, is of no consequence: under the influence of the popular opinion, plastics are evil. Plastic = oil, oil = pollution + energy crisis, pollution = health scourge.

Such a context also reactivates the prejudices about the aesthetic mediocrity of plastics with, as a corollary, substituted materials. Therefore, an "upmarket" yogurt, originally in a terracotta pot, is today placed in a colored glass jar imitating a terracotta pot, but certainly not in polystyrene like the others – a question of image!

Plastics industries have a number of problems to be cleared: the mediocrity of the materials, proven toxicity or not of certain substances and harmful influence on the environment. We will return to them. After assessing the issues, we will weigh the benefits against the harm.

4.1.2. *The world as it is ... 2030*

In the collective European imagination and especially that of the French, plastic materials today symbolize pollution, wastage of fossil fuels and toxicity. In reality, they contribute to sustainable development in various capacities. To better understand the difference between a dark vision and positive facts, we must bear in mind the view of the future such as that drawn by scientists, statisticians and futurists.

There are now seven billion people on the planet. Two billion of us do not have direct access to potable water and electricity distribution; 925 million live in a state of undernourishment. In 2030, we will be about nine billion. Bringing water and producing electricity and food will accordingly multiply our energy consumption even as fossil fuels dwindle. We know that we need to find new resources. In addition, we also know that energy consumption generates greenhouse gas emissions which affect the climate; therefore, we have to guard ourselves against extreme cold and against heat waves, which may devastate agricultural production.

Given this, there is much data likely to break such a vicious circle. We will start with globalization. We see the other side of the coin – delocalization and job losses in the West – we forget that every year, because of the export of industrial technologies, 45–50 million people on Earth gradually get out of poverty and even worse states. For a long time, centered on Europe, the plastics industry has been creating many jobs in Asia. It responds to the growing power needs in this region: equipment, transportation, communication tools, etc. With the consumption growth of plastic materials reaching two digits per capita, while in Europe it is between 1.5 and 2%, it was logical to install production factories locally. If we place ethical considerations as a priority, the plastics industry is a factor of economic development that provides access to better living conditions. This is the positive side of globalization.

Other positive developments are looming. They are related to the computational power of computers, which generates an exponential development in innovations. The amount of technological tools created post-war up to today is huge compared with the previous period, 1840–1940. And when we focus only on the last 25 years, we are even more amazed. The speed of miniaturizing the computer, and its consequences in all areas, constantly opens new perspectives. In 2007, concerning the satellite that the Russians launched to redirect sunlight onto some farmland in Ukraine, the futurologist Ray Hammond noted that what seems to be science fiction 25 years away into the future is not:

> just go back and be convinced. 2007 minus 25 equals 1982. What was the weight of your cell phone in 1982? It did not exist. How many e-mails did you received in 1982? There were none. [...] I started my career as a futurologist in 1968, based upon a story that was published and where I predicted one day worldwide computers would be connected together to form a grid, etc. At that time, it was anticipated that copper will no longer be available on Earth in the mid-1990s. The reason was the absolute necessity of making cables bigger and bigger to carry information. This was predicted without counting on the human genius and the utilization of the optical fiber which was invented in 1973.

The optical fiber is a very good example to highlight the role of plastics in high technology. Taking into account the deadlines in 2030, their physical qualities will be invaluable.

4.1.3. *Vital qualities of plastics*

4.1.3.1. *Participation to the development of food resources*

We already know the benefits offered by mulching films and films covering the frames of greenhouses. We know that, in this form, plastics boost agricultural productivity, promote multiple cropping and optimize seed development. We ignore, however, the role of plastic packaging in food preservation, the fact that they prolong its life – a fundamental quality in the near future when we will have no right to waste. As an illustration, the case of cucumbers is perhaps the most striking. In the supermarket, when we discover them packaged in a plastic film, we are surprised and frustrated by the unnecessary protection. Like for bananas, it seems to us that their skin is sufficient to protect. This is an error. A cucumber without packaging rots in three days while the micro-porosity of the film allows the evaporation of saturation gas without letting oxygen enter, a rotting factor. Result: 1.5 g of packaging film prolongs the freshness of cucumber for two weeks[1].

Generally, the lifetime of food which has been transported and stacked on the shelves in plastic packaging is extended by 20% or by 30% in the case of fresh produce. In countries whose economic and industrial development remains embryonic, 50% of the food produced does not reach its place of consumption. Without proper packaging, it rots on the way. In developed countries, through packaging and especially plastic packaging, we lose only 2% of food between the producer and the consumer's table. (That said, in developed countries, we spoil it in a different way. Statistics show that in France 25% of foods end up in the trash, 7% without being unpacked.)

4.1.3.2. *Conserving water resources and creating more*

Saving water is first and foremost simply about avoiding leaks in the routing ducts. Whether they are intended for habitat or irrigation, pipes made of PVC or polyethylene have demonstrated their watertight properties. In London, where 50% of the drinking water supplied will never reach the Londoner's tap, pipes are old, therefore made of cast iron, therefore corroded and therefore leaking. The analysis of the lifecycle consistently shows the supremacy of plastic, its better resistance compared with cast iron. For proof, only plastic pipes demonstrate resistance for settlements in extreme environments: high mountains, deserts and ice sheets.

---

1 Source: *Advisory Committee on Packaging*, 2008, United Kingdom.

Fighting water scarcity also involves figuring out how to produce it from waste water, sea water and even clouds. Desalination by reverse osmosis, purification by membrane filters, mist collectors in desert environments: all of these methods involve the use of plastics, or even the creation of new products. The first fog sensor, 4 m high and 12 m wide, made of a hundred "nets" or polypropylene panels, was installed in the desert of Chile in 2010. The fine particles of water (0.05 to 0.5 g per m^3 of fog) slide along the nets toward gutters which direct them into tanks. A completely passive process, using neither pumps nor electricity, it can collect 15,000 l of water per day, on average.

4.1.3.3. *Reducing energy needs*

We often imagine that plastics are major consumers of oil. The reality – largely ignored – is quite different. The proportion of oil which enters, every year, into the composition of all plastics produced in the world is equivalent to only 4% of the annual production of barrels. As for the part of energy burned to produce these plastics and process them at the end of their life, it represents only 1–2% of the oil and gas extracted annually. In total, the production of plastics (namely 260 million tons in 2011) consumes between 5 and 6% of fossil fuels per year. This low percentage becomes even more significant when compared with other uses for fossil fuels: 42% for heating and electricity production, 45% for transport. In other words, 87% of oil is purely and simply burned.

But the most interesting in terms of the future is the ability of plastics to reduce energy consumption through the valorization of their waste. It is developing recycling possibilities[2] beyond existing industrial processes, such as those already in place for bottles and flasks or for PVC window frames for example, and encouraging the establishment of recycling circuits in areas where plastics are already significantly present. Plastics will occur even more in the future and will constitute in fact an important part of waste: vehicles which no longer work, waste from electrical and electronic equipment among others. And when recycling is not possible, it is about recovering the plastic waste in other ways[3]: either by exploiting its energetic power to produce heat and electricity, and get especially solid recovered fuels (SRF); or by going back to the raw material, the monomer, or even to fuel or synthesis oil and thus closing the loop. This is not science fiction. The first experiments date back to 2006. The first production of fuel from plastic waste was installed in Ireland at the initiative of Cynar, a company specializing in the new technologies of converting waste.

2 Transforming objects out of use or their fragments belonging to the same family of plastics in a "new matter".
3 This is the case of the waste mixture, i.e. plastics unsorted

4.1.3.4. *Decreasing emissions of greenhouse gases*

Plastic materials again play a positive role, a reversal of received ideas. If they did not exist, if plastic products were made out of other materials, the amount of greenhouse gases emitted throughout their lifecycle would increase by 61%, or 124 million additional tons per year. These figures come from a study by the Austrian Research Institute, Denkstatt [DEN 11]. PlasticsEurope asked it to assess the carbon footprint of plastic during their utilization phase; Denkstatt proposed establishing a comparative product in 173 cases utilizing, with regard to plastics, the most appropriate substitution materials in each case. To do this, Denkstatt relied on the work of the research institute of the German market GVM because it has an inventory of exceptional data on different materials.

When we enter the details of the study, our prejudices on plastic packaging continue to collapse[4]. Fifty-seven types of packaging were analyzed, comparing 8 kinds of plastic to 8 kinds of alternative materials: iron/steel, aluminum, glass, corrugated cardboard, linerboard, fibers and nonwovens, paper and wood-based composites. The results speak for themselves. Replacing plastic packaging with their substitutes multiplies by 2.2 the energy consumption, the annual equivalent of heating 20 million houses, and multiplying by 2.7 the emission of greenhouse gases, or 60.8 million additional tons per year. The latter figure takes on its full significance when it is compared with the 21 million tons of CO_2 emitted by all of Denmark in a year.

4.2. Engineering polymers: what is wonderful, what is reassuring?

4.2.1. *Plastics and their ignored positives effects on the preservation of the environment*

Today when we think of pollution, our collective imagination is against plastics; we think it would be good to replace plastic bottles with glass, bags with paper, etc. In fact, there are only 16% of plastic applications that are not substitutable, and 84% are those we can go without or we can replace. Nevertheless, it is the lightweight of plastic that plays a dramatic role in saving energy and reducing greenhouse gas emissions.

For example, we take the case of a tomato sauce bottle delivered by truck to supermarkets. The glass is 36% of its total weight. The plastic is 3.56% only. If the plastic packaging did not exist, delivery trucks would travel 50% more, which would

4 The term "packaging" must be understood in a broad sense of "containing" pouches, bags, bottles (of household cleaners to cosmetics to syringes), films, flexible food packaging, rigid boxes, cases, etc.

require an additional 27 million tons of fuel. And if food plastic packaging did not exist, the 20% of resulting food waste would emit a further 190 million tons of greenhouse gases. It is clear that there is a gap between our prejudices and facts: among all the uses of plastics, it is effectively the food packaging that best contributes to preserving the environment.

In fact, in the 173 plastic applications studied by Denkstatt, we noted that only three substitute materials achieved a better performance. The first case, the bag shop. Paper manufacturers have demonstrated that it is an anomaly to offer in store bags that support a content of 10 kg and more, when, most of the time, we only have to carry 250 g. It would be better to put into circulation paper bags. Indeed, they are more and more on the market. Second case, where the plastic does not offer the best solution: disposable syringes in hospitals. Analysis of the lifecycle is much more favorable than that of glass – provided, of course, a perfect sterilization is possible (on the criterion of asepsis, the disposable plastic syringue remains essential, especially outside hospital when sterilization conditions are not guaranteed). Last case, certain uses of flooring carpet in the suburban habitat. They may sometimes be advantageously replaced by wooden flooring, more expensive but more durable.

4.2.2. *Lightweight plastic, a quality that induces environmental performance*

In a sector other than packaging, the automotive sector for example, plastics take priority over steel: bumpers, wings, cockpits, under-the-hood components and so on. The structure is lighter, which reduces the fuel consumption.

We think that plastics are flimsy, but the technological reality is different. Because of their natural elasticity, they absorb shocks better than steel, which helps in the recovery of the original shape. In addition, plastics have such thermal and mechanical resistance that they constitute, in part, the structure of aircraft. Wings, tailplanes, fuselages are made of plastic, and also the interior structure, compartments, overhead lockers and so on. This again results in a reduction of energy consumption. 22% of plastics constituting the weight of an Airbus A380 already saves 15% of kerosene. With the Boeing 787 Dreamliner and the A350, the part of plastics will reach 50%.

In the field of wind turbines, plastic is irreplaceable. Given the huge dimensions of a blade, only the lightweight of composites (epoxy or unsaturated polyester resins reinforced with glass fiber) allows us to obtain the correct yields. Blades constructed of other materials, necessarily heavier, react less to the wind and therefore produce less energy.

In general, if we examine the 84% of applications – all fields included – where we can replace plastics with other materials, the statistics are clear:

– the weight of substitute products would be multiplied by 3.7;

– the amount of energy required to produce them, use them and process them at the end of their life would increase by 57%;

– the quantities of CO_2 emissions would increase by 61%.

In fact, for a ton of CO_2 emitted during production, plastics emit seven times less than other materials during their lifecycle.

4.2.3. *When plastics protect us…*

Against earthquakes, typhoons, floods and others, plastics are valuable materials, both to prevent collapse and facilitate rescue. They enter into the seismic design of buildings and bridges, to strengthen the defenses with elements like kneecaps. They are in emergency facilities, tents, houseboats, artificial islands to bring the victims to shelter.

In terms of security, they are often irreplaceable. This is the case of the airbag in automobiles. Without polymers, how can we achieve this protective ball? With a pig's bladder? But then, how to integrate onto this skin the very sophisticated technology which allows the liberating explosion of the ball without injuring the driver? There is no solution outside the plasticity of polymers. So, today, 50% of plastics applications in the automotive sector are no longer substitutable.

This is almost the same in the medical field. The current quality of care could not be achieved without plastic to which no one pays attention: blood bags, heart valves, prostheses, artificial corneas, capsule medication, etc. Once again, if plastics are the most effective materials, it is because they possess, unlike other materials, all the properties required in medicine: exceptional sealing, perfect hygienic barrier, lightweight, flexibility, plasticity, durability, transparency, compatibility with other materials and low cost.

4.2.4. *How plastics will prevail in the future energy solution?*

If their lightweight is a big factor for saving energy, their insulation quality is equally important, which closely concerns the building industry. In France, for example, a third of the total energy is absorbed by domestic use, of which 72% is for heating[5]. Such consumption has become absurd given the possibilities offered by current plastic technologies. They make it possible to build positive energy buildings[6]: roof insulation in polyurethane, wall insulation in expanded polystyrene, triple-glazed windows in polyvinyl chloride (PVC), special paint which absorbs heat, solar panels, heat recovery ventilation, fuel cells, etc.

Energy saving going hand in hand with a reduction of greenhouse gas; it has been calculated that in Europe[7], if the best technologies of plastics were used to isolate half of the buildings, we would avoid emitting 340 million tons of CO_2. That is 35% of the current emissions of the building industry. And we would reach 45% of Kyoto targets.

Plastics which are involved in saving energy are also players in the production of renewable energy. We have seen it with wind turbines. In the near future, they will definitely replace the silicon hitherto necessary for the production of solar energy. Photovoltaic cells made of 100% plastic are under development. Currently, they offer a yield of 8–9%, which will soon reach 10–12%. Certainly, it remains lower than that of silicon cells: 19%, but the new polymers offer notable compensation. With 100% plastic transistors, which produce light and conduct electricity, the lower yield will be offset by the savings in manufacturing, ease of use and especially the multiplicity of suitable location. Given their low price, photovoltaic films can be installed on all surfaces exposed to light, including transparent glass. They offer us huge resources.

4.2.5. *Plastics at the heart of technological advancement*

Plastics are an integral part of nanotechnology and of the latest developments in embedded digital communications. Futurologists foresee the popularization of nanoplastic implants that are placed behind the ear, integrating a mobile set of amenities such as a telephone, GPS and the like. Certainly, science fiction always raises a fright when it is about to cross the barrier of reality and enter our daily lives. This is the case with nanotechnology. In contrast, invention seems harmless when it saves lives. Today, plastic stents that sit in arteries contain an active ingredient which acts to harden the walls to prevent them from weakening by friction in contact with the spring. This is the paradox of the collective imagination: we panic in front of the supposed toxicity of the packaging, but we are delighted to be saved by an implant inside our body.

5 ADEME (2006).
6 A building which produces more energy than it consumes.
7 Source: Badische Anilin- und Soda-Fabrik (BASF).

We disregard mainly the inexhaustible applicability of polymers. Artificial blood, "plastic blood", is under consideration in some French hospitals to compensate for the lack of plasma in stock. Self-healing polymers have also been created.

Miniaturization, nanocomposites, nanotubes, nanopolymers used as vectors for active ingredients, micro-processors and photovoltaic cells based on electrically conducting polymers: plastics are each time vectors of news. Since their origin, they have constantly revolutionized the way we live, whether it is in terms of comfort, safety, health, communication, transportation, architecture, design, sports or leisure.

But ... they also created problems which still remain to be resolved.

4.3. Plastic industries: progress to be made

4.3.1. *Environmental issues, the European plastics industrial acts*

Our actions are part of an ambitious program: to valorize all plastic materials at their end of life, to treat all the waste from our industries. This program is based on the study of the impact of polymers, from their production to their treatment after use. Each year, PlasticsEurope invests 250,000 euros to update and complete the lifecycle analysis of the production of polymers. We share data and collaborate closely with the *Joint Research Center* of the European Commission which works on calculating the environmental impact of products throughout their lifecycle.

Therefore, whether we measure the footprint we leave on the environment or whether we implement solutions to reduce it, we are at the forefront of the industry. The *Joint Research Center* considers us to be a privileged partner. But owing to a lack of communication on the subject, our citizens imagine us devoid of environmental awareness and a sense of responsibility. It is true that in France, the interest of our industry to the environment is recent. Without doubt this is the cause of our delay in recycling and waste treatment compared with Northern Europe. With respect to the media environment, unfavorable to the plastics, it increases the lack of visibility of our actions.

It is also true that recycling and treatment of waste – not an interesting subject – does not interest journalists. Therefore, they avoid the link between cause and effect of marine pollution, for example. Since 2009, they have dramatized the existence of "vast continents of plastic waste" on the ocean surface, recalling that in 1997 we discovered the largest of these bins in the Pacific. Illustrations of the phenomenon, *The Great Eastern Pacific Garbage Patch*, show groups of floating objects which could easily be caught in fishing nets. False images were taken near the coast where, unfortunately yes, the spectacle of some shorelines is distressing. But regarding the "bins" identified in the ocean, it consists essentially of plastic disintegrated into billions of small particles whose largest density at the location of the highest concentration is 5 kg per km^2. NGOs ask us how to clean these traces. Faced with the amount of energy it would take to do it, we propose instead to work together to dry up sources of pollution. It should be noted that 80% of the waste comes from land and 20% from maritime activities: transport, cruises, boating and especially fishing (the number of nets abandoned or thrown over board is considerable).

We will not clean up the surface of the ocean. However, we will develop ways not to pollute it. We are committed globally by mobilizing 47 professional associations related to plastics. We announced our goals in Honolulu at the 5th Conference on the Protection of the Oceans. Text appeared under the title *Declaration of the Global Plastics Associations for Solutions on Marine Litter*[8].

4.3.2. *From polluting plastics to non-disposable plastics*

To return to the bad image of plastics in France, it is fed with the spectacle of littering in the open air, especially on the beaches or in landfills, whether wild or public. It is true, up to now, that the performance of our collection systems has been poor, that landfills are misused and that our behaviors still leave something to be desired though, in rich countries, we have made tremendous progress.

But at the origin of our problems in France, there is an almost morbid disinterest, especially for plastics at the end of its life. It is identified almost as carrion. We cannot imagine that recycling is a largely mechanical process developed across Europe, nor that it can recreate the raw material which other manufacturers will use again. We consider it to be only about objects recovered in poor condition and processed into produced gadgets, such as pen. As for the valorization of the plastic waste mixture[9], we ignore it even more. We do not know that we can synthesize SRF which is more economical than heavy fuel, or that its chemical treatment allows it to go back to its monomer state or to produce synthetic oil.

In any case, the abandonment of plastics in landfills is absurd. That is why PlasticsEurope defends along with public authorities a clear and ambitious goal: zero plastics in landfill by 2020. This is not a utopia when the figures are examined.

With regard to recycling, the Germans are the leaders, closely followed by the Swedes, Estonians, Czechs and Belgians. The French are squarely in the rear

8 See: *www.marinelittersolutions.com.*
9 Non-sorted waste.

platoon of European countries. A particularly unfortunate observation since recycling creates jobs which cannot be outsourced, promotes both economic development and the reduction in energy consumption. As an indication, Germany consumes 11.5 million tons of plastics per year and recycles 34%. France consumes 4.6 million and recycles 16%[10].

Concerning the recycling of mixed waste, the Swiss and the Germans have already reached a level of almost 100%. The Belgians 93%. The French are still only at 54.6%. When we examine things more closely, we see that if Germany took such an advance, it is because of the TASI law that prohibits the landfilling of products still having a heat content. The result: Germany produces 11 million tons of SRF of which 25% comes from plastics (the rest coming from wood, paper, cardboard, etc.). In France, our SRF production in 2011 was only a few hundred thousand tons. The economic consequences are significant. As an example: since the French cement industry still uses heavy fuel oil, which is non-competitive in terms of price, they cannot export their cement. Accordingly, Lafarge Poland and Lafarge Germany have gained market share as the expense of Lafarge France.

4.3.3. *Recycling and valorization: the French cultural handicap*

Our environmental awareness is recent, even very recent. French manufacturers have for a long time remained disinterested in what was happening beyond their borders. What have they found? Plastics in Germany – produced by the same companies – sell better, are more expensive, with better margins. What does it mean? A much better picture for plastics in Germany.

In France, as the longevity of plastics contributes to their poor image – at least 40 years of life for a PVC frame – this quality is less important to us than the downside: the feeling of being invaded by plastics. As consumers, we blame them for not deteriorating like vegetable peelings. Continuing to exist when no more in use. But what do the figures say about the feeling of lethal flooding? On the one hand, 4.6 million tons of new products in plastics were placed on the French market in 2011. On the other hand, 3.2 million tons of end-of-life plastics were collected of which the vast majority was packaging. The difference between these two figures is that which we are using and, in part, which we no longer use, but which we accumulate without knowing what to do with it. Because we keep in our closets, garages or attics old phones, old toys, old TVs, old VCRs without ever thinking they could be introduced into the system for valorization of end-of-life plastics. The part that obsesses us, plastic waste abandoned in the wilderness is tiny.

10 Source: *Plasticseurope/Consultic Marketing und Industrieberatung GmbH.*

Our groundless impressions and our non-coherent behavior can be explained: the sorting instructions for our individual bins are unclear; the recycling center with well-organized sorting of waste is still rare, and as we do not know that our waste represents real economic and energy resources we do not care for it, swearing to boot. Unlike the countries of Northern Europe, we are actually misinformed and ill-equipped to get the exponential benefits of plastics at the end of their life. The plastics industry has an important role to play, in terms of both communication and implementation of the optimal management of plastics at the end of their life.

4.3.4. *Bisphenol A or how to spread anxiety and misinformation*

We have entered into a culture of zero risk. Our role of industry is to integrate and even anticipate this evolution and its impact on the composition of our products. And the media's role is to alert us. Provided, however, that this does not become overblown. However, in the case of bisphenol A, from dramatization to dramatization, the media have come to say that the mere touch of a drink, food or our skin with a polycarbonate containing bisphenol A poisoned our blood – and no matter that it enables the manufacture of a very transparent plastic, a quality particularly necessary in incubators for premature babies, for example.

First point, the chemical reality, bisphenol A in a pure state is actually a product to be handled with care. Injected as such in laboratory to rats and at high doses, it is clear that it has negative effects on their body. But the chemical polymerization reaction traps the bisphenol A and neutralizes it. Once entered into the composition of a polycarbonate material, for example a baby bottle, it becomes inert. It is necessary to heat the polycarbonate to above 100°C to observe the release of bisphenol A molecules. Therefore, what can be proved by experiments performed on rats which have been injected with pure bisphenol A?

In any case, with objectivity no longer being in place for the subject, the dismissal of plastics has increased. Today, with the questioning of phthalates, all polymers become subjects of anxiety. More seriously, it is impossible to stop the misinformation. When PlasticsEurope strives to tell reporters that phthalates no longer enter the composition of food packaging, nor have they for the last 20 years, they either refuse to hear it, or they refuse to admit it, or they refuse to report it. Misinformation is fierce. There is also the fact that people often lack a minimum background in chemistry. In this respect the case of "dihydrogen monoxide" is exemplary. Since 1997, the Internet has relayed dangers that it caused and denounced its presence across the globe, a "colorless solvent in its gaseous form, it can cause severe burns – prolonged contact with its solid form causes tissue damage – It is the main component of acid rain – it is released in large quantities by nuclear power plants, directly into rivers or seas, or into the atmosphere – it is used in cases of torture or by police against protesters – most young offenders consume it – it is found in blood plasma". In the United States, as in France, millions of signatures were collected against the use of "dihydrogen monoxide". A U.S. senator has even written to the Department of Health asking whether it had any intention to ban this drug. Among the millions of signatories, nobody has questioned the meaning of the formula "dihydrogen monoxide", "colorless solvent" was enough. However, this formula is the exact synonym of H_2O. Yes, water. This gigantic hoax by American students remarkably illustrates the problems currently faced by industrial plastics

4.3.5. *Bioplastics: from advertising to reality*

The term "bioplastics" covers two different concepts that we tend to confuse. The first categorizes the so-called "bio-sourced" plastics from agricultural resources, *a priori* renewable, they are not necessarily degradable. The second includes biodegradable plastics: they are not necessarily bio-sourced; they are also polymers derived from fossil resources and chemistry.

The bio-sourced plastics sparked media interest toward 2006–2007. They were endorsing the role of the good against the wicked guzzling fossil fuels. At the time, the European association *European Bioplastics* published provisional figures announcing a rapid development. PlasticsEurope[11] had tempered the illusions of the media and especially the political media. The latter was already asking when all plastics would be replaced by bioplastics. But the solution which seems ideal hid in turn problems which are worth looking closely at. Particularly those of arable lands. Given the imminent 9 billion human beings on the planet, how much land should be devoted to food and how much to produce plants for the creation of materials? In any event bioplastics account today for only 0.2% of annual global production. They are obviously expected to develop but in niche markets.

Biodegradable plastics have also a good press. But we give them magical virtues that they do not necessarily have. Or not yet. An example: on our Website, a person from Chambery pointed out that in her compost at the bottom of her garden, biodegradable plastic bags were not biodegradable. Lack of information – the advertising message tends to oversell – she did not know that biodegradation occurs under certain conditions, including temperature. However, at Chambery, in the open air, it is too cold.

Work remains to be done from a communication point of view on biodegradable plastics, including municipalities. Those who set up a coherent system for collecting

11 PlasticsEurope represents also the manufacturers of bioplastics, such as Novamont, BASF and Arkema.

waste are still few in number. There are, however, some exemplary cases such as Lorient. Separate collections of organic waste are carried out in biodegradable plastic bags that the municipality has previously distributed to residents. Therefore, bags of organic waste end up directly in the system for industrial composting. It sounds obvious, but most of the time, this is not the case. Organic waste is collected in non-biodegradable bags which must have their contents poured out.

Do not oversell, inform our citizens well – that is to say show them respect – is a prerequisite for the proper development of bioplastics.

4.4. Conclusion

Three intrinsically linked phenomena maintain the dark image of plastics in France: the lack of information, a socio-cultural tendency to be wary of chemistry, and media dramatization. Unlike in Germans, chemistry is not an area that attracts us. Polymer engineering is not understood by us and we concern about it. No matter the comfort and security that plastics constantly bring in all areas, regardless of the future solutions they represent – industrial plastics are seen as black sheep. Spontaneously, we do not give them any credibility.

On the other side of the mirror, the plastics industry has long slept on its laurels without bothering to communicate. When environmentalists have sounded the warning bell, rightly on fundamental issues of the environment and health, but also indiscriminately, it was a little late to be heard. Or to listen to reason. Recognize too: we were an industry which cared very little about the use of its products and their impact on the environment.

That said, under the cultural influence of northern Europe in the late 1990s, we began to implement ways to eliminate our sources of pollution. This is what allows us to say "zero plastic in landfills for 2020".

For the rest, polymer engineering is ready for anything, and also to do its best.

4.5. Bibliography

[DEN 11] DENKSTATT CABINET, L'impact du cycle de vie des plastiques sur la consommation d'énergie et sur les émissions de gaz à effet de serre en Europe, Summary report, French translation, February 2011.

5

Lifecycle Assessment and Green Chemistry: A Look at Innovative Tools for Sustainable Development

5.1. Contextual element

5.1.1. *The chemical industry mobilized to deal with global upsets*

Very recently, our society – indeed the whole human race – has become aware that we have been mortgaging our collective futures to satisfy our appetites for individual riches. Although only a few hundred millions of us on the planet shared between us the vast amount of wealth and therefore generated the vast amount of anthropic pollution, an undeniably critical balance was maintained. Yet with the arrival, in the last few decades, of nearly three billion individuals – Indians, Chinese and others who, quite rightly, expect a high level of consumption, and with the projected swelling of the world population in years to come, the international community is calling for sustainable development, so as to be able to establish a new balance, which is truly sustainable.

Thus, the 20th Century was marked by population growth, economic development and unprecedented environmental change. Between 1900 and 2000, the world population exploded from 1.6 billion people to 6.1 billion [UNI 00]. However, while the world population quadrupled, the actual global GDP increased between 20- and 40-fold [BRA 98], thus making it possible for the world not only to support this quadrupling of the population, but also, in doing so, to move toward significantly better living conditions. However, this rapid demographic and economic growth was very unequal from one country to another, and not all regions

Chapter written by Sylvain CAILLOL.

benefited equally from the economic growth. In addition, the population explosion and the economic development, which occurred first, have given rise to an increasingly untenable usage of the Earth's physical environment.

Analysis of the reciprocal links between population, the environment and economic development dates back to long before the work of Thomas Malthus (late 18th Century) [MIN 11]. Indeed, since Antiquity, statesmen and philosophers have expounded their opinions about questions such as the optimal population level and the drawbacks to excessive demographic growth. Thus, Plato and especially Aristotle have fanned the flames of numerous reflections on the balance between population and natural resources, defined as the means of subsistence and, in more concrete terms, as food and water [MOR 49]. This reflection was also put forward by Montesquieu in the 18th Century [THO 1798]. The United Nations Organization's thinking and activity relating to population, environment and development are as long standing as the Organization itself. In the 1960s, we saw increasing awareness of the fact that the world population growth had reached unprecedented levels – a situation which was judged to be of serious concern in a large number of studies and debates. A report by the Secretary General entitled *Problems of the Human Environment* cites the "explosive growth of human populations" as being one of the signs heralding a global crisis concerning the relationship between humans and their environment [UNI 69]. This report constituted an essential step in the process which led the UN to hold the United Nations Conference on the Human Environment, in Stockholm in June 1972. This was the first worldwide intergovernmental conference devoted to environmental protection. Thus, the 20th Century was marked by an extraordinary growth of the world's population, from 1.6 to 6.1 billion; 80% of this increase occurred from 1950 onward. The world's population is projected to continue to swell. Based on the average birth rate, the UN predicts that the world's population will reach 9 billion in 2043 and 9.3 billion in 2050. However, small but sustained differences in birth rates could influence the world's actual population in the long term. Thus, a high birth scenario where the birth rate is half a child higher than the average fertility scenario would lead to an actual population of 10.9 billion in 2050.

Urbanization is also another important trend. Indeed, as the world population is predicted to double in the next 40 years, the urban population, which is currently at a high of 3 billion, will likely reach 6 billion – i.e. a doubling of the urban population, with a considerable increase in energy requirements. In addition to these neo-urbanites, in 2050, there will also be nearly a billion climate refugees, displaced from their homes by large-scale mining or damming projects, or by the effects of global warming and the conflicts inherent to the upsets thus caused [CHR 07].

This demographic growth expected in the coming years will be accompanied by an increase in net consumption per individual. Greatest among these consumptions will be energy consumption. The estimations of the International Energy Agency (IEA) suggest a worrying *scenario* corresponding to a doubling of the energy demand by 2030 [INT 10] in relation to the 11 Gtoe (gigatons of oil equivalent – in other words, 11 billion tons of oil – 1 ton of oil corresponds to 41.9 GJ) consumed in 2007! If this evolution were to come about, it would double the production of CO_2, which is the main greenhouse gas! In parallel to this, the population increase would necessitate a 70–100% increase in agricultural production to support its food needs, by 2050. However, our planet's resources are not inexhaustible, and some of them are beginning to be in short supply. Indeed, the last two reports on the "environment" from the US inter-armed forces *Joint Operating Environment* (JOE) occupy an important position, among the recent analyses which recognize the eventuality of a decline in world oil extraction by the middle of the current decade. Indeed, these reports [JOI 10] all give a similar diagnosis which is now more pessimistic than ever regarding the question of a possible structural oil shock by 2015: "In 2012, the surplus in oil production capacity could disappear completely, and by 2015, the deficit in production could be nearly 10 million barrels per day", or the equivalent of the daily extraction from Saudi Arabia. The IEA's 2010 report states that nearly 30% of the production of the oil wells currently active will have disappeared in 10 years, dropping from 68 to 48 million barrels per day (mb/d) by 2020. Within a generation, by 2035, the currently active oilfields will only produce 17 mb/d, which is a fifth of future demand. Even if new resources are discovered (e.g. oil shale and shale gas), the cost of these will be a crucially important issue. Indeed, depending on their cost, we may or may not be able to afford to use them to produce energy and consumer goods – the effect of the cost of them, thus, is felt in terms of supply. Other non-carbon-based resources are also being drastically depleted, such as sources of terbium, hafnium or silver [USG 11]. We can now recall the famous quotation from Paul Valéry [VAL 45], "The age of a finite world is beginning"; we are experiencing that finite world on a daily basis.

5.1.2. *New stresses being exerted on industrial chemistry*

Thus, our society is currently reliant on a near-exclusive use of fossil resources – particularly for its energy supply and consumer goods. The question is no longer *whether* there will be peaks in the production of fossil resources, but rather *when* these peaks will occur. Indeed, almost all experts are in agreement as to the quantity and duration of our global reserves of oil, coal, gas, nuclear fuel, etc., in view of our current consumption rate. Thus, by the end of this century, we will have exhausted all of the Earth's reserves, which took millions of years for nature to create. However, this exploitation of fossil energy resources – of fossil carbon – is accompanied by a transfer of matter – a transfer of carbon – which, when burned, is returned to the form of CO_2 of fossil origin in our atmosphere, accumulates and contributes to the increase in concentration of the infamous "greenhouse gases", which are reputedly responsible for the increase in average temperatures around the world.

These questions are many stresses on industry, and particularly the chemical industry – the industry of industries – in that over two-thirds of its products are destined for downstream industries [UIC 08]. In view of these stresses, the chemical industry is experiencing a revolution, hinged on:

– the anticipation of the exhaustion of raw materials derived from fossil resources, accompanied by significant volatility of prices. The unequal distribution of these fossil resources – oil, in particular – gives rise to significant speculations which compromise a stable supply chain;

– an obligation to drastically reduce pollutant emissions from chemical processes – particularly the release of greenhouse gases (CO_2, NO_x, etc.). The evolution and the level of supply of fossil resources have considerably increased the quantities of fossil CO_2 emitted into the atmosphere each year. Chemistry produces relatively small quantities of CO_2, but it is very much in demand to identify new processes to help reduce CO_2 emissions of energy-based origin, to store CO_2 or recover and thus create value from it by remediation (use as an input element[1]);

– significant legislational pressure with regard to the toxicology and eco-toxicology linked to the use of raw materials, intermediary products in synthesis procedures and end products of the chemical industry, notably including the legislation REACH [REA 06], the water framework directive [JOU 00a], but also numerous European directives concerning the end of life of the materials (end-of-life vehicle directive [JOU 00b], waste electrical and electronic equipment direction [JOU 03], directive on volatile organic compounds [JOU 04] emitted by varnishes, paints and vehicle-refinishing products, etc.).

While the earliest legislation on industrial activity dates from 1810, with a Napoleonic decree imposing the conformity to a distance around production sites [JOU 10], the first European directive governing the toxicity of chemical production dates from 1967, with the Directive 67/548/EEC on "classification, packaging and labelling of dangerous substances". Since that time, the number of European directives relating to the environment has considerably increased – particularly since the end of the 1990s, with a very sharp upturn. All of these regulations represent severe constraints for the chemical industry, but they may also prove to be springboards for innovation.

Thus, catering to humanity's needs (food, energy, care, etc.) while respecting our environment is the challenge which lies before us, which chemistry is going to need

1 Input element: a substance which is taken in at the start of a production process.

to solve in years to come. Chemistry has already succeeded in winning battles in the last century – fighting epidemics, increasing agricultural, agro-alimentary and industrial productivity to serve the increasing needs in terms of food and consumer goods, etc. Today, the demand is different, but *it is still chemistry which holds the keys to sustainable development* [AZE 07]. Indeed, from a strictly etymological point of view, chemistry is the science – the art – of the Earth. Thus, it is chemistry which is called upon to deal with these issues, as a science, but also as an industry, insofar as chemistry is upstream of all the other industries (60% of the production of the chemical industry is returned to industry[2]). Thus, chemistry is at the heart of the process of production, but it cannot, alone, handle the choices and decisions relating to the arbitrations concerning the production process, and the resources, in particular with a view to the reduction in environmental impacts. Lifecycle assessment (LCA) can be a useful tool in illuminating these choices. Yet in any case, we shall see that the choices can only be made after interpretation of the results of an LCA. Indeed, chemists, biologists and toxicologists are all called on for interpretation of these results of impacts. Which criteria must we use, though, to guide these choices? Which goals should be prioritized? It is perhaps in this interpretation that we are in need of illumination in view of the work of philosophers and historians.

5.2. Lifecycle assessment, as an eco-design tool: definitions and concepts

5.2.1. *Eco-design: some definitions*

The production of goods and services is now under stress. Thus, it will no longer be sufficient to satisfy a set of specifications technically in respect of the costs set out; here onwards we will need to integrate respect for humans and for the environment. This means reducing consumption of fossil resources, limiting greenhouse gas emissions, conforming to environmental constraints – in short, *limiting the environmental impacts*. However, taking account of all environmental impacts during the manufacturing process, rather than only measuring the carbon footprint or CO_2 emissions, entails integrating *eco-design* with conventional design processes, and therefore with the innovation process as well. Thus, this innovation process is experiencing dramatic changes. No longer do we simply expect a rapid response, but rather we allow ourselves the time to reflect to provide an exhaustive response concerning the environment – an "eco-designed" response.

In addition, eco-design is an integral part of the recommendations of the environmental Grenelle held in 2007. Indeed, commitment no. 217 encourages environmental analysis techniques for product and eco-design.

---

2 Source: *Union des industries chimiques* (UIC), 2010.

Commitment no. 217: generalize the environmental information present about the products and services: energy label applied to all products with heavy energy consumption, with a unique referential framework; development of eco-labels; accompaniment of voluntary approaches in terms of displaying information about the ecological impacts, with a progressive obligation to display this information; examination of the generalization of the ecological cost (double cost to inform the consumer about the environmental footprint of the goods he is buying), eventually leading to a concerted ecological contribution. [GRE 07]

Finally, eco-design now stems from a legislational obligation with the framework directive on eco-design which, for energy-consuming products, stipulates that "eco-design of the products in an essential axis of the community strategy regarding integrated product policy. As a preventative approach, intended to optimize the environmental performances of products whilst preserving their quality of use, it presents new and real opportunities for the manufacturer, the consumer and society as a whole" [JOU 05]. This directive was reinforced by a directive establishing the requirements of eco-design for the following products [JOU 08]: water heaters fed by liquid or gaseous fuel, refrigerators, freezers and electro-domestic combination devices and filers for fluorescent lighting. Thus, eco-design is beginning to become an obligation. It is also a response to consumers' expectations. Indeed, end-users now expect products which respect the environment. According to the IRSN[3] data, since 2006, environmental damage has been among the three main concerns of French consumers [IRS 09]. Hence, eco-design is an all-encompassing approach centered on the product. In principle, it consists of taking account of the environmental and human criteria even from the product design phase. These criteria generally relate to all of the phases a product goes through: product, distribution, use and end of life – that is, the *lifecycle* of that product. In its objective, eco-design is a multicriterion preventative process which aims to identify and reduce environmental impacts at their source.

The concept of eco-design is based on a powerful tool for identifying environmental impacts: LCA.

5.2.2. Lifecycle assessment: definitions and concept

In actual fact, the notion of LCA first appeared in the SETAC workshop in Vermont in 1990, which stressed the need to expand the eco-balance founded on the materials/energy balances to a real LCA – the notion of impact evaluation emerged. Thus, the first LCA was performed in France on the steel-coating products made by

---

3 *Institut de radioprotection et de sûreté nucléaire* – Radioprotection and Nuclear Safety Institute.

the company SOLLAC. In 1993, SETAC proposed a best-practices code which served as the reference point for subsequent developments. In 1997, the ISO – International Standardization Organization – published the first international standard on lifecycle analyses – ISO 14040: environmental management – LCA – principles and framework. Other standards are to be published in the next few years. LCAs continued to expand in France in the 2000s, with the conducting of LCA by specialized task forces and the organization, in 2005, of the first symposium on eco-design and chemistry in France, by the *Fédération française des sciences pour la chimie* (FFC) and *Chaire ChemSuD*.

Hence, LCA is an assessment method which consists of quantitatively evaluating all the potential environmental impacts of a product or service, considering the whole of its lifecycle.

This assessment applies to all of the lifecycle, in what is called a "*cradle-to-grave*" approach, given that at each stage in the lifecycle, there is consumption of energy and resources, and generation of environmental, social and economic impacts.

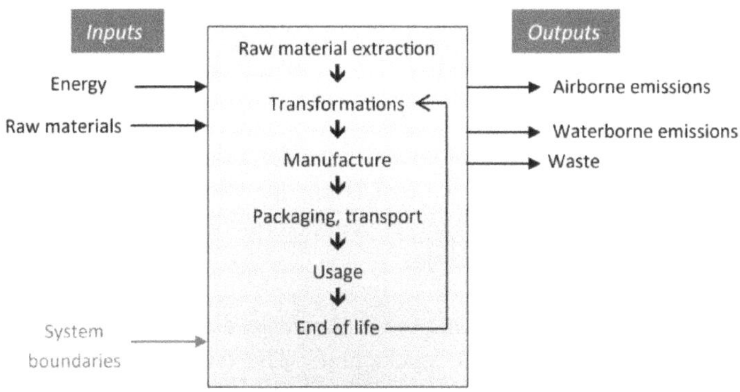

Figure 5.1. *Principle of LCA*

LCA thus consists, within a system defined by *boundaries*, of evaluating the impacts due to the *inputs* – consumption of natural resources – and to the *outputs* – emissions into the air, water and soil and other problems (Figure 5.1).

In reality, this assessment hinges on four clearly defined phases: the *definition of the goals and scopes* of the LCA; lifecycle *inventory*; *evaluation* of the impacts of the lifecycle; and finally *interpretation* of the lifecycle. The assessment is based on a scientific methodology, which employs software toolkits, within the framework laid down by the ISO 14040 and 14044 standards.

5.2.3. *Definition of the goals and scope of the lifecycle assessment*

The definition of the goals and scope for an LCA is the subject of a reference document which is updated at every stage of that LCA. In general, LCAs are carried out to answer specific questions about the environmental impacts in comparison to those of different products or services. In such cases, they are *attributional* comparative assessments. They can be conducted in order to answer questions relating to the consequences of the massification of a process – e.g. the consequences of the generalization of a restricted or localized practice: in this case, we are dealing with *consequential* LCAs.

In order to define the *goals* of the study, we first need to define the intended application, the reasons for conducting the study and the target audience – i.e. the people to whom the results of the study are intended to be communicated. Secondly, we need to define the *scope of the study*, which entails restricting the study to specific *boundaries* and establishing the limits of the system under examination, defining the activities and impacts which are included in or excluded from the study. To this end, we define the *temporal* scope (the lifetime of the system), the *geographical* scope, the *technological* scope, the coverage of the *processes* (the limits of the system), the *environmental interventions* (inputs and outputs) and the *potential impacts*.

The definition of the system also includes the definition of the *functional unit* and of the *reference flow*. The functional unit is a value which can be used to quantify the function of the product system under examination, and compare different systems which serve the same function.

Example.– In the context of an LCA to evaluate different forms of packaging, the function under examination is "packaging". The functional unit to be defined, therefore, is a particular volume V packaged (in m^3) rather than a mass (kg) packaged or a mass of packaging materials.

This definition stage is absolutely crucial, in that the results of an LCA depend closely on the goals and scope determined previously (but usually not on the client order or the manufacturer). Thus, the LCA of a plastic yoghurt pot made by a particular American manufacturer, with precise knowledge of the distances the products are transported and the composition and different means of energy production that are employed, will not yield the same results as the LCA of a European yoghurt pot, performed on the basis of averages representative of European production. Consequently, in order to avoid inappropriate interpretations or subsequent damaging generalizations in the use of the results, the goal and scope of the study must clearly and explicitly define the issue being looked at.

5.2.4. *Lifecycle inventory analysis*

This is the phase which, on a methodological level, is best developed. It has benefited from the methods stemming from the raw material/energy balance studies performed in the 1970s. The definition of lifecycle inventory analysis as given in the international standards is as follows: "the phase of lifecycle assessment involving the compilation and quantification of the inputs and outputs for a given product system throughout its lifecycle". Inventory is the most basic goal of LCA, because it is made up of elementary processes which obey the physical laws of conservation of mass and energy. This type of inventory, however, is not absolute. Indeed, this approach entails a data collection phase connected to the forming of working hypotheses. The data can be collected from production sites, but we can also use all-encompassing data from professional unions or organizations.

The goal here is to gather data or assemble existing data, and perform computations with those data using a precise method: flow diagram, description of each elementary process, collection and validation of data. The quantitative input and output data for each elementary process, calculated in relation to the reference flow, are correlated against the functional unit.

All environmental interventions (resource use and pollutant emissions) for the product system, for each of the unitary processes at every stage in the lifecycle, are thus summed in an inventory table and expressed in relation to the reference flow of the system. At this stage, when this aggregation takes place, the spatial characteristics (place of emission) and temporal characteristics (time of emission) are generally lost. This may prove damaging for certain uses to be made of an LCA, in that the inventory is very widely spatially distributed throughout our global economy (e.g. oil production in the Middle East, refinery in the United States, manufacture of intermediary synthesis products in Europe and ore production in Asia), and also differ over time, because technologies are evolving rapidly, and therefore so too are their environmental impacts.

The data for inventories, therefore, include flows of materials (resources such as iron ore, bauxite, water and so on) and energy (oil, gas, coal, etc.) entering into the system at hand and the corresponding outgoing flows (solid waste, gaseous or liquid emissions, etc.). Lifecycle inventory databases are available – particularly relating to common raw materials, energy and transport. These data are accessible for a reasonable price in the form of public or published databases (e.g. the *Ecoinvent* database set up by the *Swiss Centre for Lifecycle Inventories*).

5.2.5. Lifecycle impact assessment

The impact assessment phase consists of explicitly stating and interpreting the results obtained during the inventory, in terms of environmental impacts, in a sufficiently summarized form as to be comprehensible for a non-specialist. This phase should enable the analyst to prepare the communication of the elementary data relating to the product's environmental impact. It is particularly sensitive.

The previous two phases – inventory and assessment – are those which are most closely linked to chemistry, in that the whole of the manufacturing process is broken down into a material balance and an energy balance, and is divided into primary inputs: oil, gas, electricity, etc. At each stage of this process, the by-products, effluents, etc. are identified, and the fate of those sub-products is assessed in terms of potential direct or indirect pollution.

In order to bring the phase of impact assessment to fruition, we must first choose the *categories of impacts* to look at (see Table 5.1), define the *impact indicators* and characterizing models, and *classify* the results of the inventory into the different categories of impacts. For each of the category indicators, we have to calculate the results (*characterization*), calculate the amplitude of those results in relation to the reference values (*normalization*), perform grouping and classification of the indicators and finally *weigh* those indicators.

Basic categories of impacts	Additional categories of impacts
Exhaustion of abiotic resources Land surface occupation Climate change Destruction of stratospheric ozone layer Human toxicity Eco-toxicity Formation of photo-oxidants Acidification Eutrophication	Loss of biodiversity Impacts of ionizing irradiations Odors Smells Drying out

Table 5.1. *Categories of impacts*

The methods upon which the impact indicators are based are drawn from various sources.

It is vital that the chosen indicator gives an appropriate representation, i.e. the weak indicator should correspond to a slight impact and the high indicator should indicate a relevant environmental phenomenon.

Therefore, the number of indicators needs to be limited – the indicators must be able to be determined on the basis of the existing data and models, and it must be possible to perform the computations in a limited amount of time and for a reasonable cost.

In reality, the category indicators represent the amount of potential impact. They are distinguished into two major types, depending on their position in the causal chain between emission and impacts: *midpoint* and *endpoint* indicators.

Midpoint indicators correspond to aggregation for each type of impact (acidification, destruction of the ozone layer, etc.). They are more easily accessible, in that they represent measurable values (pH, surface area of the ozone layer, etc.), with limited uncertainty, but their environmental relevance is low. Indeed, we do not know with precision (in an impact study or epidemiological study, say) the consequences of a change in the pH of a lake or the hole in the ozone layer. *Endpoint* indicators correspond to the effect on target criteria (loss of biodiversity, harm to human health, etc.). However, they are difficult to gain access to, in that they represent realities (disappearance of species, loss of human lives, etc.). In any case, these indicators reflect cross-disciplinary impacts. Thus, they must be exploited by chemists, toxicologists, eco-toxicologists, biologists, etc. However, it may also prove highly relevant to bring in the interpretations of philosophers, historians or epistemologists, particularly with a view to the survival or sustainable development of civilization. This issue is even more interesting in that it is often complex to choose between various, non-prioritizable indicators. Analysts tend to (falsely) agglomerate indicators in order to artificially assist in making choices, but to the detriment of the information which had been capitalized upon.

5.2.6. *Lifecycle interpretation*

The previous two phases – inventory and impact assessment – represent the domain of expertise of LCA. Indeed, the approach here is very technical, and the data voluminous. In the interpretation phase of LCA, it is the user, the responsible party, the decision maker, who uses the results of the impact assessment in order to determine the main actions to be taken, which must take other dimensions (research and development, marketing, production, finance, etc.) into consideration.

The results of an LCA are expressed in the form of a series of data which presents both the potential impacts (e.g. *X* kg_{eqCO_2} *for the greenhouse effect*) and physical flows (e.g. *Y MJ of non-renewable energy*). They are the subject of a report and, if published, a summary document for the public.

For an LCA comparing two products, A and B, the results concerning each individual impact can be expressed for each of the steps of the lifecycle, in order to compare them and identify which steps have the highest impacts. This can also help compare the contribution of each of the products to each stage of the lifecycle. From the results illustrated below (Figure 5.2), we can establish that product A has a higher contribution to the greenhouse effect than does product B in the stages of raw material extraction and production, but its recyclability at the end of life means it will actually absorb CO_2. *This representation allows us to see which stages require the most concentrated effort in order to reduce the impacts.*

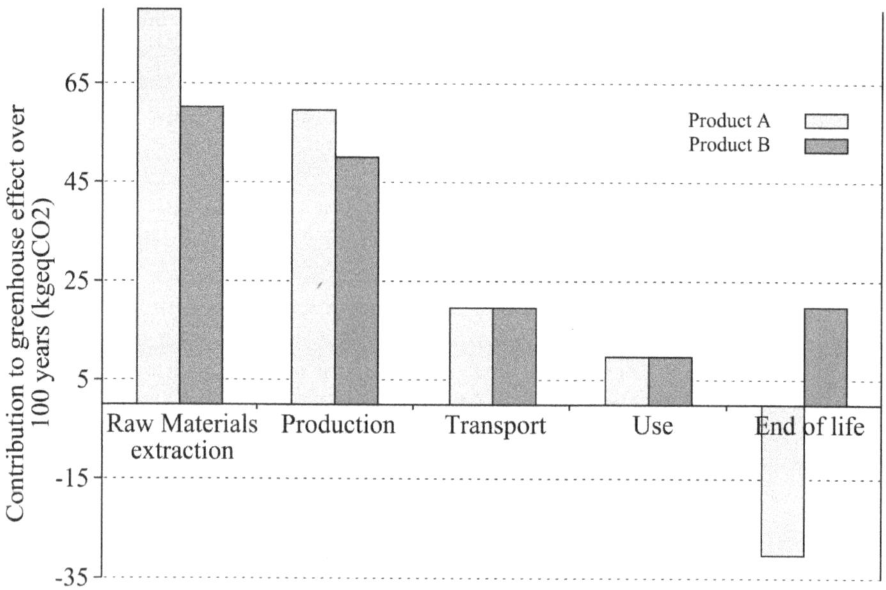

Figure 5.2. *Contribution to greenhouse effect of the lifecycle of products A and B*

The aggregation of the results concerning the contribution to the greenhouse effect per product enables us to discern macroscopic trends (see Figure 5.3) but eliminates the differences between the lifecycles of the two products. This representation enables us to make an overall choice regarding the product whose impact is lesser.

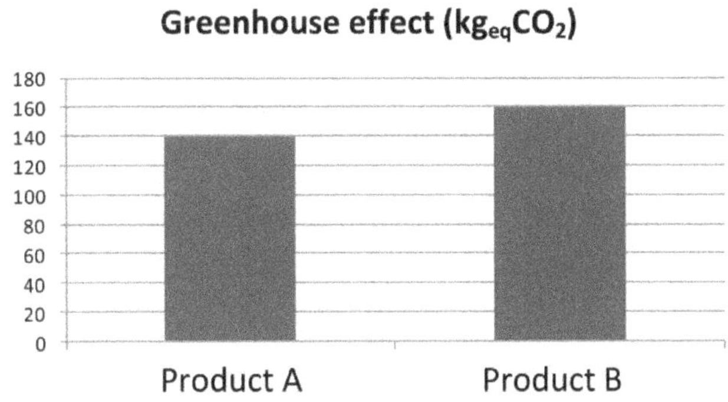

Figure 5.3. *Aggregation of the impact "Greenhouse effect"*

The essential step in the interpretation phase is the drafting of the report, which needs to contain the main elements of the LCA: a reminder of the scope and goals of the LCA, cogently argued definition of the functional unit chosen, methodology used for the LCA, sources and document bases used and the limitations inherent therein, and the technical, methodological and scientific difficulties encountered. This report necessarily must include a critical review – i.e. an examination of the study by an expert not involved in its realization. This expert may act alone or as part of a critical review board bringing together specialists in the sector in question and the main stakeholders: *what is essential is to guarantee the experts' impartiality regarding the LCA in the sector in question*. The comments and responses to the recommendations arising from the critical review must be included in the final report.

The results of an LCA can also be expressed in the form of an environmental declaration, known as a "Type-III environmental declaration", or an *eco-profile*, which can be printed on the product label. The international standard ISO 14025, published in 2006, establishes the principles and procedures for developing Type-III environmental declarations and for using the ISO 14040 series of standards to develop Type-III environmental declarations. Type-III environmental declarations as defined by ISO 14025 are primarily intended for communication between corporate organizations, but it is also possible to use them for communication between an organization and individuals, in certain conditions. The quest for improvement is the component of LCA in which the options for reducing the system's environmental impacts are identified and evaluated. This stage includes the identification, evaluation and selection of the options to improve the environmental ramifications of the products or processes.

At present, LCAs are usually used to serve certain needs, such as analysis of the environmental performances of an industrial process, environmental publicity discourse, comparison of the environmental impacts of two or more products and finding the total environmental balance.

In addition, the initial characteristics of the product generally determine the possibilities for recovery of the product at end of life. Finally, this approach presents strategic interest in terms of communication. Indeed, the results obtained using this type of approach can be communicated to the customers and render a competitive advantage which helps differentiate the product from its competitors.

5.3. Green chemistry and eco-design

The concept of green chemistry first appeared in a publication by Anastas and Warner, citing 12 principles for chemistry which is more respectful of the environment [ANA 98]. In reality, these 12 principles can be grouped together to give two rules: affordable and safer chemistry. In his later work, Anastas published the 12 principles of *Green Engineering* [ANA 03] which actually underlie an eco-designed approach. In addition to these publications, green chemistry is defined as chemistry which is respectful of sustainable development and brings together existing practices in order to reduce the environmental impacts of its products and procedures.

In this "green" chemistry, a significant place is accorded to the use of renewable resources. Indeed, the tensions relating to oil supply, in terms of prices and access to the resource, and the issues of reduction of greenhouse gas emissions, contribute to the development of chemistry based on renewable resources. In addition, industrial actors in the field of chemistry have committed to increasing the proportion of renewable resources used in chemistry in years to come. In France, through the *Union des industries chimiques* (UIC), industrialists have committed to use 15% renewable resources by 2015 (i.e. three times more than current usage). In the United States, the goals are to use 25% renewable resources by 2030.

Biomass production on Earth corresponds to around 120,000 Mt per year. However, only 5% of that biomass is captured and used by humans, and of those 6,000 Mt, only 5% is *not* destined for energy or food usage. Thus, 300 Mt are already used for chemistry. Yet these quantities are fairly close to the 500 Mt of fossil resources used annually for chemistry, so we can see the significant potential for substitution of fossil resources by increasing use of renewable resources. The resources used essential are starch and its derivatives – sugar, ethanol, etc. – cellulose, vegetable oil and glycerol, plant fibers, etc.

With regard to polymers, there are many types of bio-sourced polymers: natural polymers (cellulose, starch, etc.) which can be modified physically, either alone or in combination with other polymers; polymers produced by micro-organisms or genetically modified crops (PHAs, etc.) and polymers obtained by polymerization of bio-based *building blocks* (synthons) by chemical means or by fermentation. The renewable carbon content of bio-sourced polymers is measured by the yardstick of the ASTM D6866 standard, which is based on the presence of carbon-14 isotopes in carbonate molecules of renewable origin. Indeed, the half-life of ^{14}C, which is 5,730 years, implies the disappearance of ^{14}C isotopes in carbonate molecules of oil or coal. The presence of renewable carbon is now attested by labels such as the *Ok Biobased* label.

Also, the use of renewable resources to synthesize polymers results in the development of new scientific challenges, such as reduction chemistry, because biomass is highly oxygenated (whereas earlier, petrochemistry was based on the oxidation in air of the heavily reduced fossil resources); depolymerization of natural polymers with high molar masses; polycondensation in that the bio-sourced building blocks are polyfunctional and may not necessarily have polymerizable double bonds; and use of LCA as a decision-support tool to identify pathways with the least environmental impact.

In addition, the mere fact that the building blocks are bio-sourced does not mean that they are not dangerous, or that they necessarily produce biodegradable polymers. Thus, the substitution of molecules already synthesized from oil with identical bio-sourced molecules does not solve the problems of toxicity. Epichlorohydrin which is bio-sourced from glycerol has the same level of toxicity as does petro-sourced epichlorohydrin. Hence, for certain polymers, we need to envisage new means of access based on new, bio-sourced, non-dangerous molecules, with a view to double substitution. Thus, catechin or tannin derivatives can be advantageously used to synthesize bio-sourced aromatic epoxy resins containing no bisphenol A [NOU 10].

5.4. Limitations of the tool

LCA is, as we have seen, a particularly useful tool, because it enables us to carry out a multicriterion analysis on the whole of the lifecycle, rather than limiting ourselves to just one stage (e.g. end of life) or just one impact (e.g. carbon balance). In addition, this tool is standardized, which is because of the ISO standards describing it. Also, the same impact, evaluated at each stage, can be "added" to yield a clear vision and support decision-making. Furthermore, nothing is published without first having been reviewed by a group of experts. Finally, LCA, used as a design guide, can prove an excellent tool in assisting eco-design. While – in the hands of experts – LCA can be shown to be a very powerful tool, it does have its limitations, particularly in view of how complex it is to implement. Indeed, the required use of expert software and often costly databases means the technique is only used by a few multinationals, and that its contribution is restricted to the assessment of pre-existing products and processes. Thus, it is only usable for experts in LCA, rather than for any and all designers or creators, which imposes limits in terms of innovation. In industry, SMEs with limited resources may not be able to afford the software, experts or access to the level of expertise required to fully exploit LCA. Thus, a large portion of the production sector is excluded. Above all, LCA is based on the use of existing data. Thus, it is very risky to work on non-industrial procedures for which no data yet exist, which again proves a limiting factor.

However, while LCA enables us to identify the stages, in a given process, which have the most significant environmental impacts – e.g. the stage of raw material extraction – it gives us an *a posteriori* view, but cannot guide the path of the innovation process. Indeed, LCA is an examination performed on a product which is already industrialized or commercialized, and identifies the stages during the manufacture of that product which have the heaviest impact on the environment. The objective of an eco-design process is, in a second phase, to produce solutions in order to reduce the most significant impacts during the stages previously identified. Hence, innovation is born of the quest for solutions in a second generation, or for improvement in the manufacturing process for the product. Therefore, the primary objective of such LCAs tends to be compilation of data, drawing of an environmental balance, and production of results with a view to communication about the manufacture of a product; support for eco-design is required, secondly, because LCA is more able to evaluate the final impact of a product than to guide in its design. In addition, other limitations of this tool lie in the steps of definition of the hypotheses, in the choice of allocation rules followed or the end-of-life scenarios envisaged, as we intend to demonstrate through the examples given below.

Thus, LCA provides no solutions to help design products or processes with low environmental impact, but constitutes a guide in selecting which steps can be improved. Hence, LCA is a tool that enables us to make choices, and thereby orient our research to favor innovation. Indeed, the results of an LCA on a given initial product (see Figure 5.4) enable us to identify that the greatest *impact X* is caused during the raw material production stage. Thus, it is on this stage that research needs to be carried out, to identify avenues for access or raw materials with less of an impact, to create a new, *eco-designed* product. It is in this process of research that the innovation lies, and it is in that sense that eco-design represents a genuinely useful tool for innovation [CAI 08].

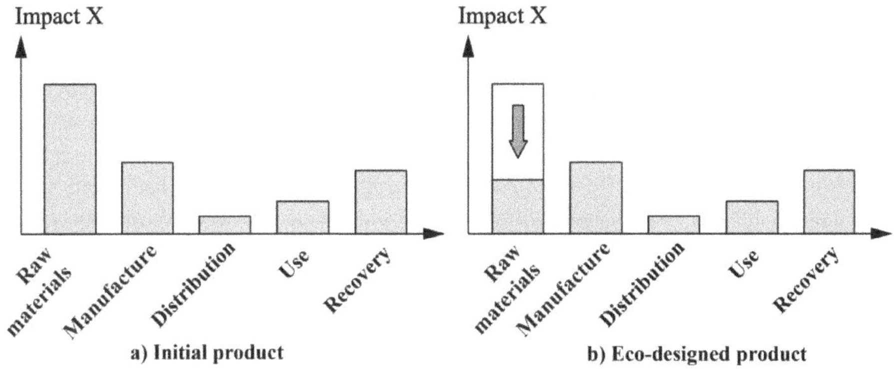

Figure 5.4. *Innovation process*

5.4.1. *Importance of the hypotheses*

The example of comparative LCA of shopping bags at supermarket checkouts is most enlightening. This study was performed in 2004 by a panel of experts in LCA, for a supermarket chain. It was supplemented by a critical review organized by ADEME [ECO 04]. The *goal* of this study was to quantify and compare the environmental impacts of four types of shopping bags available to customers at supermarket checkouts: a single-use polyethylene (PE) bag, a reusable soft PE bag, a single-use paper bag and a single-use biodegradable bag. The *inventory* was performed using data collected from the distribution company's suppliers of bags, and supplemented with data from Ecobilan's own database.

The *functional unit* defined for this study corresponds to the service rendered by the bags – i.e. packing customers' purchases in the stores. The *hypotheses* defined in the context of the study predict an average of 45 visits per year for each customer to the stores, with 200 L of purchases per visit (a trolley filled to 80% capacity), which corresponds to 9,000 L of merchandise per year. The functional unit chosen is therefore "package 9,000 L of merchandise in the stores of the Carrefour Group".

The *boundaries* of the system take account of the production and transport of the materials used to make the bags, the manufacture and printing of the bags, transport of the bags, the usage phase and the different end-of-life scenarios envisaged. The *inclusion threshold* defined is 5%: this means that the sum of the *inputs* whose production is not included in the system represents less than 5% of the total mass of inputs to the system. The inclusion thresholds are set at similar levels, depending on usual and pragmatic practices, but they may also introduce biases into the analyses: a very significant toxicological impact due to an imperfectly known impurity could alter the results.

The LCA related to the *lifecycles* of the four types of bags envisaged. By way of example, the lifecycle of the single-use high-density polyethylene (HDPE) bag first takes account of the exploitation and refining of the oil to synthesize ethylene, the polymerization of the ethylene by Ziegler–Natta catalysis, the production of the HDPE pellets and their transport. These data were taken from databases of inventories performed in 1999 by APME (averages taken over 24 European sites producing 3.87 Mt HDPE/year, which is 89.7% of Western Europe's production). The lifecycle also takes account of titanium dioxide production (data taken from a production site – 1992), calcium carbonate (data from the Swiss Ministry of the Environment) and linear low-density PE (averages from APME) which are fillers in the bag. The production of glue and ink is also taken into account. In these lifecycles, energy production is modeled on the basis of the energy ratios of electricity production in the bag-producing countries – France, Italy, Spain and Malaysia. The differences are particularly significant for the impacts concerning natural resource consumption and greenhouse gas production. Seventy-five percent of France's electricity is nuclear in origin, while 75% of Malaysia's is produced by hydrocarbon combustion. In addition, the computations performed to evaluate the impacts relating to the transport phases are based on the consumption of diesel oil by the trucks. The model takes account of the average consumption of a fully-laden truck (38 L/100 km), weighted to a third by the mass of the filler, including the influence of the return trip of the *empty* truck. Finally, the end-of-life stage was modeled on the basis of data provided by ADEME regarding household waste. Thus, 51% of waste is sent to landfill and 49% is incinerated. Eighty-eight percent of the waste incinerated is put to profitable use in terms of energy, 5% in the form of steam sold and 22% in the form of electricity sold.

The main outlines of the conclusions show that reduction in the bag's weight and reuse of the bag are two factors of prime importance which help to reduce environmental impacts. The *conclusions* of this study are thus very favorable to the use of soft LDPE bags – assuming the hypothesis that these bags are reused at least three times. Yet at no time was it hypothesized that the other types of bags might be reused (particularly the HDPE and biodegradable bags, which could withstand at least a second use, or be reused as bin liners). In addition, the study compares bags with very different volumes, which implies that rather different quantities of materials can be packaged in the different bags. However, given that the weight of the bag is a crucially important factor, it was relevant to envisage solutions which, for the same bag, make possible the larger volumes to be packaged. Thus, we can clearly identify that the LCA evaluates the selected products and helps us choose the best candidate *from among* that selection. However, experience tells us that the solutions to a problem can often be found outside of a pre-selection. "The relevance of a problem sometimes lies beyond the bounds of the data used to formulate that problem. We need to imagine the reality beyond the data we have to help grasp the situation; we need to imagine that which we do not know. To formulate or model a problem is often to misrepresent it and open the way to the wrong solutions" [BOT 08]. Thus, we have to display innovation to truly find eco-designed solutions.

Similarly to this study, another LCA [ECO 08] was performed, with a view to "rehabilitating" paper bags and showing that, for certain applications, paper bags prove to be the candidates with least environmental impacts. Thus, a study was commissioned by *Sacs papier de France* and *Procelpac* and conducted by *Ecobilan* in 2008, with the following functional unit: "a paper or plastic shopping bag intended to transport and provide good protection of the purchases, excluding high-consumption products, while constituting a reliable means of transporting the goods" – (packaging is not enough). The criteria used were the following: containment; presence of a pleat (protection); strength; surface/area for advertising; rigidity and quality of printing. In the context of this study, indeed, for all categories, the bag with least environmental impact is the paper bag. However, in this study, we envisage that the paper will be recycled, but not the plastic bag.

The example taken from the study by Kim and Dale [KIM 05] is also highly illustrative of the importance of the choice of hypotheses for the LCA. Indeed, in their study, that team compared the environmental impacts of two types of polymers: polystyrene (*PS*) – produced by the polymerization of styrene, an oil-based monomer – and a polyhydoxyalkanoate (PHA) – a polyester produced by bacterial fermentation of sugars extracted from maize grains, agricultural in origin. The comparison was performed for identical masses, in spite of any differences in properties. The results presented in this study show firstly that the impact on the greenhouse effect is lower in the case of the manufacture of polystyrene (2.9 kg_{eqCO_2} for PS and 3.5–4.4 kg_{eqCO_2} for PHA). However, secondly, the authors add into the lifecycle the production of sugars from the co-product straw, and also the recovery of energy by profitable exploitation of the straw. In this case, the production of PHA becomes an absorber of CO_2, as the indicator is –1.2 to –1.9 kg_{eqCO_2}. This example perfectly illustrates the importance of the choice of hypotheses and boundaries for the final result.

5.4.2. *Relevance of the inventory data*

Let us now return to the LCA of the shopping bags described in section 5.4.1. The LCA examined the *lifecycles* of the four types of bags envisaged. For example, the lifecycle of the single-use HDPE bag takes account of the extraction and refining of the crude oil for synthesis of ethylene, polymerization of the ethylene by Ziegler–Natta catalysis, production of HDPE pellets and their transport. However, these data were drawn from databases from inventories by APME in 1999 (averages over 24 European sites producing 3.87 Mt of LDPE/year, which accounted for 89.7% of Western Europe's production). Yet the study reveals that HDPE is made in France, but also in Asia and Brazil. Not only do the APME's inventory data on HDPE production date from five years prior to the study, but also they are in no way representative of production in Brazil or in Asia – thus, we can no longer use them to evaluate the environmental impacts. Similarly, the lifecycle of the HDPE bag also takes account of the manufacture of those bags. In this case, the data are taken from APME's inventory databases (averages from 1993 on 8 production sites in the UK); however, these HDPE bags are made in France. Hence, these data are not only outdated (they are over 10 years old) but they also only partially represent the situation in the UK and can under no circumstances be considered the representative of the situation in France – even from the point of view of energy consumption. Indeed, while France's electricity production is chiefly nuclear (78%), followed by thermal production (11%) and renewable sources (11%), this proportion is completely different to the UK, where the majority of electricity comes from thermal production (75%), followed by nuclear (20%) and renewable energy (5%) [DIR 05]. Hence, the environmental impacts are completely different (in terms of contribution to global warming) and the data cannot be used from one country to another.

5.4.3. *Influence of rules of allocation*

The rules of waste allocation can also play a crucially important role in the results of an LCA. Thus, if we compare two LCAs on bio-ethanol as a biofuel – the first conducted by ADEME in 2002 [ADE 02] and the other by EDEN in 2006 [SAD 06], the results are very different. Indeed, the energy yield of bio-ethanol produced from wheat (energy produced/renewable energy mobilized) varies two-fold between the two studies: 1.10 for EDEN and 2.05 for ADEME. These differences are attributable, in particular, to the differences in the choices made for the allocation of the waste products of the bio-ethanol manufacturing process. EDEN chose to integrate into its LCA all of the impacts generated by waste from the bio-ethanol process, favoring a systemic approach, while ADEME allotted only 43% of impact from waste. For this study, ADEME made the choice of allocation by mass – bio-ethanol, the target product for the process, represented only 43% of the dry material used. Both these options can be chosen, but it is important to understand that the results may be radically modified as a result.

5.4.4. The choice to recycle

With regard to the rules for comparison on the end-of-life stages, *Centre interuniversitaire de recherche sur le cycle de vie des produits, procédés et services* (CIRAIG) draws our attention to the relevance of comparing different recovery scenarios. Indeed, it is normally impossible to directly compare the environmental impacts generated by two means of recovery of a product. If we want to process

1 ton of old papers, we cannot directly compare *recycling* – which will produce *Y* kg ($Y < 1{,}000$) of recycled paper – and *thermal valorization* – which will produce *X* MJ of electricity. In the first case, we will still need electricity production, and in the second, we will still need paper – these two systems do not provide the same service. Hence, we need to supplement this comparison but adding to each system the process which is avoided by adopting that option (see Figure 5.5).

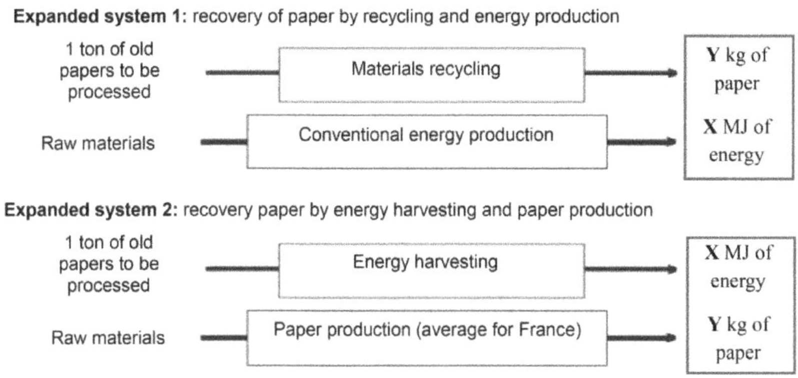

Figure 5.5. *Comparison of two expanded systems*

Thus, in the first case – that of paper recycling – in evaluating the impacts, we must also take account of the production of *X* MJ of electricity in the local conditions for conventional electricity production. In the second case – that of thermal energy harvesting – we also need to consider the conventional production of *Y* kg of paper from wood, when evaluating the impacts. We can then perform a valid comparison of two systems which have the same products: *Y* kg of paper and *X* MJ of electricity.

5.5. Conclusions: the future of eco-design

The design processes in industry, and in particular the chemical industry, are in the full throes of mutation at present. They now need to respond to an overarching need to reduce environmental impacts at every stage in the production process. They have to integrate eco-design of the product and of the process. In this approach, LCA is a crucial tool in evaluating the environmental impact. In doing so, LCA, which identifies areas for improvement in terms of environmental impacts, energy and resource consumption, becomes a strategic innovation tool. Indeed, this tool helps to guide our efforts in R&D to identify innovative solutions with lesser environmental impacts, leading to new products which are "greener", eco-designed, serving the ever-more-stringent requirements of the market and of legislation. However, the process of LCA is carried out *a posteriori* on an existing product or process, and enables us to analyze the environmental impacts of that product or process. The results of that LCA show the stages whose impact on the environment is greatest. It is these stages which we must then focus on to reduce the environmental impact, but only in the phase of product improvement, or during the design phase for the so-called "second generation" of that product. At first glance, LCA is only able to compare the environmental impact of two products to help find the best compromise. However, as we saw earlier, the solution to a problem may lie outside of the pre-selected elements, as may the eco-designed solution. Thus, the tool must be made to evolve and become more usable in an innovation approach.

Yet the most restrictive limitations of this tool are qualitative and methodological. We have qualitative limitations because the relevance of the data is of fundamental importance in evaluating the impacts, and these data are not always relevant or updated in the databases – they are not always representative of the local reality. We also see methodological limitations, because this tool only offers us a comparison at a given moment in time, is only capable of evaluating relative impacts and cannot take account of the margin for progress of the technologies which it is comparing. In addition, the definition of the hypotheses, the boundaries, the functional unit, the allocation rules obeyed or the end-of-life scenario envisaged, can noticeably modify the results. Thus, it is desirable that this LCA tool should evolve, to remedy these limitations and better evaluate certain impacts relating to toxicity and nuisances. Furthermore, in order to satisfy the new constraints on the processes of innovation, to be located at a very early point in the phases of projects, to accompany the design of the products and processes in the chemical industry and to take account of the new regulatory aspects, we need new tools which give orientations to follow to guide researchers and chemists in their choices. It becomes more important to assist the innovation process with a control support tool, *gate to gate*, rather than conclude it with an exhaustive *a posteriori* assessment. Also, it is essential to extent this environmental design to all projects in the chemical industry, in order to make eco-design a reality in the industry. To this end, the designers of products and processes in the chemical industry need an appropriate tool, which is easy to use rather than being the preserve of experts in environmental balances; a tool which can guide them as to the final environmental impact, from the very beginning of the process when the access pathways are chosen. Industrialists also need to expand the collection of inventory data, and share these inventory data to contribute to keeping the inventory databases up to date, which is the real "Achilles' heel" of LCA. In addition, it is crucial to link the inventory databases with those classifying dangerous substances. In today's world, how can we envisage identifying a chemical access pathway without anticipating the constraints imposed by the regulations, and in particular, the REACH legislation? It is also of crucial importance to be able to render the use of such a tool a viable option for smaller enterprises, so as to improve their competitiveness. For this purpose, we need to give them a tool which is usable at every stage of the project, from very early on in the innovation phase, including guiding the choices made in terms of R&D. The existing tools are not necessarily able to serve this objective, particularly for SMEs in the chemical industry, who want a simplified frame of reference to help them integrate the concept of sustainable development into the design of their products. Thus, while quality management began being implemented around 1992 by the major corporate groups, it was not until 2004–2006 that smaller structures began using these techniques. Today, in the industrial sector, only a quarter of companies envisage end-of-life scenarios for their products, and eco-design practices are only in place for 40% of them [REY 08]. Finally, it is of prime importance to relay the commitment of companies in research and teaching, and help to mobilize public research teams in this domain.

In any case, eco-design is a crucially important concept, which gives rise to change in chemistry. Indeed, it forces producers to adopt a holistic assessment of their processes and an all-encompassing approach, and avoid any shift in pollution. It also enables us to justify the choice of a pathway with lesser environmental impact, or steer away from a path which seemed, at first glance, to be better; thus, it is a decision-making tool, weighing up the choices in search of a reduction in environmental impact. In that sense, while chemists, mechanics experts, biologists, toxicologists and eco-toxicologists are inevitably involved, it is also important to bring in historians, philosophers and epistemologists when interpreting the results of LCAs with a view to sustainable development of civilization. Indeed, the notions of negative external factors which are the environmental impacts need to be envisaged in light of the progress made from a historical and philosophical perspective. Thus, LCA is merely an environmental analysis, and should be accompanied by a social component – by way of a social LCA – in which the place occupied by historians, economists and philosophers could be still more significant. Finally, eco-design causes a change in processes by justifying recourse to re-use and recycling, and the circular economy insofar as prolonging the lifecycle is the preferred means of reducing environmental impact. As Antoine de Lavoisier's maxim tells us, essentially stating the first law of thermodynamics, "nothing is destroyed, nothing is created; everything is transformed", eco-design aims to reduce losses and new creations, but encourages (re)transformations. However, this circular evolution of the economy, of the resources used by chemistry (which, having once been based on renewable resources, then coal and then oil, is now turning back toward renewable resources) also need to be supported by the thinking of historians and philosophers.

5.6. Bibliography

[ADE 02] ADEM, Etude ADEME/DIREM, 2002.

[ANA 98] ANASTAS P.T., WARNER J.C., *Green Chemistry: Theory and Practice*, Oxford University Press, New York, 1998.

[ANA 03] ANASTAS P.T., ZIMMERMAN J.B., "Design through the Twelve Principles of Green Engineering", *Env. Sci. Tech.*, vol. 37, no. 5, pp. 94A–101A, 2003.

[AZÉ 07] AZÉMAR M., *Les sciences chimiques au service de l'environnement*, Pollutec, Paris, 2007.

[BOT 08] BOT L., "A quoi la philosophie peut-elle servir dans une formation professionnalisante?", *Actes du colloque Questions de pédagogies dans l'enseignement supérieur, Enseigner, étudier dans le supérieur: pratiques pédagogiques et finalités éducatives*, Brest, Telecom Bretagne, ENSIETA, Naval College, University of Western Brittany, June 2008.

[BRA 98] BRADFORD DE LONG J., *Estimating World GDP, One Million B.C.–Present*, 1998.

[CAI 08] CAILLOL S., Analyse de cycle de vie et écoconception: les clés d'une chimie nouvelle, The annals of Mines industrial realities, November 2008.

[CHR 07] CHRISTIAN AID, *Human Tide: The Real Migration Crisis*, May 2007.

[DIR 05] DIRECTION GÉNÉRALE DE L'ÉNERGIE ET DES MATIÈRES PREMIÈRES, Observatoire de l'énergie 2005.

[ECO 04] ECOBILAN PWC, Evaluation des impacts environnementaux des sacs de caisse Carrefour, 300940BE8, 2004

[ECO 08] ECOBILAN PWC, Evaluation des impacts environnementaux des sacs boutique, 2008.

[GRE 07] GRENELLE ENVIRONNEMENT, ministère de l'Ecologie, de l'Energie, du Développement durable et de la Mer, 2007.

[INT 10] INTERNATIONAL ENERGY AGENCY, *World Energy Outlook* 2009 and 2010.

[IRS 09] IRSN, Baromètre de perception des risques et de la sécurité 2006, 2007, 2008 and 2009.

[JOI 10] JOINT OPERATING ENVIRONMENT, JOE 2008 and 2010.

[JOU 10] JOURNAL OFFICIEL, Article 8 du décret du 15 October 1810 relatif aux manufactures, 1810.

[JOU 00a] JOURNAL OFFICIEL DE L'UNION EUROPÉENNE, Directive 2000/60/CE, L 327, 22/12/2000.

[JOU 00b] JOURNAL OFFICIEL DE L'UNION EUROPÉENNE, Directive 2000/53/CE, L269, pp. 34–43, 21/10/2000.

[JOU 03] JOURNAL OFFICIEL DE L'UNION EUROPÉENNE, Directive 2002/96/CE (modifiée 2003/108/CE), L345, pp. 106–107, 31 December 2003.

[JOU 04] JOURNAL OFFICIEL DE L'UNION EUROPÉENNE, Directive 2004/42/CE, L143, pp. 87–96, 30 April 2004.

[JOU 05] JOURNAL OFFICIEL DE L'UNION EUROPÉENNE, Directive 2005/32/CE, L 191/29, 6 July 2005.

[JOU 08] JOURNAL OFFICIEL DE L'UNION EUROPÉENNE, Directive 2008/28/CE, L81/48, 11 March 2008.

[KIM 05] KIM S., DALE B., *Int. J. LCA*, vol. 10, no. 3, pp. 200–210, 2005.

[MIN 11] MINOIS G., *Le poids du nombre, L'obsession du surpeuplement dans l'Histoire*, Perrin, Paris, 2011.

[MOR 49] MOREAU J., "Les théories démographiques dans l'Antiquité grecque", *Population*, vol. 4, no. 4, pp. 597–614, Press Editions 1949.

[NOU 10] NOUAILHAS H., BURGUIERE C., CAILLOL S., *et al.*, Procédé d'élaboration de nouvelles résines thermodurcissables à partir de composés phénoliques naturels, FR 0902589, 2009, WO 2010136725, 2010.

[REA 06] REACH, Règlement (CE) no. 1 907/2006 du Parlement européen et du Conseil du 18 December 2006.

[REY 08] REYES T., Report, integrating the environment into product design, UTT, 2008.

[SAD 06] SADONES P., Les agrocarburants, Report EDEN 2006.

[THO 98] THOMAS-ROBERT M., *Essai sur le principe de population*, J. Johnson, London, 1798.

[UIC 08] UIC, Bilan de l'industrie chimique en France, 2008.

[UNI 69] UNITED NATIONS, Problems of the human environment: report of the Secretary-General – Forty-seventh session of the Economic and Social Council, 1969.

[UNI 00] UNITED NATIONS, Long-Range World Population Projections: Based on the 1998 Revision, 2000.

[USG 11] US GEOLOGICAL SURVEY, 2011.

[VAL 45] VALÉRY P., *Regards sur le monde actuel*, Gallimard, Paris, p. 23, 1945.

6

Are Bioplastics "Green" Plastics?

6.1. Introduction

In recent years, the image of plastics in the media and therefore in public opinion, particularly in France, has become extremely poor. An opinion poll conducted by the plastic producers union *PlasticsEurope* shows that, indeed, the image has considerably suffered in France, where, between 1997 and 2011, it declined from 67 to 35% positive opinion; this decline is less dramatic in other European countries (in terms of overall European perception of plastics, the same survey revealed that over 40% of European citizens expressed a desire for a decrease in the use of plastics, and 28% actively spoke out against plastics). These materials which, only a few decades ago, were a symbol of modernity because of their durability, excellent performances and low cost, have now become a symbol of the most damaging excesses of the consumer society, and are held to be the cause of a great many problems, in terms of both environment and public health. We need only type "plastics" into a search engine and look through the first few pages of results to see the dreadfully low popularity of plastics at present, and the controversies surrounding them: "Plastic pollution is harming increasing numbers of seabirds"; "Should we do away with plastic bags?"; "Plastic waste pollution reaches record levels"; "Plastic bottles – mounting concerns"; etc. Television reports show the stomachs of cormorants filled with large pieces of plastic rubbish, or genital deformities in both animals and humans suspected to be caused by residual additives or monomers present in certain plastics which act as endocrine disruptors, and we frequently hear talk of the "seventh continent of Plastic".

In this chapter, we shall avoid adopting an entrenched position between denouncing a media firestorm which is perhaps exaggerated, and at the very least

---

Chapter written by Olivier TALON.

simplistic (because it too often skips over the different impacts which would be caused by certain alternative options), and denouncing plastics in general, as do the media and public opinion. Rather, we shall simply note that there are a certain number of very real problems, which need to be dealt with. Increasingly, in their conclusions, reports are mentioning bioplastics (or rather "bioplastic", as it is common simply to speak of "plastic" rather than "plastics") as a ray of hope, a miraculous solution to all the problems with conventional plastics. However, this obviously idyllic presentation of bioplastics should inevitably raise a certain amount of reasonable doubt, and it is worthwhile analyzing things properly, asking how these bioplastics could actually solve some of the environmental problems, but also in what aspects they still leave room for improvement in terms of the paradigm of sustainable development. This reasoned discussion is necessary so as to avoid arousing excessive hopes by presenting these materials as a panacea – hopes which would ultimately be dashed, which would inevitably, and undeservedly, take the wind out of the sails of the development of such materials.

Thus, in this chapter, we shall attempt to answer the question "Are bioplastics 'green' plastics?", firstly by way of a brief literature review on the solutions that lifecycle assessment (LCA) methods are able to bring, also examining the limitations of these methods, and then by putting forward an additional reference framework – one which is more qualitative – established on the basis of the 12 principles of green chemistry. In the interests of clarity, it should be noted that the term "bioplastic" is to be understood here in terms of nature rather than properties. Thus, for our purposes, a bioplastic is a plastic material created, at least partially, from or with renewable resources, exclusively. Hence, we shall not look at biodegradable polymers derived entirely from petroleum, though certain authors consider these products also to belong to the family of bioplastics.

6.2. Bioplastics and LCA – some basic points

6.2.1. *Overview of methods used and results obtained*

Table 6.1 shows data published by Landis and Theis in 2006 [LAN 06], on the impact categories examined in a representative number of publications dealing with LCA of bioplastics, now supplemented by a similar number of works published since then. For the latter, the impact categories examined were not systematically indicated by the same terminology, so we have attempted to amalgamate the terms in order to make the table consistent.

Reference	GW	FF	E	HH	WU	EQ	LU
Gerngross and Slater [GER 00]	■	■					
Kurdikar *et al.* [KUR 00]	■						
Slater *et al.* [SLA 02]	■	■					
Akiyama *et al.* [AKI 03]	■	■					■
Dornburg *et al.* [DOR 03]	■	■					■
Lynd and Wang 2003 [LYN 03]		■					
Vink *et al.* 2003 [VIN 03]	■	■			■		
Kim [KIM 05]	■	■	■	■		■	
Vink [VIN 07]	■	■			■		
Murphy *et al.* [MUR 08]	■	■	■	■		■	
Madival *et al.* [MAD 09]	■	■	■	■		■	■
Häkkinen and Vares [HAK 10]	■	■					
Hermann *et al.* [HER 10]	■	■	■		■	■	■
Groot and Borén [GRO 10]	■	■	■			■	■
Hermann *et al.* [HER 11]	■	■					
Kendall [KEN 12]	■	■	■			■	

Table 6.1. *Evolution of impact categories examined in lifecycle assessments of bioplastics (GW: global warming; FF: fossil fuel use; E: eutrophication; HH: human health; WU: water use; EQ: ecosystem quality; LU: land use)*

Looking at this table, we can clearly see that up to a period which began in around 2005, environmental impact studies on bioplastics focused almost exclusively on global scale impact categories such as global warming and fossil fuel use. The trend thereafter has turned toward the development of multi-criterion analyses. There are a number of reasons behind this change in direction. To begin with, it should be understood that LCA is a relatively new discipline, and that studies on the carbon balance were the first to be carried out (a) because they were simpler to perform and required easier data collection, and (b) because their results corresponded directly to society's expectations, as global warming was a concept of which there had been public understanding for a number of years (which is not true of eutrophication, for instance). Another reason, less clearly defined and less certain, but which cannot be discounted *a priori*, relates to the fact that such studies are often carried out in conjunction with industrial actors whose objective is to publicize the qualities of their products. Thus, it is not hard to see that these industrial actors would have given preference to studies limited to the points with regard to which the environmental benefit of their products was most probable and evident. Indeed, the bio-sourced nature of bioplastics derived from agrarian resources lends them an intrinsic added value in terms of carbon balance, in that they were not created using fossil resources and (even better) constitute a means of storage of atmospheric CO_2. On the other hand, the more local impacts relating to ecosystem quality or human health are likely to be affected by the agricultural practices leading to the production of the raw material (use of fertilizers, pesticides, etc.). However, these ulterior motives are probably no longer true of more recent studies, which no longer gloss over this problem of potential shifting of impact when we need to compare a bioplastic to a petro-sourced plastic.

Thus, we can see that the studies published on bioplastics generally look at different impact categories, and rarely with the same computational methods. In addition, we need to take account of the fact that the goals and scopes of the studies are also variable, as are the functional units defined (simple lifecycle inventories as well as complete LCAs can be found; some studies are performed from cradle to factory gate for pellets or finished products, whereas other studies also integrate one or more end-of-life scenarios). In such conditions, it is understandable that it is not easy to write a coherent summary of publications relating to LCA of bioplastics in order to provide a simple response to the question "Are bioplastics 'green' plastics?"

A few choice examples can be used to illustrate the diversity of the studies, and to highlight the type of results they produce.

Vink, in 2003 [VIN 03], published an early version of a cradle-to-grave LCA of poly(lactic acid) (PLA) produced by NatureWorks, from maize (certain data on this product were updated in 2007 [VIN 07] and then in 2010 [VIN 10] in view of the actual and predicted evolutions of the processes used). This comparative study (see Figure 6.1) shows a lesser consumption of fossil energy in the lifecycle of PLA than in that of traditional polymers taken as reference points. On the right-hand side of this figure, we can see the significant difference between the results for "PLA1" and "PLA BioWP". This difference is due to the effect of predicted process improvements, such as use of co-products and of wind energy. Here Vink introduces a question which repeatedly appears in studies comparing the environmental impact of bioplastics, which are relatively new materials for which the processes are not yet optimized and which present significant room for improvement, to that of traditional plastics for which the production and transformation processes have, some considerable time ago, reached a certain level of maturity, and whose environmental balances are not, *a priori*, expected to evolve significantly: should we make decisions on the basis of the result obtained for a given technology at time *t*, or integrate the probable evolutions of the emerging technologies into the decision?

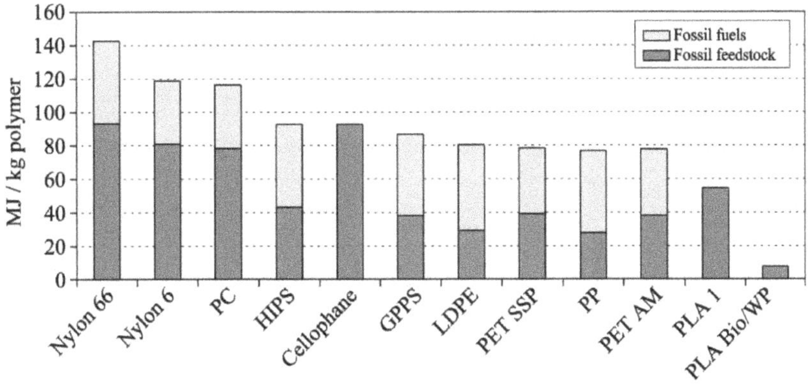

Figure 6.1. *Fossil energy required for PLA and for a number of petroleum-based polymers [VIN 03]*

Kendall, in 2012 [KEN 12], used the LCA method to evaluate the environmental impact of a potential industrial installation using municipal waste as a raw material for the production of poly(hydroxyalkanoates) (PHAs) and compared that impact to that of a PHB produced from maize material. This is what is known as a *cradle-to-gate* study – up to the production of the plastic pellets. The approach used consists of using data taken from previous studies on processes similar to each step in the process envisaged (waste processing and separation, fermentation and post-fermentation stages). Figure 6.2 shows that, both in terms of primary energy and of contribution to global warming, the impact of such a technology could be significantly lesser than that of production of PHAs from maize expressly cultivated for this purpose. It should be noted that here, we touch on the solution to one of the controversies often facing bioplastics: food competition (consider, for example, the relative impact of the nuanced proposals of the designer Philippe Starck who, not so long ago, compared biomaterials to a "crime against humanity" [EGO]) which weighs against their development, and led to an effort to no longer use noble fractions of a plant cultivated primarily for that purpose (sugar or starch) as a fermentable raw material for biotechnological processes of production of biopolymers or biomonomers, but instead to use by-products or waste.

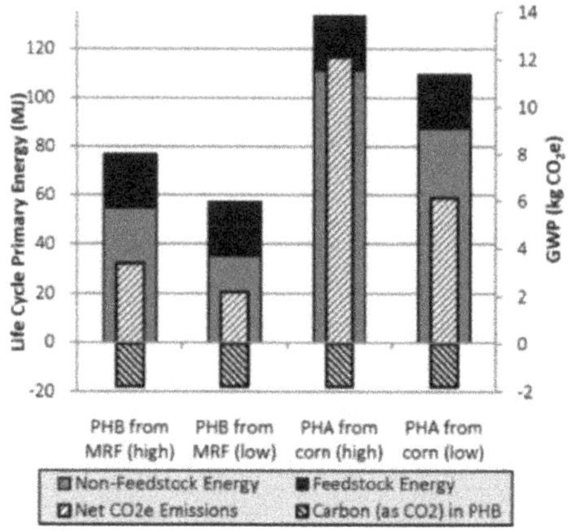

Figure 6.2. *Environmental profiles of PHAs produced from waste products (MRF) and maize [KEN 12]*

The question of transportation steps in the lifecycle of a plastic article is clearly dealt with by Madival *et al.* [MAD 09], who compare the impacts of strawberries depending on whether they are transported across the USA in packaging made of PLA, PET or PS. Figure 6.3 shows the results obtained for different impact categories. It emerges from this study that the transport phase for the packaged strawberries is the phase with highest impact in all cases (and that impact is obviously invariable or almost invariable). Hence, the values here are presented in order to set this phase apart from the rest, and the results are normalized in relation to this phase.

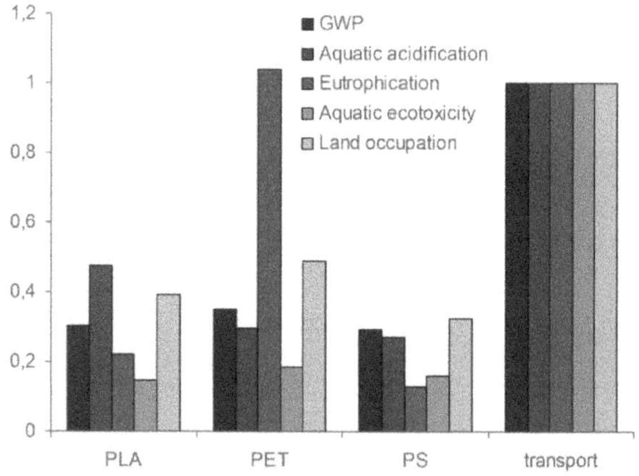

Figure 6.3. *Environmental impact of strawberries transported in clamshell packaging in the United States [MAD 09]*

Here, we can clearly see the importance of defining the system boundaries, and choosing an appropriate functional unit in an LCA. Indeed, if the question is "Is it better to package strawberries in PLA, PET or PS?", the response given by these results seems to be to avoid PET, but a more detailed analysis may need to be carried out in order to decide between PS and PLA. If, however, the question is "Which packaging should be used for strawberries which, after conditioning, are to be shipped the length and breadth of the United States?", the response appears to be "it makes little difference".

6.2.2. *Limitations of the LCA methodology for the study of bioplastics*

As we have seen, the diversity of LCA studies, both in terms of the form and their objectives, makes it difficult to give a simple summary. However, this diversity is not the only limitation to the LCA method – particularly so in the context of a study of bioplastics. As previously mentioned, bioplastics fall down because of the problem of lack of maturity of the technologies used for their production. In case of a comparison between a "bioplastic" solution and a "conventional plastic" solution, this lack of maturity may partially or totally offset the benefit we expect to be able to draw from using materials from renewable sources. The choice of input data for the computation of the lifecycle is of crucial importance, and it is important for future technologies to ask the question of whether to use real and current data or projected future data.

Apart from the question of the choice of data, there is also the question of the quality of those data. From this point of view, bioplastics are plagued by a host of problems. Firstly, the modeling of end-of-life scenarios in LCA is, as yet, often tainted with major uncertainty, particularly in terms of the fate of gas emissions in landfill, but also in terms of biodegradation. Yet we often observe a fairly strong dependency of the results on this part of the lifecycle [HAK 10]. In addition – and this point is particularly relevant for bio-sourced materials – it is very difficult to integrate the local characteristics of the territories for production of the resources (agricultural practices, nature of soil, weather, etc.) into the computations (because these details are usually absent from the databases, and are not easy to collect in the form of primary data). Certainly, it is possible to find average data – say, on a European scale – but this is not always appropriate for projects under consideration at a more regional scale [BOI 12]. However, the agricultural phases, in the case of bioplastics, account for a significant proportion of some of the impacts. By way of example, the use of SimaPro to compute the impact of PLA of the database using the CML 2 baseline 2000 method shows that the maize production phase accounts for 53% of eutrophication or 40% of acidification due to the lifecycle of the PLA granulate.

6.2.3. *Advantage to an additional qualitative approach*

While LCA of bioplastics is capable of providing us with interesting data in terms of evaluation of their environmental impact, as we saw earlier, there are certain limitations, the list of which could still be added to. In addition, we must consider that LCAs are quantitative studies, which often need to be boiled down to a unique score, or aggregated to a lower level. This method is very useful for identifying the stages which produce the greatest impact in a series of processes, and therefore those to which most effort needs to be devoted with a view to eco-design. In the context of a comparative study, the method also helps to identify impact shifts when a product or process is substituted (e.g. a decrease in greenhouse gas emissions counterbalanced by an increase in soil acidification). However, it is sometimes difficult to tell whether this shift in impact should lead to the adoption of the substitution or the maintenance of the previous technique; computational methods do exist, but all of them are open to debate. As we can see, while LCA is an undeniably useful tool, it does present the danger – in that it presents figures (which are what remain once we have ironed out the uncertainties, the hypotheses of approximation and of weighting, in order to draw conclusions) – potentially leading to imperfect or even partially erroneous conclusions and decisions. Therefore, it seems useful to accompany LCA with a more qualitative approach, with no aim to express the results in numerical form, which – precisely because it is not quantitative – will enable us to take account of certain aspects which are overlooked by an LCA, either *a priori* or because of lack of data, or as a consequence of the propensity to concentrate on a unique figure, or simply on the figure which is considered the most important. With this in mind, the next section looks at bioplastics within the framework of the 12 principles of green chemistry.

6.3. Bioplastics in light of the 12 commandments of green chemistry

6.3.1. *The twelve principles of green chemistry: a reference framework*

In 1998, Anastas and Warner [ANA 98] compiled a list of what they called the 12 principles of green chemistry. They can be transcribed as follows:

1) Thou shalt avoid waste.

2) Thou shalt design safe products.

3) Thou shalt design safe processes.

4) Thou shalt use renewable feedstocks.

5) Thou shalt use catalysts rather than stoichiometric reagents.

6) Thou shalt avoid temporary modifications.

7) Thou shalt incorporate all material into thy product.

8) Thou shalt use safe reaction conditions.

9) Thou shalt optimize energy efficiency.

10) Thou shalt keep end of life in mind.

11) Thou shalt prevent pollution.

12) Thy products shall be safe.

Note that, if we do not go into the finer details of their meaning, these "12 commandments" can be reduced to a low number of meta-rules (limiting harmful emissions or waste products, optimizing yields in relation to energy resources or raw materials, limiting the number of stages and limiting the risks) into which all of these fit. Note, also, that the term "commandments" used here, partly as a play on words, may ultimately correspond more closely to what Anastas had in mind than does the term "principles", because what he is proposing here is more a set of guidelines defining a conceptual framework than a list of principles in the scientific sense of the term.

The aim of this conceptual framework is to guide the chemist in his approach to design; the more faithfully these principles are respected, the more benign the developed product can be considered. In order to determine whether a product can be considered "green", it thus seems consistent to follow an inverse approach and use this framework as a referential guide to check which of these 12 commandments the product actually obeys, and envisage how it might be possible to improve its performances in relation to other principles from this list.

6.3.2. *Examples of use of this referential framework*

Our aim here is not to give a complete overview of all bioplastics viewed in the light of all these 12 commandments, but rather to illustrate the approach using a number of examples.

The first principle stipulates that waste should be limited when making products. In the case of PLA, it is known that during the phase of production of lactic acid by fermentation, the process leads to the formation of gypsum (calcium sulfate) in large quantities (one mole of $CaSO_4$ per mole of lactic acid, or approximately 1 kg of gypsum per kg of PLA) [VIN 10]. This waste product can be viewed instead as a co-product, which can be used in cement-making, for example; it can also be used as a filler in a PLA matrix, which is an elegant method [MUR 09]. Nevertheless, though marketable or reusable, it is still a waste product, of which it is useful to minimize production. Thus, in 2007, NatureWorks announced that for the next generation of its PLA, it was expecting to achieve an 85% reduction in the coproduction of gypsum [VIN 07].

The third commandment encourages the chemist to develop safe procedures, meaning that the substances used and produced should be the least harmful (toxic) as they can be, both for humans and for the environment. In 2012, Alvarez-Chavez *et al.* drew up a list of points of concern in terms of environmental danger and danger to health for numerous bioplastics [ALV 12]. A number of these dangers are not specific to bioplastics (the toxicity of isocyanates, for example, would equally be a point of concern for a bio-sourced polyurethane as for a petro-sourced polyurethane); the rest are linked mainly to the agricultural practices of the production of the raw material, such as exposure to pesticides or use of GM crops. The toxicity and potential for accumulation of certain reagents used in the synthesis of PLA (for example, octanol to initiate polymerization or tin-based catalysts) are also highlighted, as are the steps of extraction and purification in biotechnological processes (in the case of production of PHAs by fermentation, the use of pyridine, methanol or diethyl ether is cited). The extraction and purification of PHA during production by fermentation indeed require 50–150 L of solvents per kg of plastic produced [NOD 98].

Of the 12 principles of green chemistry, that which is most often used to advertise the merits of bioplastics is, quite rightly, the fourth, which recommends using renewable resources wherever possible. The justification of the environmental added value of biomaterials in light of this principle lies essentially in what Narayan [NAR 11] calls the "intrinsic carbon footprint" of the products. The intrinsic carbon footprint of a plastic can be defined as the mass of "old" carbon given off into the atmosphere at the end of life (e.g. in a scenario of incineration) per 100 kg of that plastic. By looking at Figure 6.4, we can easily see that this footprint will be zero for an entirely bio-sourced product, and directly proportional to the mass fraction of the carbon for a product made from fossil resources. In fact, it is on the basis of a determination of the amount of carbon 14, i.e. "new" carbon, that we determine the bio-sourced nature of a product made with renewable resources in the sense of the ASTM D6866 standard.

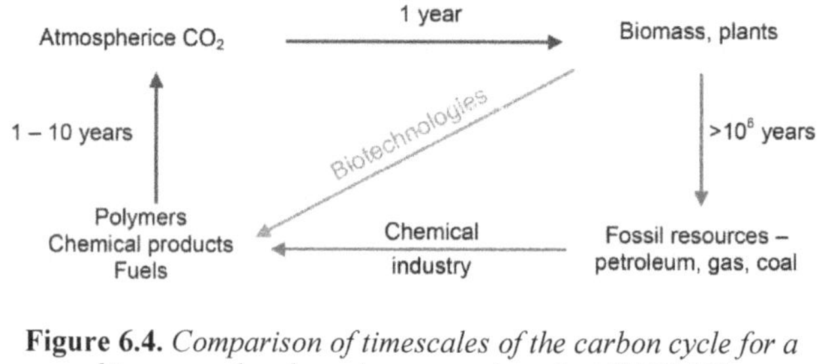

Figure 6.4. *Comparison of timescales of the carbon cycle for a bio-sourced and a petro-sourced product [NAR 11]*

The seventh commandment stipulates that we avoid atomic wastage. The mass ratio of incorporation of input ingredients into the final product thus needs to be optimized. It is possible, using biotechnological and chemical processes, to obtain several bioplastics from sugar. Figure 6.5 gives a simplified view of the path followed to produce PE, PHB, PLA and PBS from sugar. In this diagram, of particular note is the fact that the bio-ethanol path taken to produce bio-PE involves a release of CO_2, while the production of succinic acid by fermentation requires an input of CO_2, and therefore constitutes one possible means of storage of atmospheric carbon. The production of PLA, for its part, appears to be carbon-neutral. Thus, the same molecule of glucose ($C_6H_{12}O_6$) can be used to obtain several bioplastics, including PE $(C_4H_8O_0)_n$, PLA $(C_6H_8O_4)_n$, PHB $(C_4H_6O_2)_n$ or PBS $(C_8H_{12}O_4)_n$. Based on these arrangements, we can therefore establish the theoretical atomic yields which are possible, at most, to achieve for these four polymers (Figure 6.6).

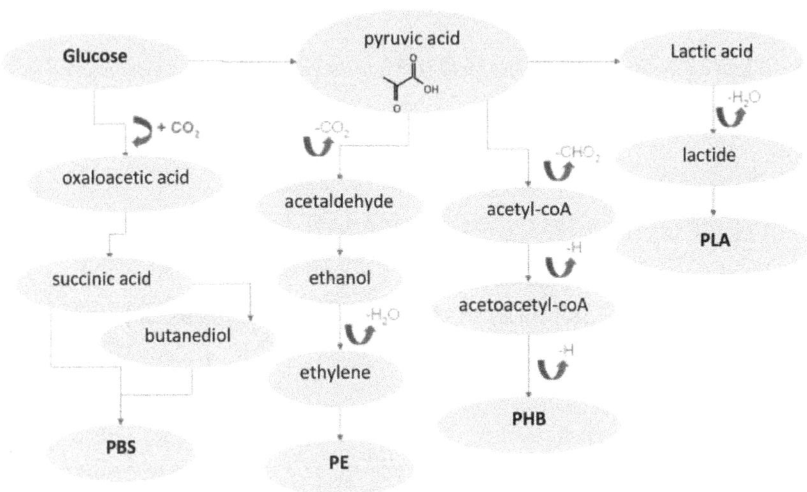

Figure 6.5. *Pathways for synthesis of different bioplastics from glucose*

A microbiologist would, rightly, protest that these yields are merely hypothetical and in no case achievable. In fact, these yields are those which are obtained on the condition that all chemical processes take place with no losses, and also that the bacteria used in the fermentation phases use all of the raw material for the production of the chosen metabolite, which is never the case in reality. To illustrate this point, we can take the example of the production of succinic acid by *Escherichia Coli*. As shown in Figure 6.7, the actual yield of succinic acid (calculated in percentage of carbon taken from the sugar being converted) is a long way from the theoretical value of 133% announced above – instead it is only 66%. This is, to a large extent, due to the fact that the metabolism of that bacterium produces not only succinic acid, but also formic, acetic and lactic acid, as well as ethanol and CO_2. One technique used to improve the metabolic yields of bacteria in biotechnological processes is to genetically modify the strains used in order to block certain metabolic pathways so as to favor those which are of interest to us. Figure 6.7 also shows how, in the case of *E. Coli*, it is possible with this technique to increase the yield from 66 to 113% [SAN 10].

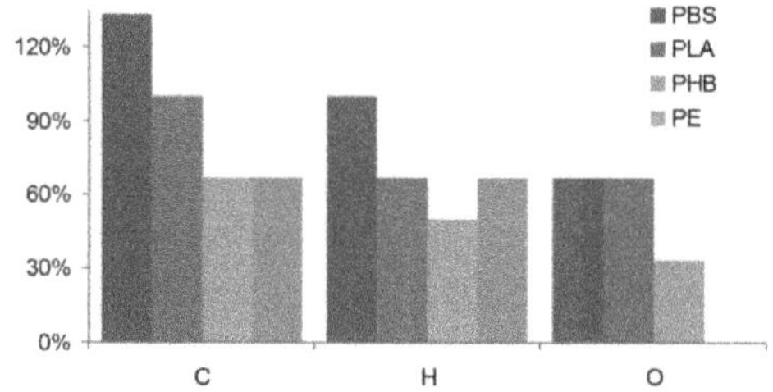

Figure 6.6. *Maximum atomic yields for four bioplastics derived from sugar*

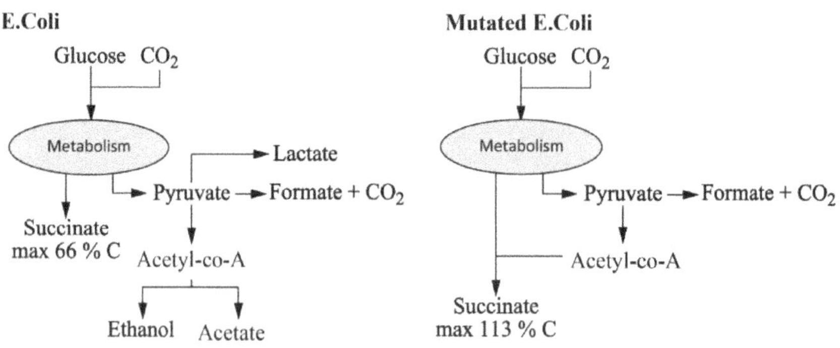

Figure 6.7. *Improvement of succinic acid yield from fermentation by a mutated strain of E. Coli. (data taken from US7790416 [SAN 10])*

The ninth commandment stipulates that we seek to optimize the energy efficiency of the processes. If we look, this time, not at the production of bioplastics but at another stage in their lifecycle – the implementation of the finished products – we can see that PLA, in the applications for which it is likely to be in competition with PET, can offer significant energy savings because of its lower transformation temperatures. Figure 6.8 illustrates this point in the case of production of bi-oriented films. The same type of profile can be established for injection-stretch blow molding of bottles, where we can achieve energy gains of around 30% in the blow-molding stage by switching from PET to PLA.

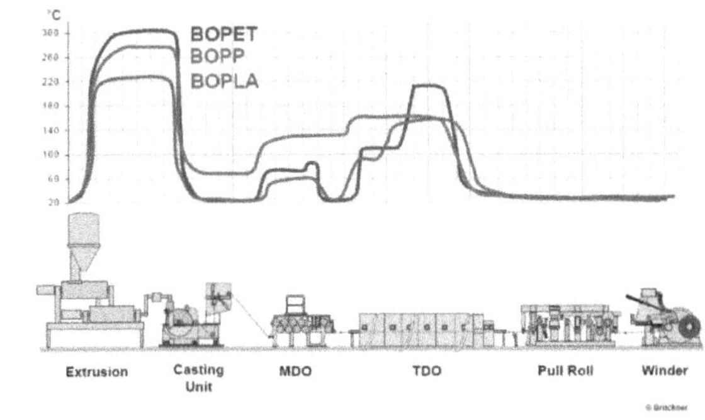

Figure 6.8. *Temperature profiles on a production line for bi-oriented film. Comparison of PLA, PP and PET, by Zimmermann, 1st PLA World Congress, Munich, 2008*

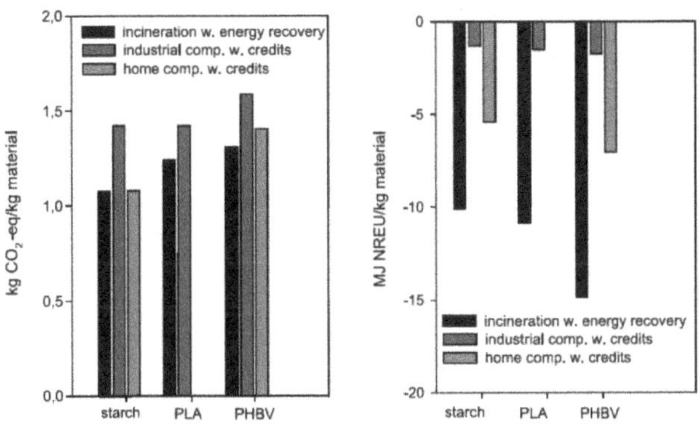

Figure 6.9. *Comparison of environmental impacts of three end-of-life scenarios for three bioplastics [HER 11]*

The tenth commandment recommends that account be taken of the end of the products' useful life at the stage of their design, and that care be taken that their degradation products are non-harmful and non-persistent in the environment. From this point of view, a large number of bioplastics, which present a biodegradable nature (polyesters such as PLA or PHB, starch, etc.), seem to be able to be distinguished. Note, however, that biodegradability is only ever one more of the properties of a material, which is only useful if we are able to exploit it, and that it cannot be defined as an intrinsic characteristic of a material, except contextually and normatively [TAL 11]. One way to exploit the biodegradability of a material is to subject it, at the end of its life, to degradation in an industrial composting facility. European norm EN 13432 applicable to packaging, for example, enables us to determine whether a product can be degraded in such a facility without reducing the quality of the compost produced. However, we can see that certification as conformant to this type of norm does not systematically guarantee that composting will take place as planned. Indeed, Hailey [HAI 10] presents the results of a study conducted on a facility in the United States during which 105 certified products were tested. Of these, only 37 (all based on PLA) actually exhibited total biodegradation, while more than half were only degraded by 20% or less. A study such as this highlights an imperfect correspondence between norms which need to be revised and the reality of certain composting installations. In addition, LCA studies have shown that composting is not always the end-of-life scenario with the best environmental impact for bioplastics. Hermann *et al.*, in 2011 [HER 11], for example, compared the environmental performances of several end-of-life scenarios, including incineration with energy recovery, industrial composting and domestic composting for PLA, PHBV and starch (Figure 6.9).

6.3.3. *Practical case study: Bio-PET, decryption of a communication and avenues for improvement*

In 2009, the Coca-Cola Company released its first communications regarding a new bottle, renewable in origin, known as the PlantBottle. The following phrases were to be found in the communiqué: "The new bottle is made from a blend of petroleum-based materials and up to 30% plant-based materials. [...] is currently made through an innovative process that turns sugar cane and molasses [...] into a key component for PET plastic. [...] it can be processed through existing manufacturing and recycling facilities without contaminating traditional PET". Even without any further information, it was possible to deduce, on this basis, that what was being used was a PET synthesized using bio-based ethylene glycol taken from the bio-ethanol pathway and conventional terephthalic acid, which was confirmed by subsequent communications.

On the basis of this composition of the product, it is possible to perform an analysis of it, using the guiding framework presented above. The first point of note in this communication is the announcement of a 30% plant-based material. This is true if we take account of the mass fractions of the monomers making up the molecule, but as we saw previously, bio-sourced nature is quantified in terms of carbon atoms present in the molecule, which is not exactly the same thing. In the case of such a PET, indeed, the bio-sourced nature as determined according to the standard ASTM D6866 cannot be 30%, but rather only 20%.

The ethylene glycol in the composition of this PET is derived from sugar cane, by way of the synthesis process described earlier for ethylene (Figure 6.5), followed by two stages of chemical transformation – first into ethylene oxide and then into ethylene glycol. As we saw earlier, the bio-ethanol pathway does not completely fulfill the seventh commandment, because the yield from fermentation will, at best, transform a sugar (C6) into two molecules of ethanol (C2). Also, the synthesis of ethylene glycol by this technique involves numerous stages, which contravenes the sixth principle. Other processes can be envisaged for producing bio-sourced ethylene glycol. Thus, in 2008, Ji *et al.* [JI 08] proposed to obtain ethylene glycol from cellulose in a single reaction step. The use of cellulose as a raw material also constitutes an advantage in terms of the question of food competition. In the same vein of ideas, we can also envisage following the ethanol pathway to produce PET, but on the basis of by-products or waste products rather than sugar; Novozymes and Shengquan recently announced the opening, in China, of a factory capable of producing bio-ethanol from cellulose on a commercial scale.

In addition, if we consider bio-PET from the point of view of the tenth commandment, which ordains that we take account of the end of life of the products, it is worth noting that this partially bio-based plastic presents the advantage of being strictly identical to PET, and consequently being perfectly able to be integrated into the recycling channels which are already firmly established for this material.

6.4. Conclusion

In the clearly defined context of a comparative study, the method of LCA helps determine whether, for a given application, a bioplastic can be considered to be "greener" than a traditional plastic. The qualitative approach advocated here, using the guiding framework of the 12 principles of green chemistry, enables us to determine the points with regard to which a bioplastic can be deemed green, and the points of view in relation to which it appears possible to make it greener still. These two approaches must be considered to be complementary, but are not – either separately or together – capable of providing a simple, binary response to the question "Are bioplastics 'green' plastics?", given that the environmental question is largely incompatible with Manichaeism.

Other similar guiding frameworks can be envisaged, such as that proposed by Plastics Scorecard [CLE], which attributes a score in relation to a series of environmental, health-related or social criteria, to various stages of the lifecycle of a product (production of raw material, manufacture of plastic, use and end of life). Notably, this approach was employed by Alvarez-Chavez *et al.* [ALV 12] in putting forward a classification for various bioplastics.

6.5. Bibliography

[AKI 03] AKIYAMA M., TSUGE T., DOI Y., "Environmental lifecycle comparison of polyhydroxyalkanoates produced from renewable carbon resources by bacterial fermentation", *Polymer Degradation and Stability*, vol. 80, no. 1, p. 183, 2003.

[ALV 12] ALVAREZ-CHAVEZ C., EDWARDS S., MOURE-ERASO R., *et al.*, "Sustainability of bio-based plastics: general comparative analysis and recommendations for improvement", *Journal of Cleaner Production*, vol. 23, p. 47, 2012.

[ANA 98] ANASTAS P., WARNER J., *Green Chemistry: Theory and Practice*, Oxford University Press, New York, 1998.

[BOI 12] BOISSY J., GODARD C., GABRIELLE B., "L'analyse du cycle de vie: Spécificités de l'application de cette méthode d'évaluation environnementale aux matériaux agro-sourcés", *Proceedings, Ecobat Conference 2012*, Paris, 2012.

[DOR 03] DORNBURG V., LEWANDOWSKI I., PATEL M., "Comparing the Land Requirements, Energy Savings, and Greenhouse Gas Emissions Reduction of Biobased Polymers and Bioenergy", *Journal of Industrial Ecology*, vol. 7, nos. 3–4, p. 93, 2003.

[GER 00] GERNGROSS T., SLATER S., "How green are green plastics?", *Scientific American*, vol. 283, no. 2, p. 37, 2000.

[GRO 10] GROOT W., BORÉN T., "Lifecycle assessment of the manufacture of lactide and PLA biopolymers from sugarcane in Thailand", *International Journal of LifeCycle Assessment*, vol. 15, p. 970, 2010.

[HAI 10] HAILEY P., "compostables trial at municipal yard trimmings operation", *Biocycle*, 28, August 2010.

[HAK 10] HÄKKINEN T., VARES S., "Environmental impacts of disposable cups with special focus on the effect of material choices and end of life", *Journal of Cleaner Production*, vol. 18, no. 1, p. 458, 2010.

[HER 10] HERMANN B., BLOK K., PATEL M., "Twisting biomaterials around your little finger: environmental impacts of bio-based wrappings", *International Journal of LifeCycle Assessment*, vol. 15, p. 346, 2010.

[HER 11] HERMANN B., DEBEER L., DE WILDE B., *et al.*, "To compost or not to compost: Carbon and energy footprints of biodegradable materials' waste treatment", *Polymer Degradation and Stability*, vol. 96, no. 1, p. 159, 2011.

[JI 08] JI N., ZHANG T., ZHENG M., *et al.*, "Direct Catalytic Conversion of Cellulose into Ethylene Glycol Using Nickel-Promoted Tungsten Carbide Catalysts", *Angewandte Chemie International*, vol. 47, no. 44, p. 8510, 2008.

[KEN 12] KENDALL A., "A lifecycle assessment of biopolymer production from material recovery facility residuals", *Resources, Conservation and Recycling*, vol. 61, p. 69, 2012.

[KIM 05] KIM S., DALE B., "Lifecycle Assessment Study of Biopolymers (Polyhydroxyalkanoates) Derived from No-Tilled Corn", *Int. J. LCA*, vol. 10, no. 3, p. 200, 2005.

[KUR 00] KURDIKAR D., FOURNET L., SLATER S., *et al.*, "Greenhouse Gas Profile of a Plastic Material Derived from a Genetically Modified Plant", *Journal of Industrial Ecology*, vol. 4, no. 3, p. 107, 2000.

[LAN 06] LANDIS A., THEIS T., "Comments on Workshop Report on the Economic and Environmental Impacts of Biobased Production", *International Journal of LifeCycle Assessment*, vol. 11, no. 3, pp. 213, 2006.

[LYN 03] LYND L., WANG M., "A product-nonspecific framework for evaluating the potential of biomass-based products to displace fossil fuels", *Journal of Industrial Ecology*, vol. 7, nos. 3–4, p. 17, August 2003.

[MAD 09] MADIVAL S., AURAS R., SINGH S., *et al.*, "Assessment of the environmental profile of PLA, PET and PS clamshell containers using LCA methodology", *Journal of Cleaner Production*, vol. 17, p. 1183, 2009.

[MUR 08] MURPHY R., DAVIS G., PAYNE M., Lifecycle assessment (LCA) of biopolymers for single-use carrier bags, Research report for NNFCC, 2008.

[MUR 09] MURARIU M., DA SILVA FERREIRA A., BONNAUD L., *et al.*, "Calcium sulfate as high-performance filler for polylactide (PLA) or how to recycle gypsum as by-product of lactic acid fermentation process", *Composites Interfaces*, vol. 16, p. 65, 2009.

[NAR 11] NARAYAN R., "Carbon footprint of bioplastics using biocarbon content analysis and life-cycle assessment", *MRS Bulletin*, vol. 36, p. 716, 2011.

[NOD 98] NODA I., Solvent extraction of polyhydroxy-alkanoates from biomass facilitated by the use of marginal nonsolvent, Patent US5821299, 1998.

[SAN 10] SAN K.Y., BENNETT G., SANCHEZ A., Mutant E. Coli strain with increased succinic acid production, Patent US7790416, 2010.

[SLA 02] SLATER S., GLASSNER D., VINK E., GERNGROSS T.,"Evaluating the environmental impact of biopolymers", in STEINBUCHEL A. (ed.), *Biopolymers*, Wiley-VCH, Weinheim, p. 173, 2002.

[TAL 11] TALON O., "Les plastiques biodégradables dans l'emballage alimentaire", *Food Science and Law*, vol. 6, p. 7, 2011.

[VIN 03] VINK E., RABAGO K., GLASSNER D., *et al.*, "Applications of lifecycle assessment to NatureWorksTM polylactide (PLA) production", *Polymer Degradation and Stability*, vol. 80, no. 3, p. 403, 2003.

[VIN 07] VINK E., GLASSNER T., KOLSTAD J., *et al.*, "The eco-profiles for current and near-future NatureWorks polylactide (PLA) production", *Industrial Biotechnology*, vol. 3, p. 58, 2007.

[VIN 10] VINK E., DAVIES S., KOLSTAD J., "The eco-profile for current Ingeo® polylactide production", *Industrial Biotechnology*, vol. 6, no. 4, p. 212, 2010.

6.5.1. *Websites*

[EGODESIGN] http://www.egodesign.ca/fr/article.php?article_id=534&page=2&switchLang=fr ou http://www.floornature.eu/architectes/interviews/philippe-starck-chapter-2-3019/.

[CLEANPRODUCTION] http://www.cleanproduction.org.

7

Environmental Characterization of Materials for Product Design

7.1. Introduction

The goal of environmental characterization of materials is to facilitate a measurement of the environmental quality of the materials to assist in the choice during the process of eco-design. Thus, rather than putting forward an absolute classification of the various materials on the basis of their environmental impacts, the intention is to provide information to support decision-making during an eco-design process. Indeed, from the point of view of eco-design, it is of little use to attempt to compare materials except in the context of a reflection about the use of the product and the functional unit (FU) of the system at hand.

In addition, environmental evaluation and characterization of materials raise specific issues – some of them peculiar to lifecycle assessment (LCA); others to the handling of multicriterion information as part of a decision-making process.

Using a concrete example (characterization of a selection of materials for the design of a bottle), this chapter illustrates the issue of environmental characterization of materials, and offers a discussion of the limitations of LCA and the functional approach required.

Chapter written by Stéphane LE POCHAT and Arnaud ROQUESALANE.

7.2. Environmental characterization for a drink container

7.2.1. *Description of the case study*

The case study chosen for our example here is that of a 330 cm^3 bottle of water. This is a theoretical example, based on hypotheses deemed to be probable (the masses of aluminum, glass and plastic have been measured). An initial selection, based on the mechanical characteristics of a variety of materials, leads us to shortlist those shown in Table 7.1.

Thermoplastics	Metals	Minerals
– Poly(ethylene terephthalate), PET	– Aluminum	Glass
– Poly(lactic acid), PLA	– Steel	
– Poly(hydroxy alkanoate), PHA, [poly(hydroxy butyrate) PHB, poly(hydroxyl valerate) PHV]		
– Poly(vinyl chloride), PVC		
– Poly(amide 6.6.), PA 66		
– Poly(propylene), PP		
– Poly(carbonate), PC		
– Poly(methyl methacrylate), PMMA		
– Poly(styrene), PS		
– Poly(ethylene), PE		
– Cellulose acetate, CA		

Table 7.1. *Categorized list of possible materials selected for the example of a water bottle*

7.2.2. *Characterization of materials by LCA*

7.2.2.1. *Perimeter*

The environmental evaluation will relate to two different FUs:

– FU1: 1 kg of "factory output" material, ready to be transformed;

– FU2: lifecycle of a material as defined in Figure 7.1 (material, shaping processes, transport and end of life). The respective masses of the different materials for FU2 are given in Table 7.2. The masses of the bottle with the different types of plastics are considered to be identical (within 1 g). The data for the end-of-life scenario (household waste, in France) are taken from a study conducted by ADEME in 2009.

Masses measured for a 330 cm^3 bottle		
Aluminum	Glass	Plastic
27 g	242 g	22 g

Table 7.2. *Mass of material for a 330 cm^3 bottle*

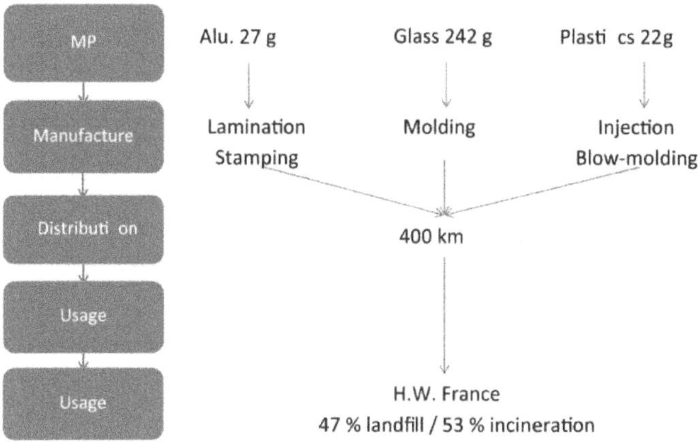

Figure 7.1. *Representation of lifecycle of the 330 mL bottle for each of the three categories of materials*

7.2.2.2. *Method and indicators*

The environmental evaluations are performed using the software SimaPro v7 and the database Ecoinvent v2.2.

The computation method employed is the ReCiPe method, with midpoint, endpoint and single score (World H/H).

The LCA indicators chosen are shown in Table 7.3. For the purposes of the example, the results are presented successively in multi-indicator form (*midpoint* and *endpoint* indicators) and then in the form of a unique *single-score* indicator.

It is worth noting that the selection of the indicators to use must conform to the rules set out by the ISO 14040:2006 standard, which stipulates that the selected indicators must be consistent with the objectives defined and be representative of the system under examination (ISO 14040:2006).

Midpoint indicator		Endpoint indicator		Single Score	
Indicator	Unit	Indicator	Unit	Indicator	Unit
Climate change	kg eq. CO_2	Human Health Quality of ecosystems Resources	DALY Species.yr US $	Single score	Pt
Ozone depletion	kg eq. CFC-11				
Photochemical oxidant formation	kg COVNM				
Freshwater eutrophication	kg eq. P				
Freshwater ecotoxicity	kg eq. 1,4-DB				
Agricultural land occupation	m^2 eq.				
Metal depletion	kg eq. Fe				
Fossil depletion	kg eq. oil				

Table 7.3. *Series of environmental impact indicators used for the evaluation of the materials*

7.2.2.3. *Environmental evaluation for 1 kg of material (FU1)*

Figures 7.2 and 7.3 give the relative results normalized with 100 full scale in multi-indicator criteria (midpoint and single score) for the following materials: aluminum, glass, PC, PET, PE, PLA, PP and steel.

It should be noted that, in relation to the initial selection made on the basis of the mechanical characteristics, the following materials are not evaluated, because they are absent from the database Ecoinvent v2.2: PHB, PHA, PHV and CA. In addition, for ease of reading, Figures 7.2 and 7.3 give the results for a reduced selection of the materials deemed most promising for the desired function (a 330 cm^3 bottle).

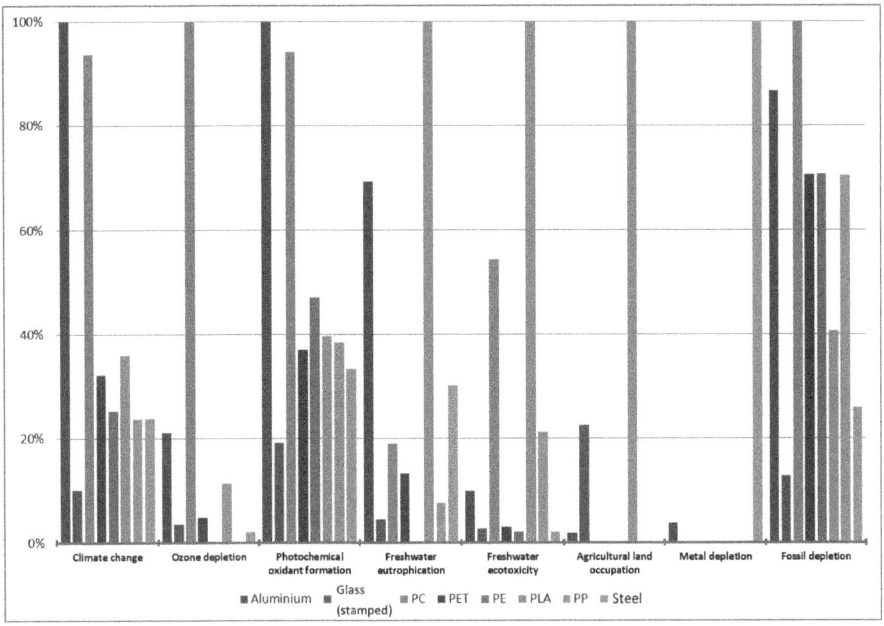

Figure 7.2. *Environmental evaluation of 1 kg of material (midpoint, ReCiPe). (For a color version of this figure, see www.iste.co.uk/hamaide/polymers.zip)*

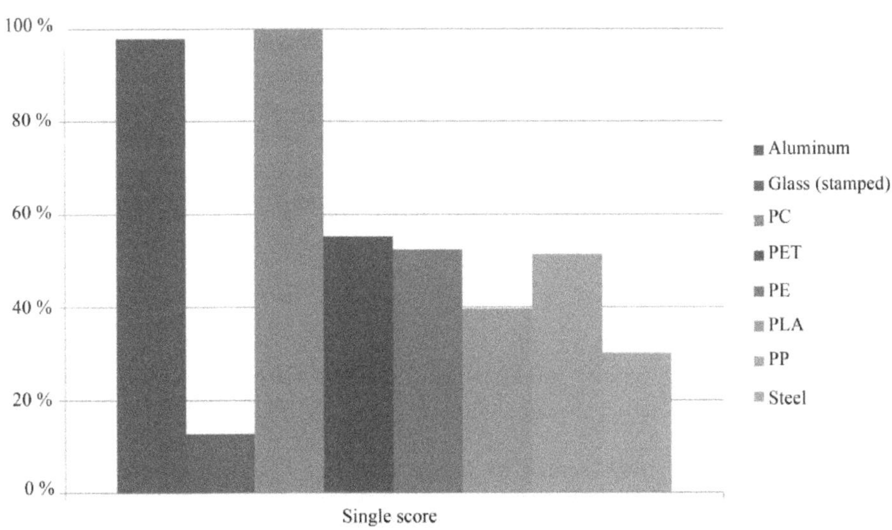

Figure 7.3. *Environmental evaluation of 1 kg of material (endpoint, ReCiPe). (For a color version of this figure, see www.iste.co.uk/hamaide/polymers.zip)*

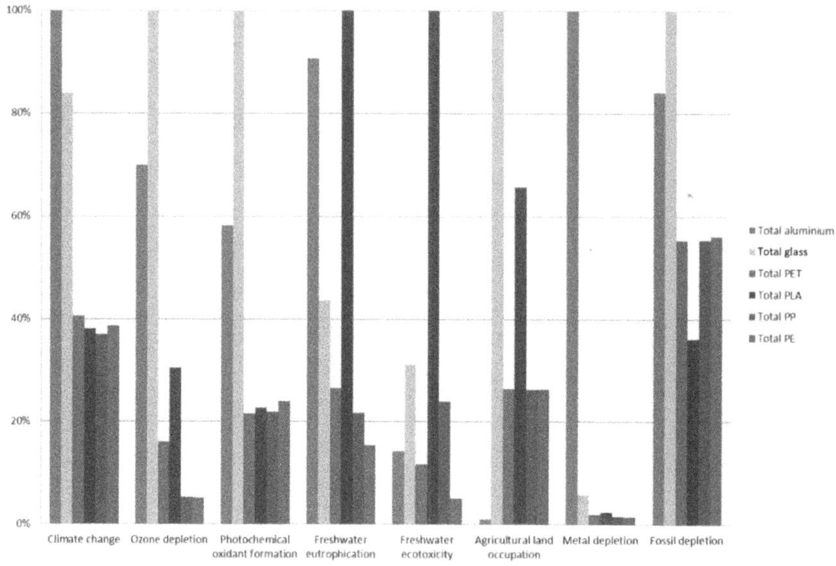

Figure 7.4. *Environmental evaluation of the lifecycle of the materials (midpoint, ReCiPe). (For a color version of this figure, see www.iste.co.uk/hamaide/polymers.zip)*

Figure 7.2 shows that aluminum, polycarbonate and PLA (with regard to the agricultural-based impact indicators) emerge as having a significantly greater impact than the other materials, but in relation to different indicators. As Figure 7.3 shows, in terms of the "single score", aluminum and PC emerge as the most damaging materials. Glass appears to have impact, and steel appears relatively innocuous. The group of thermoplastics seems to be relatively homogeneous, with a relatively slight impact on average. Within that group, PLA appears to have the least impact, in spite of its major impacts in relation to the agricultural indicators (freshwater ecotoxicity, agricultural land occupation and eutrophication).

7.2.2.4. *Environmental evaluation for the lifecycle of the materials (FU2)*

Figures 7.4 and 7.5 show the relative results with reference base 100 in terms of multi-indicators – respectively midpoint and endpoint – and single score, for the lifecycle of the following materials: aluminum, glass, PC, PET, PE, PLA and PP.

Over its lifecycle, glass appears to have a very high impact, owing to its mass, and therefore the mass transported during the transport phase. Aluminum also appears to have a high impact in terms of several indicators, while PLA stands out in terms of the agricultural impact indicators. Finally, apart from PLA, there seems to be little difference among the group of thermoplastics.

Throughout the entire lifecycle, with the hypotheses considered here, and with the single-score indicator, it becomes apparent that aluminum and glass have significantly greater impacts than the other materials. Again, there appears to be little difference between the plastics, with PLA having a greater impact on ecosystems (Figure 7.5(a)), but the least impact overall (Figure 7.5(b)).

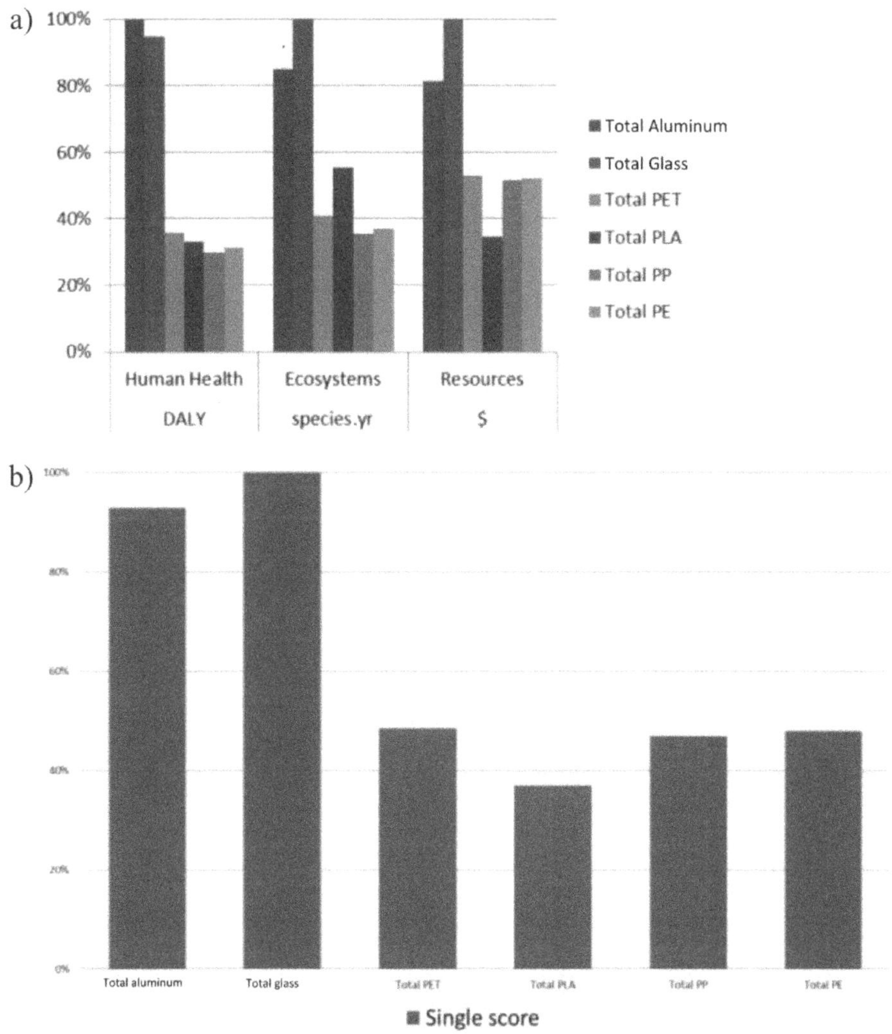

Figure 7.5. *Environmental evaluation of the lifecycle of the materials (endpoint and single score, ReCiPe). (For a color version of this figure, see www.iste.co.uk/hamaide/polymers.zip)*

7.2.2.5. *Discussion*

The respective results for FU1 and FU2 clearly demonstrate the necessity of always considering the environmental evaluation of the materials within the FU.

In addition, the joint presentation of the results in terms of multi-indicators and single score helps illustrate the compromise represented by the choice of one type of indicator to the detriment of another.

Finally, when characterizing the environmental quality of the materials, it is important to bear in mind the limitations inherent in environmental evaluation, and LCA in particular.

7.2.2.5.1. Compromise between multi-indicator and single-score criteria

Generally speaking, if we are aiming to characterize the environmental quality of materials with a view to making choices for eco-design, in comparison to multi-indicator criteria, the single-score criterion presents the undeniable advantage of being far easier to understand and use as a decision-support tool. That said, there are also certain drawbacks in using a single score.

The respective advantages and disadvantages of multi-indicator and single-score approaches are summarized in Table 7.4.

Multi-indicator		Single score	
+	Objectivity	+	Ease of interpretation and use
+	Complete information	-	Subjectivity (weighting)
+	Visibility of impact transfers	-	Partial information
+	Decreased uncertainty	-	Increased uncertainty
-	Complexity of interpretation and use	-	Non-visibility of impact transfers

Table 7.4. *Comparison of advantages and disadvantages of multi-indicatorand single-score characterizations*

7.2.2.5.2. Limitations of LCA for environmental characterization of materials

There are a number of limitations to the LCA method, which are all liable to affect the results of the environmental evaluation, and must be borne in mind when we wish to compare different materials. These limitations include the following:

– certain materials are absent from the databases used by the software packages – particularly the Ecoinvent database. Unless we are able to carry out a lifecycle inventory on these materials, it is then impossible to evaluate them;

– the results of LCA are characterized by numerous and relatively significant uncertainties (the value of these uncertainties varies depending on the indicators used). An average uncertainty of 30% for the results is generally accepted. By way of example, if we consider the results shown in Figure 7.2, for the "climate change" indicator, the materials PET, PE, PP, PLA and steel must be considered to be equivalent (insignificant differences in terms of the uncertainties), while for the freshwater ecotoxicity indicator, all the materials must be considered to be equivalent (2- to 3-fold uncertainties [HUI 10]);

– LCA does not take account of a certain number of indicators which may, in fact, be absolutely relevant in judging the environmental quality of the materials. For example, we can cite: harm to biodiversity, the various risks (risk of abandonment and dispersion into the environment, risk of dangerous substances, GM risk, etc.), degree of recyclability of the material or the social aspects.

7.3. Suggested indicators for the materials considered in this example

While it is indubitable that the choice of materials for design is facilitated by the use of a single indicator, the ideal is to have a set of environmental indicators that are as representative and complete as possible (including LCA indicators, but also other types of indicators, be they semi-quantitative or even qualitative) so that we can, on a case-by-case basis, employ those indicators which are most relevant for the particular objective (technical specifications) and the FU under consideration.

For illustrative purposes, Table 7.5 shows an example of a set of indicators that can be used for the environmental choice of a material to make a 330 cm^3 bottle. The color code white/light gray/dark gray makes for better legibility of these data.

7.4. Conclusion

The choice of a material for eco-design, on the basis of environmental criteria, is a tricky exercise – mainly for two reasons:

– first, from the point of view of eco-design, the environmental quality of a material is often dependent on the FU of the system at hand. Put differently, from an eco-design point of view, the environmental quality of a material cannot be defined absolutely;

– second, the structuring principles of any process of eco-design impose a multicriterion approach, which renders the decision-making process more complex.

In addition, it is worth mentioning that the importance of the choice of a material in a design process can often be relativized over the whole of the lifecycle of the product. Thus, the issue of choosing one material over another may not actually be a priority from an environmental standpoint.

		Aluminum	Glass	PE	PET	PLA
Indicator	***Unit***	***1 bottle, 330 mL***				
Single score	Pt	1.69	0.22	0.90	0.95	0.69
Human health	DALY, 10^{-7}	5.22	4.59	1.12	1.51	1.57
Ecosystems	Spc.yr, 10^{-9}	2.3	2.5	0.7	0.9	1.4
Resources	$	1.11	1.26	0.71	0.72	0.47
CO_2	kg eq. CO_2	8.31	0.84	1.93	2.68	2.99
Degree of recycling (number of cycles)	%	32 (infinite)	80 (infinite)	22 (1)	22 (1)	0
Danger of dispersion		Slight	None	Significant	Significant	Significant
Competition/agricultural land use		0	0	0	0	++

Table 7.5. *Set of indicators to assist in an environmentally sound choice of a material (example)*

Finally, it must be borne in mind that, in the context of any eco-design process, the primary objective is de-materialization (e.g. reduction), rather than the actual choice of one material over another.

7.5. Bibliography

[ADE 09] ADEME, La collecte des déchets par le service public en France, Données 2007, June 2009.

[ANF 06] AFNOR, NF EN ISO 14040, Environmental Management: analysis of the lifecycle, principles and frameworks, October 2006.

[HUI 10] HUIJBREGTS M., HAUSCHILD M., JOLLIET O., *et al.*, *User Manual*, February 2010.

8

Choice of Materials and Environmental Impact: Case of a Water Bottle

8.1. Introduction

In this chapter, we discuss a process of choice of materials, focusing in particular on the case of a water bottle. Commentary on the approach is provided at every step. The choice of materials is structured into three parts. The first part demonstrates the performances of the materials in relation to the properties of the use of bottle. The second part demonstrates their performances in relation to the processing of the bottle. The final part shows the influence of certain environmental criteria on the classification arrived at during the previous stages.

The choice of materials necessitates knowledge of all of the required properties for the making and use of the product. These properties play a part in the proper fulfilment of all the product's functions. These functions are identified and characterized by a functional analysis. It should be noted that the approach of choice of materials as presented here takes in only a part of the functional analysis technique used in the design. Our goal is not to improve the design of the product, but rather to exhaustively list all of its properties. The technique takes account of the shapes of a conventional bottle, and integrates the constraint of cost, which needs to be kept to a minimum whilst also maximizing the product's technical performances [DIE 97].

Chapter written by Rémi DETERRE.

8.2. Functional analysis

Functional analysis is a technique, which "consists of researching and characterizing the functions provided by a product to satisfy the needs of a user" (NF-EN 1325-1). The functional analysis used in our approach considers the functions (expressed in active form) that the product must fulfill: this is an *external* functional analysis or functional analysis of need.

8.2.1. *Functional analysis of a bottle*

The diagram given in Figure 8.1, taken from the NF-EN 12973 standard (soon to become an ISO standard), brings together all the service functions on the basis of the elements of the part's environment. With this diagram, we can visualize all of the product's interactions with its environment, and it is easier to exhaustively list the functions.

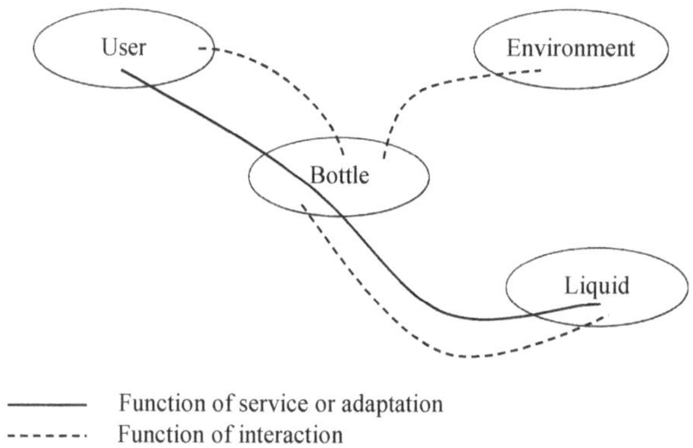

Figure 8.1. *Diagram integrating all the service functions of a bottle*

The terms *user*, *liquid* and *environment* should be understood in the generic sense. Indeed, the user may be considered to be any person likely to come into contact with the product throughout its lifetime: the manufacturer, retailer, buyer, user, urban collective, etc. Some of these elements can be considered to be elements of the environment. The term *liquid* is also general, and may relate to various types of fluids (industrial fluid, foodstuff, etc.). The term *environment* can denote elements of the physical environment, such as other bottles, the shelving units at the store or at the home of the individual, but also the environment in the broader sense, from an ecological point of view.

To begin with, it is possible to list the functions common to all bottles. Some functions may prove to be specific to certain liquids (carbonated drinks, alcoholic drinks, etc.). For simplicity's sake, we shall begin by considering a bottle of still water.

The primary function (PF) could be expressed as follows:

PF1. To manipulate (pour, transfer and transport) a certain amount of liquid

In order to serve this function, it is necessary to respect the set of service functions (SFs), which we wish to list exhaustively so as not to omit any criteria for selection of the material. The exhaustiveness of the list of functions is a gage of the quality of the material selection. To ensure this exhaustiveness, we can review, in turn, each relation between the product (the bottle) and the elements of the environment.

To begin with, let us consider the relations between the bottle and the liquid:

SF1. Receiving the liquid

SF1.1. Adapting to the pouring system (bottling factory, faucet, etc.)

SF2. Containing the liquid

SF2.1. Withstanding the pressure exerted by the liquid contained in the bottle

SF2.2. Ensuring a tight seal between the liquid and the outside of the bottle

SF3. Transferring the liquid

SF3.1. Transferring the liquid between two different containers

SF3.2. Filling the bottle

SF3.3. Emptying the bottle

SF3.4. Pouring the drink (into another container, a glass, etc.)

SF4. Ensuring the integrity of the liquid

SF4.1. Preventing leakage of the liquid (function partially performed by the bottle cap)

SF4.2. Preventing pollution or any possible contamination of the liquid by the container itself or by external elements

Secondly, consider the relations between the bottle and the user:

SF5. Handling of the bottle, by a user

SF5.1. Withstanding the handling and transport (in a truck, a car, etc.)

SF5.2. Withstanding the compressive force when the bottle is handled

SF5.3. Withstanding occasional shocks during handling

SF6. Pleasing the user (this is an estimation function – the verb *to please* is to be taken in the broadest sense, rather than in a merely esthetic sense):

SF6.1. Possessing "ergonomic" properties (in terms of weight, shape, etc.):

– lightweight;

– manageable shape;

– pleasant visual form;

– reduced cumbersomeness after use (compressible bottle, skin flasks, etc.);

– transferring all of the content (without spilling any or leaving any at the bottom).

SF6.2. Possessing the desired esthetic qualities: (shape, color/transparency, surface state, etc.)

– visually pleasant shape;

– transparency or translucency;

– desired color;

– surface state.

SF7. Protecting the user from liquid splashes

SF7.1. Preserving the properties of watertightness when being handled: shock resistance, withstanding the stresses of handling and maintaining contact with the cap

SF8. Informing the user about the bottle's contents, the level of liquid, the composition of that liquid, etc.

SF8.1. Visibility of the level of liquid to the consumer (transparent or translucent bottle)

SF8.2. Displaying legible information about the liquid contained in the bottle (bottle engraved or able to accommodate a label giving this information)

Thirdly, consider the relations between the bottle and the environment:

SF9. Being appropriate for the bottle's immediate environment: the cap

SF9.1. Attaching the cap

SF9.2. Keeping the cap in place

SF9.3. Ensuring a tight seal between the cap and the bottle

SF10. Storage of the drink in retailer aisles

SF11. Adapting to the environment

SF11.1. Withstanding exposure to varying temperatures (both hot and cold)

SF11.2. Withstanding exposure to ambient humidity, to rain, to aging

SF12. Adapting to the user's environment (supermarket, home, kitchen, cellar, refrigerator, etc.)

SF12.1. Occupying an amount of space, which is compatible with the different possible storage sites: (supermarket, home, kitchen, cellar, refrigerator, etc.)

SF12.2. Stacking bottles on top of one another

SF13. Minimizing the bottle's environmental impact

SF13.1. Minimizing energy consumption over the bottle's lifecycle

SF13.2. Minimizing greenhouse gas emissions (equivalent CO_2)

SF13.3. Minimizing water consumption

SF13.4. Minimizing ionizing gas emissions

SF13.5. Minimizing the risk of dispersion at end of life

– harnessing added value from the used bottle;

– collecting the bottle with a view to a value-added process;

– recycling the bottle as:

- an object (e.g. an igloo made out of used bottles or a recipient for a different function)

- a source of raw material (depolymerization)

- an energy source (incineration, etc.)

Not all these functions have the same importance, and it is possible to use a ranking table, establish the criteria for evaluation, and then arrange these criteria in order of importance.

Having drawn up this list of functions, with equal performances, it is advisable to select the material that generates least cost relating to its lifecycle.[1]

The standards and regulations in force (for food contact materials, etc.) are not physical elements of the product's environment. They are not expressed in a functional analysis, as they are inherent to each of the separate functions.

8.3. Choice of materials

We are using a material-choice strategy inspired by the work of M.F. Ashby and D. Cebon from the University of Cambridge [ASH 04]. To begin with, we need to express each of the functions listed above in terms of properties of the material. Then, we hierarchically rank the materials and finally make a selection on the basis of the available maximum values.

8.3.1. *Expression of the specifications*

In order to express a PF or service function in terms of properties of the material, it is necessary to identify the property of the product, which is most likely to serve each of these functions. At this stage, there is a fundamental difference between a functional analysis performed with a view to dcsign and the same analysis performed with a view to choosing a material. Indeed, the aim at this instance is not to question the original design of the product (the bottle), but rather to discover how an existing bottle fulfills each of the functions.

Table 8.1 below shows the resulting list of properties. We can classify these properties according to their nature: general, mechanical, thermal, optical and chemical properties and appropriateness for use. We shall now look at the properties concerning environmental impact in isolation, to illustrate their influence on the final result.

1 In this particular case, the cost relating to the lifecycle depends primarily on the cost of the material and the cost of its employment. The cost of employing a given material includes mainly the investment cost and labor costs.

Function	Product property	Material property
PF1. To manipulate (pour, transfer, transport) a certain amount of liquid		
SF1. Receiving the liquid		
SF1.1. Adapting to the pouring system (bottling factory, faucet, etc.)	Shape, size	Suitability for purpose (Shape, Dimensions)
SF2. Containing the liquid	Shape Tight seal	Suitability for purpose (Hollow 3D form) Hydrophobia Suitability for purpose (lack of pores)
SF2.1. Withstanding the pressure exerted by the liquid contained in the bottle	Rigidity Toughness	Young's modulus Elastic limit
SF2.2. Ensuring a tight seal between the liquid and the outside of the bottle	Tight seal Chemical resistance to the liquid	Suitability for purpose (Hollow 3D form) Hydrophobia Suitability for purpose (lack of pores) Chemical resistance to freshwater and saltwater
SF3. Transferring the liquid		
SF3.1. Transferring the liquid between two different containers	Easy-to-handle shape Shape of neck	Suitability for purpose (Easy-to-handle shape) Suitability for purpose (Shape of neck)
SF3.2. Filling the bottle	Dimensions of bottle Shape of neck	Suitability for purpose (Height, diameter, shape of neck)
SF3.3. Emptying the bottle	Manageable shape	Suitability for purpose (Manageable shape)
SF4. Ensuring the integrity of the liquid		
SF4.1. Preventing leakage of the liquid (function partially performed by the bottle cap)	Seal of bottle	Hydrophobia Suitability for purpose (lack of pores)
SF4.2. Preventing pollution or any possible contamination of the liquid by the container itself or by external elements	Tight seal Chemical resistance of the bottle to the liquid	Hydrophobia Suitability for purpose (lack of pores) Chemical resistance to freshwater and saltwater
SF5. Handling of the bottle, by a user	Easy-to-handle geometry (manageable shape and size)	Suitability for purpose (manageable shape and size)
SF5.1. Withstanding the handling and transport (in a truck, a car, etc.)	Rigidity Toughness Shock resistance	Young's modulus Elastic limit Shock resistance

SF5.2. Withstanding the compressive force when the bottle is handled	Rigidity Toughness	Young's modulus Elastic limit
SF5.3. Withstanding occasional shocks during handling	Shock resistance	Shock resistance
SF6. Pleasing the user (estimation function)		
SF6.1. Possessing "ergonomic" properties (in terms of weight, shape, etc.):		
• Lightweight	Weight of product	Density Suitability for purpose (process to reduce thicknesses)
• Manageable shape	Manageable shape	Suitability for purpose (Manageable shape)
• Reduced cumbersomeness after use (compressible bottle, skin flasks, etc.)	Shape (appropriate for compacting) Flexible bottle Squeezy bottle	Suitability for purpose (compactable shape) Young's modulus Fracture stretching
• Transferring all of the content (without spilling any or leaving any at the bottom).	Shape capable of preventing stagnation Flexible bottle	Suitability for purpose (shape without stagnation area) Young's modulus
SF6.2. Possessing the desired esthetic qualities: (shape, color/transparency, surface state, etc.)		
• Visually pleasant shape	Shape	Suitability for purpose (visually pleasant shape)
• Transparency or translucency	Transparency or translucency	Transparency or translucency
• Desired color	Tinted bottle	Suitability for purpose (Color)
• Surface state	Surface state	Suitability for purpose (Surface state)
SF7. Protecting the user from liquid splashes		
SF7.1. Preserving the properties of watertightness when being handled: shock resistance, withstanding the stresses of handling, maintaining contact with the cap	Shock resistance Rigidity Toughness Fracture strain	Shock resistance Young's modulus Elastic limit Fracture strain
SF8. Informing the user about the bottle's contents, the level of liquid, the composition of that liquid, etc.		

SF8.1. Visibility of the level of liquid to the consumer (transparent or translucent bottle)	Transparency	Transparency
SF8.2. Displaying legible information about the liquid contained in the bottle (bottle engraved or able to accommodate a label giving this information)	Engraved shape Ability to accommodate a label Ability to accommodate an adhesive label	Suitability for purpose (Engraved shape) Suitability for purpose (Shape of the bottle) Surface tension
SF9. Being appropriate for the bottle's immediate environment: the cap		
SF9.1. Attaching the cap	Shape appropriate for attaching the cap (screw system)	Suitability for purpose (screw shape)
SF9.2. Keeping the cap in place	Shape and Behavior appropriate for keeping the cap in place (fixing capsule, tightening)	Suitability for purpose (shape of neck)
SF9.3. Ensuring a tight seal between the cap and the bottle	Relative rigidity/cap Shape of joint	Young's modulus Suitability for purpose (rigid-form neck)
SF10. Storage of the drink in retailer aisles	Stackable bottle shape	Suitability for purpose (stackable shape)
SF11. Adapting to the environment		
SF11.1. Withstanding exposure to varying temperatures (both hot and cold)	Temperature resistance, hot and cold	Temperature resistance, hot and cold
SF11.2. Withstanding exposure to ambient humidity, to rain, to aging	Water resistance Aging resistance	Water resistance Aging resistance
SF12. Adapting to the user's environment		
SF12.1. Occupying an amount of space which is compatible with the different possible storage sites:(supermarket, home, kitchen, cellar, refrigerator, etc.)	Standardized shape Stackable shape	Suitability for purpose (Standardized shape) Suitability for purpose (Stackable shape)
SF12.2. Stacking bottles on top of one another	Stackable shape	
SF13. Minimizing the bottle's environmental impact		

SF13.1. Minimizing energy consumption over the bottle's lifecycle	Minimizing energy consumption over the bottle's lifecycle in relation to the functional unit (amount of water transported)	Minimizing energy consumption in relation to the functional unit
SF13.2. Minimizing greenhouse gas emissions (equivalent CO_2)	Minimizing greenhouse gas emissions (equivalent CO_2) in relation to the functional unit	Minimizing greenhouse gas emissions (equivalent CO_2) in relation to the functional unit
SF13.3. Minimizing water consumption	Minimizing water consumption	Minimizing water consumption in relation to the functional unit
SF13.4. Minimizing ionizing gas emissions	Minimizingionizinggasemis sions	Minimizing ionizing gas emissions in relation to the functional unit
SF13.5. Minimizing the risk of dispersion at end of life	Minimizing the risk of dispersion at end of life	Minimizing the risk of dispersion at end of life in relation to the functional unit
• Disposing of the used product	Degradable product Chemical degradation (decomposition) Thermal degradation (incineration)	Decomposable material Combustible material
• Harnessing added value from the used bottle	Favoring identification and concentration of waste products	Identifiable material Inclusion in a recovery sequence
• Collecting the bottle with a view to a value-added process	Minimizing the environmental impact relating to the recovery of the products Ability to withstand multiple lifecycles Ability of used bottle to be compacted (in terms of shape and rigidity)	Minimizing energy consumption in relation to the functional unit Aging resistance Suitability for purpose (compactable shape) Maximum modulus and flame resistance
• Recycling the bottle as: o An object (e.g. an igloo made out of used bottles, or a recipient for a different function) o A source of raw material (polymer or monomer)	Product adaptable for a different function Minimizing the environmental impact relating to the recovery of the material	Minimizing energy consumption in relation to the recycled functional unit Chemical degradation resistance

Table 8.1. *List of functions of the product and their correspondence to the product properties and material properties*

8.3.2. *List of properties relating to choice of material*

8.3.2.1. *General properties*

– Cost of material (C_m);

– density (ρ).

8.3.2.2. *Mechanical properties*

– Young's modulus (E);

– elastic limit;

– shock resistance (R_{shock});

– fracture stretching;

– modulus and flame resistance limited in the case of a compactable bottle. This property is tricky to evaluate: it must correspond to the ductility of the product necessary to fulfill the functions of a compressible bottle and possibly cold resistance.

8.3.2.3. *Thermal properties*

– Resistance to hot and cold temperatures. The heat resistance is often a factor in "Minimizing energy consumption in relation to the functional unit". We need to distinguish between the long-term and short-term heat resistance.

In the case of the bottle, this property could correspond to accidental exposure to hot water or to a high ambient temperature.

The cold resistance is often linked to the shock resistance. In our case, we can consider the temperature of a refrigerated container.

8.3.2.4. *Optical properties*

– Transparency or translucency. Transparency is necessary for us to be able to see the level of the liquid. It is not necessary to achieve perfect optical quality, and it is acceptable to settle for translucency.

– Color.

8.3.2.5. *Chemical properties*

– Hydrophobia. This property does not clearly describe the desired function, which is associated with impermeability to the liquid. This property is mentioned so as to prevent the use of any highly absorbent material;

– chemical resistance of the bottle to the liquid:

- freshwater and saltwater resistance.

– aging resistance;

– surface tension.

8.3.2.6. *Suitability for processing*

– Suitability for processing (hollow 3D form);

– suitability for processing (cost of use);

– suitability for processing (process to reduce thickness of product);

– suitability for processing (lack of pores);

– suitability for processing (shape and dimensions);

– suitability for processing (easy-to-handle shape);

– suitability for processing (shape of neck);

– suitability for processing (height, diameter and shape of neck);

– suitability for processing (manageable shape);

– suitability for processing (manageable shape and size);

– suitability for processing (process to reduce thickness);

– suitability for processing (visually pleasant shape);

– suitability for processing (compactable shape);

– suitability for processing (shape without stagnation area);

– suitability for processing (color);

– suitability for processing (surface state);

– suitability for processing (engraved shape);

– suitability for processing (shape of bottle);

– suitability for processing (screw shape);

– suitability for processing (shape of neck);

– suitability for processing (rigid-form neck);

– suitability for processing (stackable shape);

– suitability for processing (standardized shape);

– suitability for processing (compactable shape).

8.3.2.7. *Properties relating to environmental impact*

– Minimizing energy consumption in relation to the functional unit during the lifecycle;

– minimizing greenhouse gas emissions (equivalent CO_2) in relation to the functional unit;

– minimizing water consumption in relation to the functional unit;

– minimizing ionizing gas emissions in relation to the functional unit;

– minimizing the risk of dispersion at end of life in relation to the functional unit;

– decomposable or combustible material;

– identifiable material;

– minimizing energy consumption in relation to the recycled functional unit.

All the properties of the product regarding its shape, surface state and porosity are related to the implementation process. We need to discover whether the chosen material facilitates the use of a shaping procedure capable of giving rise to these geometric characteristics. There are few relevant databases in existence for making this type of selection.

There are also other properties, which are not available in the databases but which have an obvious impact on the choice of material (e.g. food contact conformity).

In order to perform a hierarchical ranking, it is necessary to isolate the relevant properties for this type of choice, i.e. the properties, which need to be maximized or minimized. The other properties must be selected by a process of elimination (based on whether or not the minimum or maximum value of a property is satisfied). This selection by limit value poses the problem of the quality of conversion of the product properties into material criteria with a single eliminative limit value. This conversion requires numerical evaluation of the various stresses (mechanical, thermal, etc.), which can often only be performed by simulation software. These data are valid only in the case of loading for a given geometry.

The material's performance is proportional to an index combining all of the properties; the properties needing to be maximized appear in the numerator and those needing to be minimized appear in the denominator.

If no data about the limit values are available to us, we can content ourselves with introducing the property in question into the performance index used for the hierarchical ranking.

In the present case, the most complete performance index (I_P) may be expressed as follows:

$$I_P = \frac{E\sigma R_{shock}}{\rho C m} \quad [8.1]$$

The shock resistance is not available as a measurement for all categories of materials. We find ourselves forced to replace it with a similar property: toughness (K_{Ic}), which corresponds to crack propagation resistance:

$$I_{P1} = \frac{E\sigma K_{1c}}{\rho C m} \quad [8.2]$$

The graphic visualization of the hierarchical classification of the materials based on their I_{P1} is generated by the *Cambridge Engineering Selector* (CES) software [ASH 09]. Representation in log scales renders all of the materials intelligible. On the horizontal axis, we have put the general properties, whilst all the other properties are displayed on the vertical axis:

$$Ln(E\sigma K_{1c}) = Ln(\rho C m) + Ln(I_{p1}) \quad [8.3]$$

Given that the log function is increasing and continuous, the material's performance is proportional to the ordinate value at the start of a straight line with slope = 1 passing through the material's specific coordinates.

The classification of the materials on the basis of the performance index I_{P1} is shown in Figure 8.2.

The classification of all the families of materials in relation to "universal" mechanical and economic performances shows structural materials (metals and alloys, composites), technical ceramics, glasses and even certain natural materials to be dominant over polymers. This result is not very surprising, and demonstrates the genuine necessity to integrate functions to improve the performance of polymers.

The classification on the basis of the normalized performance index I_{P1} is given in Table 8.2.

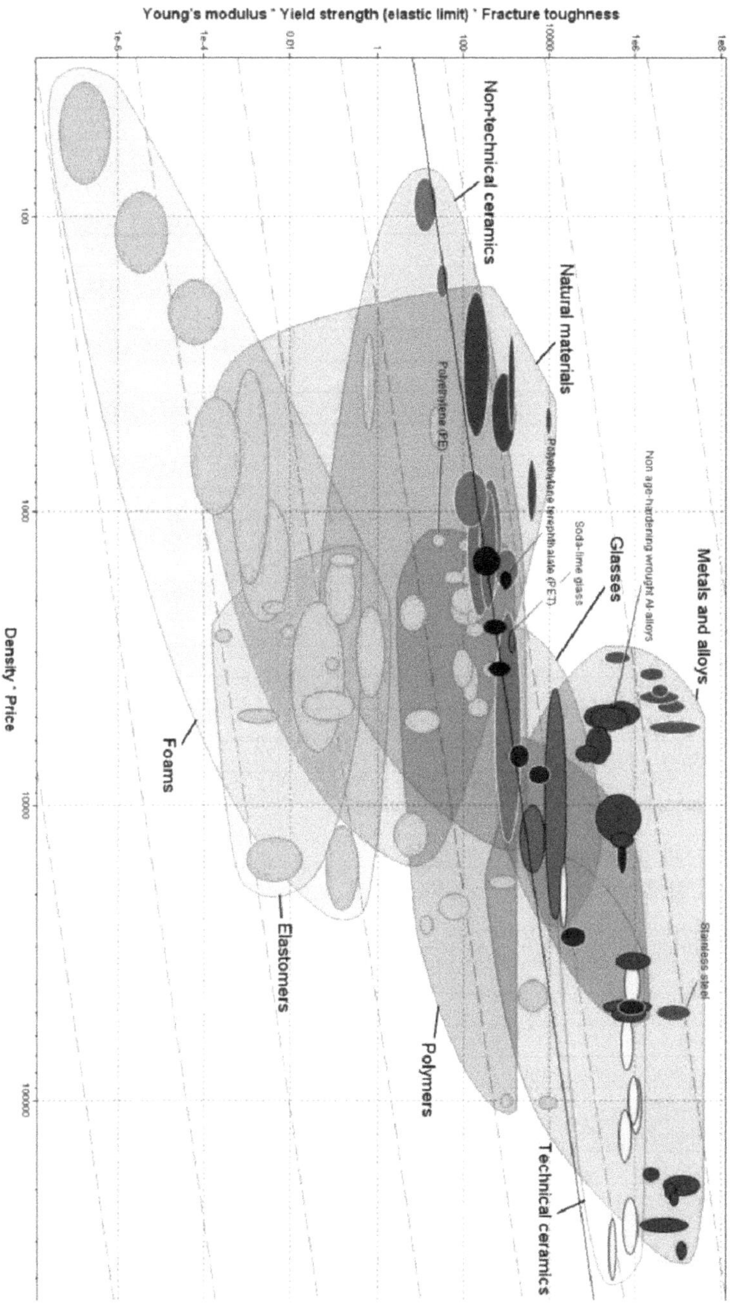

Figure 8.2. *Graphic classification of materials on the basis of* I_{PI}

Name	Performance Index
Low alloy steel	1.00^{+00}
High carbon steel	9.56^{-01}
Low carbon steel	5.86^{-01}
Medium carbon steel	5.31^{-01}
Cast iron, ductile (nodular)	4.29^{-01}
Stainless steel	9.94^{-02}
Cast iron, gray	7.88^{-02}
Age-hardening wrought Al-alloys	6.19^{-02}
Nickel-based superalloys	3.98^{-02}
Cast Al-alloys	3.20^{-02}
Non age-hardening wrought Al-alloys	2.69^{-02}
Commercially pure zinc	2.41^{-02}
Tungsten alloys	2.38^{-02}
Zinc die-casting alloys	2.36^{-02}
Nickel-chromium alloys	2.36^{-02}
Aluminum/Silicon carbide composite	2.20^{-02}
Titanium alloys	2.09^{-02}
Brass	1.72^{-02}
Wrought magnesium alloys	1.44^{-02}
Silicon carbide	1.36^{-02}
Hardwood: oak, along grain	1.29^{-02}
Nickel	1.14^{-02}
CFRP, epoxy matrix (isotropic)	1.12^{-02}
Bronze	9.00^{-03}
Copper	9.00^{-03}
Commercially pure titanium	8.38^{-03}
Cast magnesium alloys	7.56^{-03}
Alumina	6.50^{-03}
Silicon nitride	6.23^{-03}
Zirconia	5.82^{-03}
Bamboo	2.96^{-03}
Boron carbide	2.84^{-03}
Softwood: pine, along grain	2.36^{-03}
Tungsten carbides	1.84^{-03}
Paper and cardboard	1.22^{-03}
Glass ceramic	9.06^{-04}
GFRP, epoxy matrix (isotropic)	8.50^{-04}
Silicon	6.94^{-04}
Aluminum nitride	6.04^{-04}
Sheet molding compound, SMC	4.86^{-04}
Slate	4.13^{-04}
Plywood	4.13^{-04}

Polyethylene terephthalate (PET)	3.66^{-04}
Soda-lime glass	3.18^{-04}
Limestone	2.28^{-04}
Silica glass	1.98^{-04}
Dough (Bulk) molding compound, (BMC)	1.89^{-04}
Marble	1.81^{-04}
Polyvinylchloride (tpPVC)	1.48^{-04}
Polyoxymethylene (Acetal, POM)	1.43^{-04}
Granite	1.31^{-04}
Polyamides (Nylons, PA)	1.29^{-04}
Cement	1.27^{-04}
Brick	1.15^{-04}
Sandstone	1.06^{-04}
Concrete	9.13^{-05}
Polycarbonate (PC)	8.00^{-05}
Borosilicate glass	7.31^{-05}
Lead alloys	6.04^{-05}
Tin	5.78^{-05}
Polypropylene (PP)	5.76^{-05}
Phenolics	5.01^{-05}
Polymethyl methacrylate (Acrylic, PMMA)	4.79^{-05}
Polylactide (PLA)	4.68^{-05}
Acrylonitrile butadiene styrene (ABS)	3.86^{-05}
Epoxies	3.78^{-05}
Polyurethane (tpPUR)	3.12^{-05}
Commercially pure lead	2.90^{-05}
Hardwood: oak, across grain	2.69^{-05}
Polyester	2.47^{-05}
Polystyrene (PS)	2.32^{-05}
Cellulose polymers (CA)	1.74^{-05}
Polyethylene (PE)	1.33^{-05}

Table 8.2. *Classification of materials by normalized performance index* I_{P1}

We can see that the performances of metals are far better than those of the other categories of materials. It is worth noting that there are metal bottles available (cans, flasks, etc.). Now let us add a criterion of transparency or translucency into the equation, with a limit value (Figure 8.3). The aim of this criterion is to enable us to detect the level of drink in the bottle (see Table 8.1).

We can see that the criterion of transparency has eliminated metals from the classification illustrated by Figure 8.3.

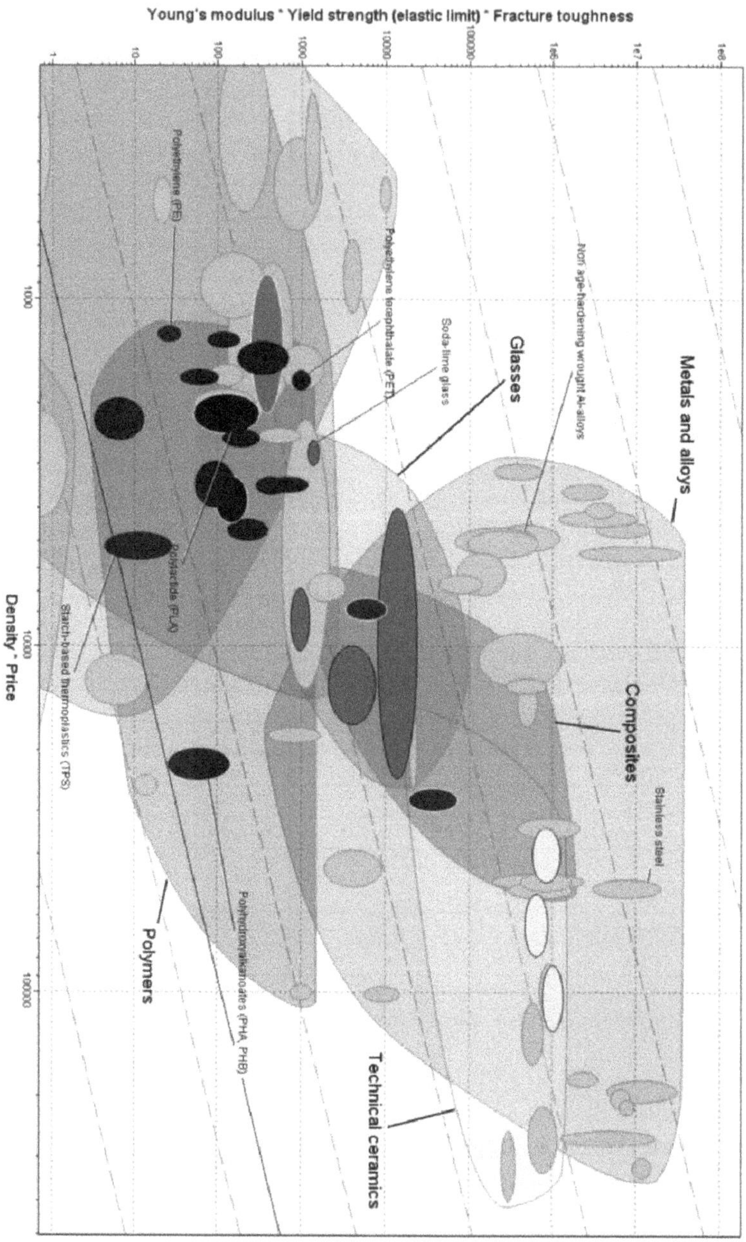

Figure 8.3. *Graphic classification of materials on the basis of* I_{P1} *with a criterion of transparency*

The table with the classification of the materials becomes Table 8.3.

Name	Stage 1: Index
Silicon carbide	1.0000
Alumina	0.4793
Silicon nitride	0.4590
Glass ceramic	0.0668
GFRP, epoxy matrix (isotropic)	0.0627
Sheet molding compound, SMC, polyester matrix	0.0359
Polyethylene terephthalate	**0.0270**
Soda-lime glass	**0.0235**
Silica glass	0.0146
Marble	0.0134
Polyvinylchloride (tpPVC)	**0.0109**
Polyamides (Nylons, PA)	0.0095
Polycarbonate (PC)	0.0059
Borosilicate glass	0.0054
Polypropylene	**0.0042**
Polymethyl methacrylate (Acrylic, PMMA)	0.0035
Polyactide (PLA)	**0.0034**
Epoxies	0.0028
Polyurethane (tpPUR)	0.0023
Polyester	0.0018
Polystyrene (PS)	0.0017
Cellulose polymers (CA)	0.0013
Polyethylene (PE)	**0.0010**
Ionomer (I)	0.0001
Polyhydroxyalkanoates (PHA, PHB)	**0.0001**
Starch-based thermoplastics (TPS)	0.0001

Table 8.3. *Classification of materials on the basis of I_{P1} with a criterion of transparency*

Certain technical ceramics are high up in the classification. The presence of these materials in that position may be surprising, given their cost. This paradoxical position stems from the introduction of criteria, which must correspond to limitations within a performance index that implies a maximization or minimization of the properties.

We must take note of the behavior of the remaining materials in terms of chemical resistance. Consider the example of a salty aqueous fluid. The database includes a qualitative appreciation criterion, which means that we cannot carry out a classification or elimination on the basis of the limit values. We propose to visualize the hierarchical classification of the remaining materials in relation to this criterion in order to determine the maximum or minimum acceptable performance on the basis of that of a reference material (Figure 8.4).

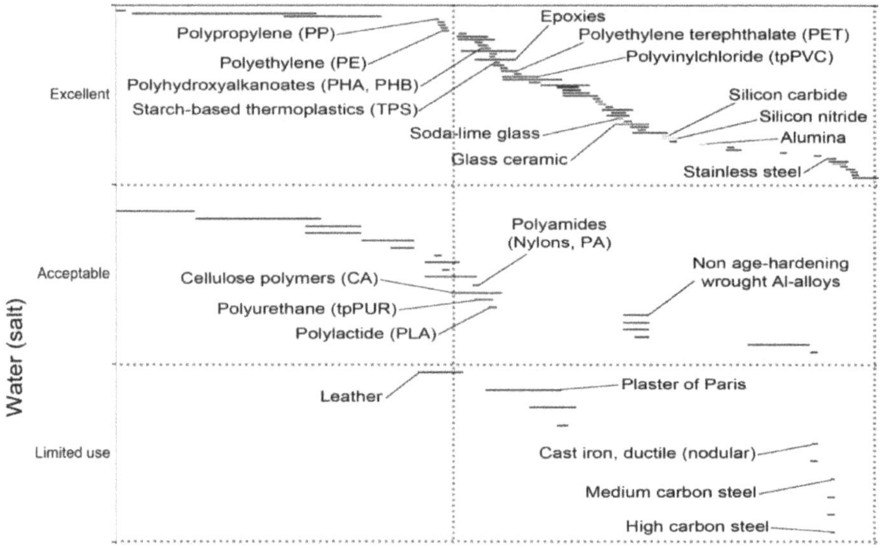

Figure 8.4. *Chemical resistance of various materials to salt water*

The chemical resistance criterion can be used to determine materials, which have an unacceptable limit property for the function of the bottle, such as non-alloyed steel.

8.4. Suitability for processing

Now, let us use the selection criteria relating to suitability for processing. The list of these criteria is given again below:

– suitability for processing (hollow 3D form);

– suitability for processing (cost of processing);

– suitability for processing (process to reduce thickness of the product);

– suitability for processing (lack of pores);

– suitability for processing (shape and dimensions);

– suitability for processing (lack of pores);

– suitability for processing (manageable shape);

– suitability for processing (shape of neck);

– suitability for processing (height, diameter and shape of neck);

– suitability for processing (manageable shape);

– suitability for processing (manageable shape and size);

– suitability for processing (process to reduce thickness);

– suitability for processing (visually pleasant shape);

– suitability for processing (compactable shape);

– suitability for processing (shape without stagnation area);

– suitability for processing (color);

– suitability for processing (surface state);

– suitability for processing (engraved shape);

– suitability for processing (shape of bottle);

– suitability for processing (screw shape);

– suitability for processing (shape of neck);

– suitability for processing (rigid-form neck);

– suitability for processing (stackable shape);

– suitability for processing (standardized shape);

– suitability for processing (compactable shape).

In order for us to be able to choose, these criteria must be expressed as quantifiable properties available in the software's database. This list of properties is given in Table 8.4.

The correspondence between the available attributes of the process and the criteria of suitability for processing required for the bottle are indicated in Table 8.5.

Technical ceramics are classed highly because of their high performance index. However, these materials are often used by sintering, and are liable to exhibit porosity, which is incompatible with the function of hermetic sealing. This criterion corresponds to a suitability for processing; yet it cannot be expressed as a material property available in the databases, but it eliminates these materials.

Physical attributes	Economic attributes
Surface state	Relative cost of equipment
Weight range	Relative cost of tools
Section thickness range	Labor intensity
Minimum pass depth	Batch size
Roughness	
Dimensional tolerance	
Cutting speed	
Cost modeling	Process characteristics
Cost of tools	Continuous
Capital cost	Discontinuous
Lifetime of tools (units)	Prototyping
Proportion of material used	
Cost index (per unit)	
Production rate (units)	
Shape	
Hollow, 3D	
Flat sheet	
Non-circular prismatic	
Solid, 3D	
Crowned plate	
Torus with prismatic cross-section	

Table 8.4. *Attributes of processes available in the CES software's database*

In order to perform an initial classification of the processes, we can plot a performance index on the basis of constant characteristics of the processes, such as the relative cost index of the product, the production rate and the batch size. Bottle production requires very high capacity and very fast production tools in order to obtain a reduced relative cost index per unit. In addition, the relation between the production capacity and the cost of the product per unit is universal [REY 98]. This being the case, the performance index for the process can be written thus:

$$Ipp = \frac{Ts \cdot Vp}{\mathrm{Pr}} \quad [8.4]$$

where Ts is the batch size, Vp the production rate and Pr the relative cost index per unit of product:

$$Ln(Ts \cdot Vp) = Ln(\text{Pr}) + Ln(Ipp) \qquad [8.5]$$

Suitability for processing	Process attributes
Hollow 3D form	Hollow 3D form
Processing cost	Relative cost of equipment Relative cost of tools Labor intensity Batch size Cost of tools Capital cost Lifetime of tools (units) Proportion of material used Cost index (per unit) Production rate (units)
Process to reduce thicknesses of product	Range of cross-section thickness
Lack of pores	
Shape, Dimensions	Weight range
Manageable shape	3D form
Shape of neck	3D form
Height, diameter	3D form
Easy-to-handle shape	3D form
Manageable shape and size	3D form
Visually pleasant shape	3D form
Compactable shape	Hollow 3D form
Shape without stagnation area	Hollow 3D form
Color	
Surface state	Surface state Roughness
Engraved shape	3D form
Shape of bottle	Hollow 3D form
Rigid-form neck	3D form
Stackable shape	3D form
Standardized shape	3D form
Compactable shape	3D form

Table 8.5. *Concordance table between suitability for processing and the attributes of the process*

Figure 8.5 shows the evolution of the processes with the batch size, production rate and relative cost index per unit.

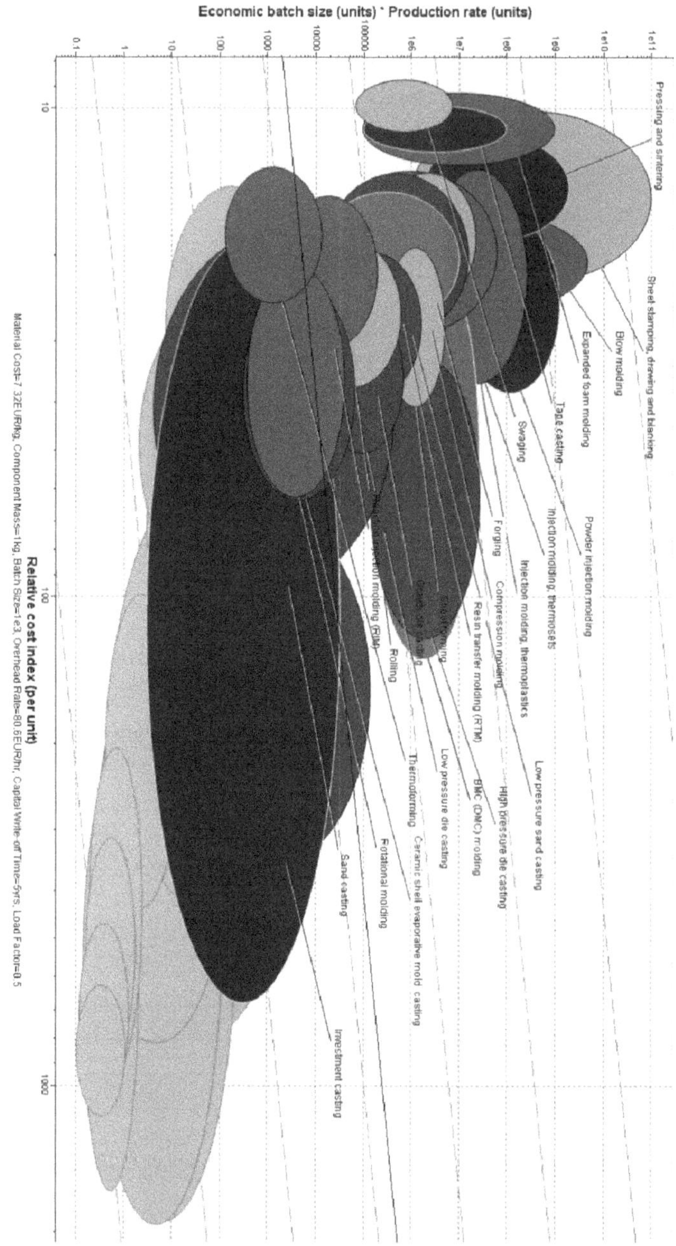

Figure 8.5. *Classification of processes on the basis of batch size, production rate and relative cost index per unit*

We can indeed see an evolution of the relative cost index, which is inversely proportional to the batch size.

Name	Stage 1: index
Sheet stamping, drawing and blanking	1.0000000
Powder injection molding	0.2277512
Pressing and sintering	0.1980861
Injection molding, thermosets	0.0531100
Injection molding, thermoplastics	0.0531100
Expanded foam molding	0.0426794
Blow molding	0.0348325
Tape casting	0.0137799
Forging	0.0057895
Swaging	0.0034354
Compression molding	0.0024354
Sheet forming	0.0021388
SMC molding	0.0018086
BMC (DMC) molding	0.0018038
High pressure die casting	0.0015311
Low pressure sand casting	0.0012010
Resin transfer molding (Rtm)	0.0007464
Gravity die casting	0.0002737
Low pressure die casting	0.0001237
Rolling	0.0000995
Reaction injection molding (RIM)	0.0000420
Ceramic shell evaporative mold casting	0.0000056
Rotational molding	0.0000051
Thermoforming	0.0000035
Sand casting	0.0000016
Investment casting	0.0000004
Hot isostatic pressing	0.0000001

Table 8.6. *Classification of processes on the basis of the performance index* I_{pp}

If what is needed is a hollow 3D form, we obtain the graph shown in Figure 8.6.

If we apply a constraint of maximum weight of 100 g (typical weight of a bottle is between 20 and 40 g) and maximum thickness of 1 mm, we obtain the result presented in Figure 8.7.

All that remains now is the processes of injection molding, powder injection molding and blow molding. Injection molding and powder injection molding necessitate additional operations to create a hollow body. The only remaining process is blow molding. The only blow-moldable materials are glasses and polymers.

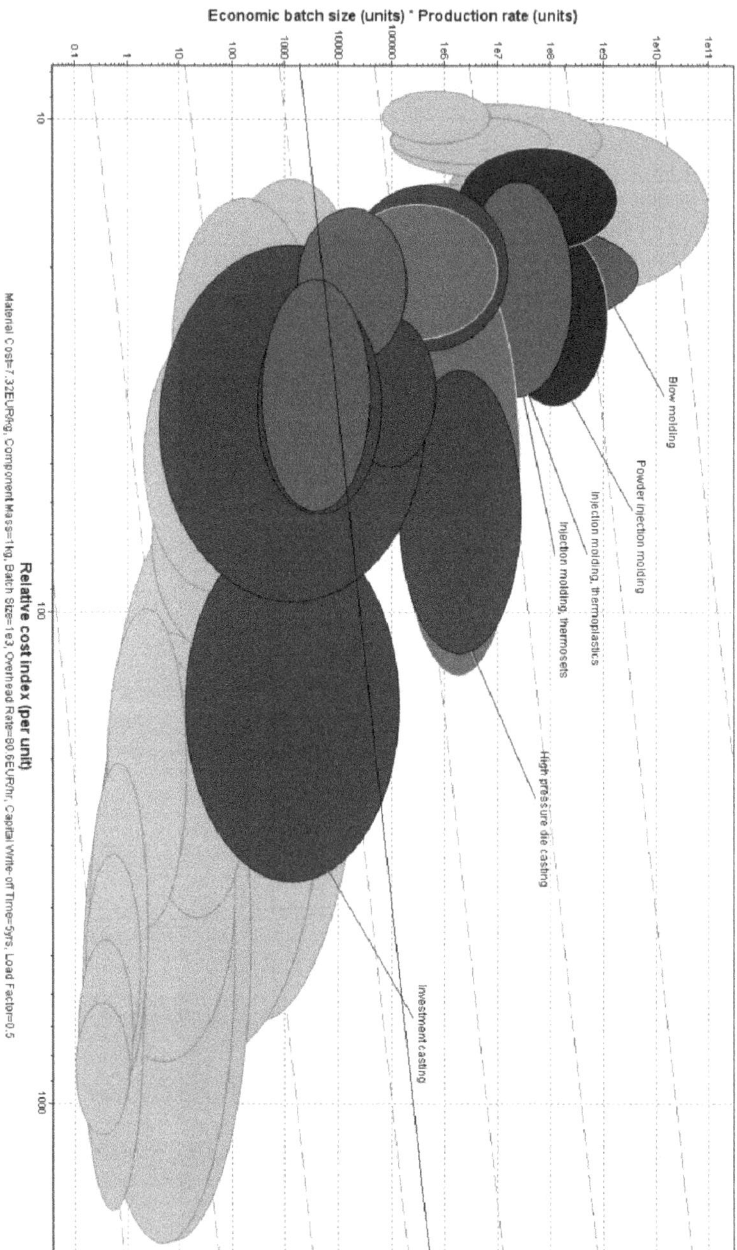

Figure 8.6. *Classification of processes on the basis of batch size, production rate and relative cost index per unit with the constraint of a hollow 3D form*

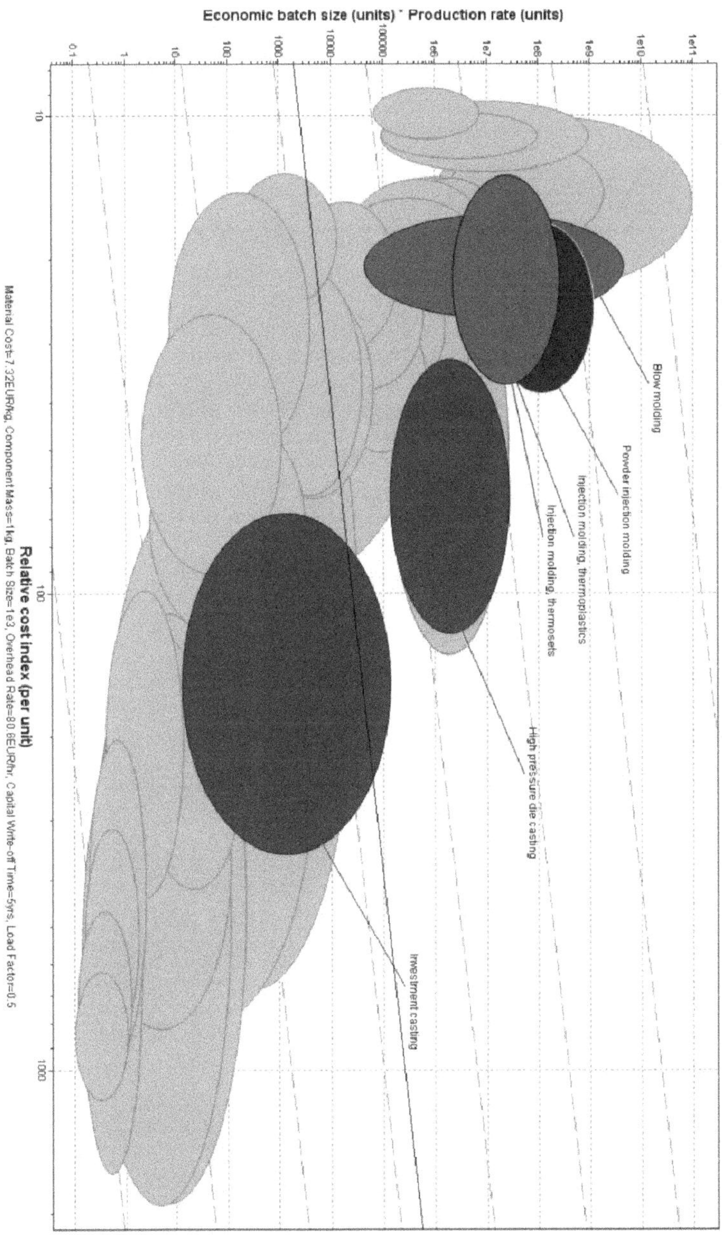

Figure 8.7. *Classification of processes on the basis of batch size, production rate and relative cost index per unit, with the constraint of a hollow 3D form and a limitation of the sizes and thicknesses of the units*

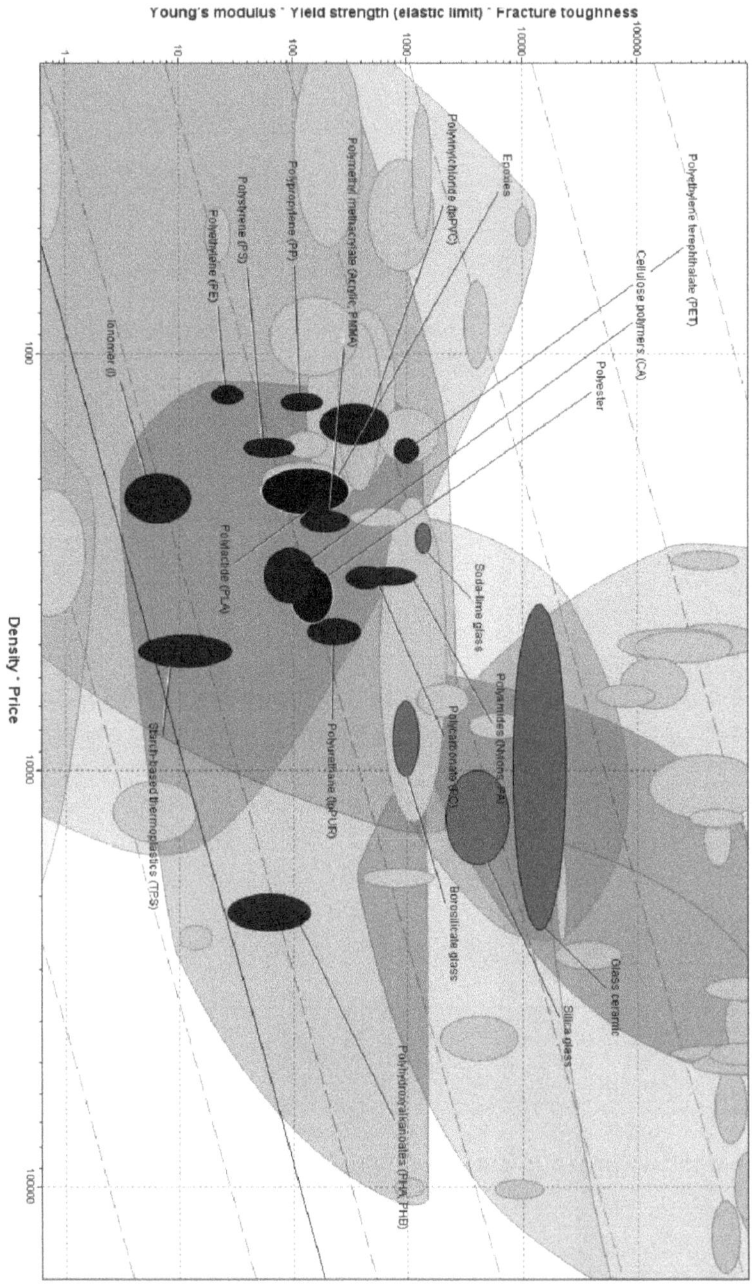

Figure 8.8. *Graphic classification of the materials on the basis of I_{P1} with a criterion of transparency and compatibility with the process of blow-molding*

Name	Stage 1: Index
Glass ceramic	1.0000
Polyethylene terephthalate (PET)	0.4034
Soda-lime glass	0.3510
Silica glass	0.2186
Polyvinylchloride (tpPVC)	0.1634
Polyamides (Nylons, PA)	0.1421
Polycarbonate (PC)	0.0883
Borosilicate glass	0.0807
Polypropylene (PP)	0.0636
Polymethyl methacrylate (Acrylic, PMMA)	0.0528
Polyactide (PLA)	0.0516
Epoxies	0.0417
Polyurethane (tpPUR)	0.0344
Polyester	0.0272
Polystyrene (PS)	0.0256
Cellulose polymers (CA)	0.0192
Polyethylene (PE)	0.0147
Ionomer (I)	0.0021
Polyhydroxyalkanoates (PHA, PHB)	0.0019
Starch-based thermoplastics	0.0015

Table 8.7. *Graphic classification of the materials on the basis of I_{P1} with a criterion of transparency and compatibility with the process of blow molding*

By integrating the result of this selection due to the process, we obtain Figure 8.8 and the list given in Table 8.7.

8.5. Integration of an environmental criterion

Let us now introduce criteria relating to the environmental impact. According to lifecycle assessments, the environmental impact of a bottle (as is the case with most forms of packaging) relates primarily to raw material production (Figure 8.9).

This idea is valid for packaging whose weight is negligible in comparison to that of the content, but it needs to be viewed relativity in the case of heavier forms of packaging, such as glass, for which the material's fragility and/or suitability for processing means its weight cannot be reduced.

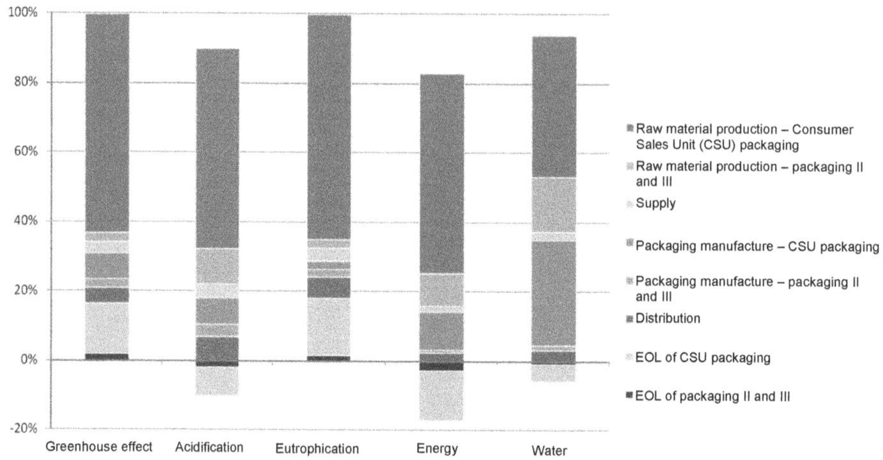

Figure 8.9. *Distribution of environmental impact of a 1.5-l PET bottle (data taken from [ECO 10])*

Given the rarity of these properties in the databases, we shall, in part, make use of those which are available in the software CES[2] – namely:

– the energy and CO_2 emission necessary for the production of the raw material. These parameters are respectively labeled *energy* and CO_2 in the future performance index. We modify the performance index as follows:

$$P_{p2} = \frac{E\sigma K_{ic}}{\rho Cm.EnergyCO_2} \qquad [8.6]$$

The classification of the materials obtained with this new performance index is shown in Figure 8.10 and Table 8.8.

If we adjust the previous performance index by the functional unit (amount of liquid transported per unit mass of the bottle), we obtain the modifications presented in Table 8.9.

2 Environmental criteria available in the CES database:
– CO_2 footprint and primary production;
– embodied energy and primary production;
– water usage;
– material processing energy;
– material processing CO_2 footprint;
– material recycling: energy, CO_2 and recycled fraction.

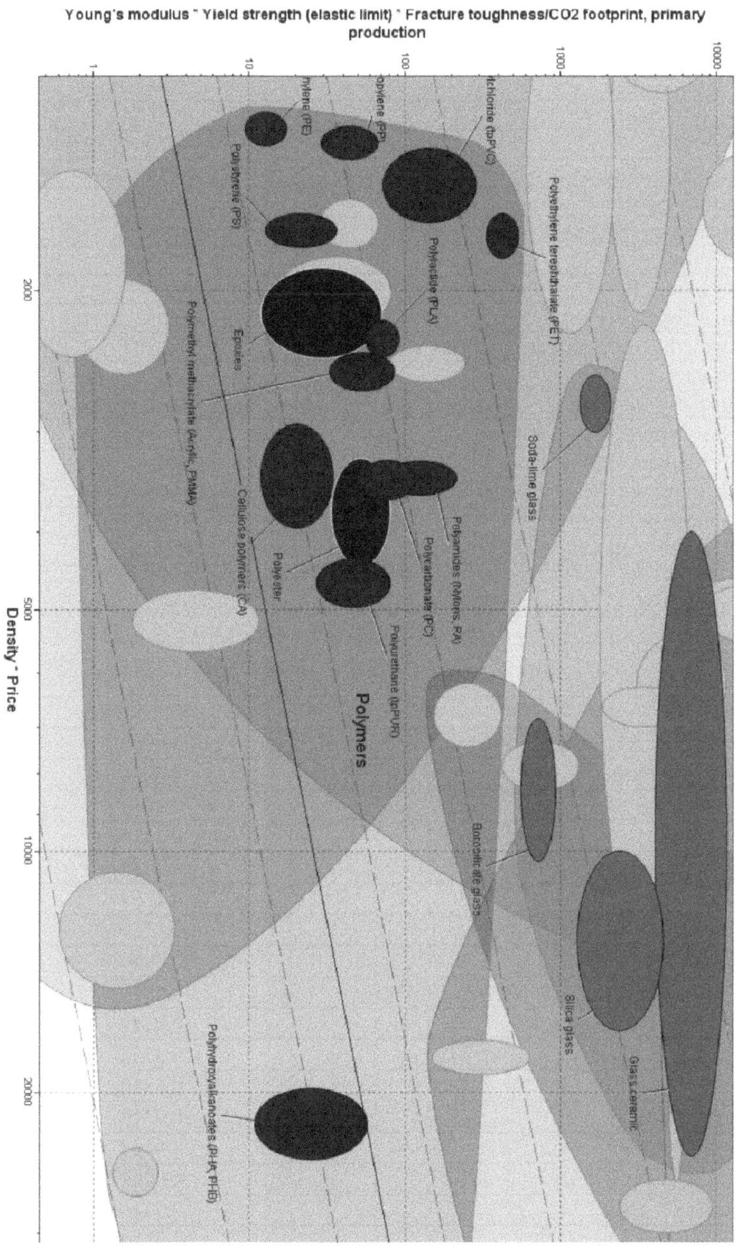

Figure 8.10. *Graphic classification of materials on the basis of I_{P2} with a criterion of transparency and compatibility with the process of blow molding*

Name	Stage 1: Index
Glass ceramic	1.0000
Soda-lime glass	0.8626
Polyethylene terephthalate (PET)	0.3555
Silica glass	0.2663
Polyvinylchloride (tpPVC)	0.1407
Borosilicate glass	0.1231
Polyamides (Nylons, PA)	0.0525
Polypropylene (PP)	0.0484
Polyactide (PLA)	0.0448
Polycarbonate (PC)	0.0332
Polymethyl methacrylate (Acrylic, PMMA)	0.0302
Polyester	0.0197
Epoxies	0.0195
Polystyrene (PS)	0.0184
Polyethylene (PE)	0.0147
Polyurethane (tpPUR)	0.0144
Cellulose polymers (CA)	0.0085
Polyhydroxyalkanoates (PHA, PHB)	0.0016

Table 8.8. *Classification of materials on the basis of I_{P2} with a criterion of transparency and compatibility with the process of blow molding*

Name	Corrected Index
Polyethylene terephthalate (PET)	1.0000
Polyvinylchloride (tpPVC)	0.3956
Glass ceramic	0.2813
Soda-lime glass	0.2426
Polyamides (Nylons, PA)	0.1478
Polypropylene (PP)	0.1363
Polyactide (PLA)	0.1259
Polycarbonate (PC)	0.0904
Polymethyl methacrylate (Acrylic, PMMA)	0.0849
Silica glass	0.0749
Polyester	0.0554
Epoxies	0.0550
Polystyrene (PS)	0.0518
Polyethylene (PE)	0.0414
Polyurethane (tpPUR)	0.0406
Borosilicate glass	0.0346
Cellulose polymers (CA)	0.0240
Polyhydroxyalkanoates (PHA, PHB)	0.0046

Table 8.9. *Classification of materials on the basis of I_{P2} corrected by the functional unit found by an LCA and standardized with a criterion of transparency and compatibility with the process of blow molding*

Here, we can see the limitations of this criterion, which corresponds to the environmental impact expressed in relation to the mass of material, whereas the environmental impact should be expressed in relation to the functional unit determined by the lifecycle assessment (LCA) – namely the amount of liquid conditioned [ASH 09]. By using plastic, we are able to condition the same amount of liquid with a mass of material up to 10 times less than glass [POT 98]. This property considerably reduces CO_2 emissions during transport of the liquids.

We have modified the final classification of the polymers by replacing the indicator of CO_2 emitted per kg of raw material with the quantity of equivalent CO_2 per kg of raw material, found by way of an LCA.

We obtain the final classification shown in Table 8.10.

Taking account of the environmental data unearthed by an LCA does not alter the top of the classification, but relegates technical polymers to the end of the classification and increases the difference in performance between PLA and PHA.

Polymer	I_{p1}/CO_2lca *normalized*	I_{p1}/CO_2lc *a*	$KgCO_2eq/Kg$ *LCA*	I_{p1}
Polyethylene terephthalate (PET)	100.0%	0.2017	2.90	0.5850
Polyvinylchloride (tpPVC)	58.7%	0.1185	2.00	0.2370
Polypropylene (PP)	23.2%	0.0468	1.97	0.0922
polylactic acid (PLA)	11.9%	0.0240	3.12	0.0748
Polyamides (Nylons, PA)	11.0%	0.0222	9.27	0.2060
Polycarbonate (PC)	8.1%	0.0164	7.80	0.1280
Polyethylene (PE)	5.5%	0.0110	1.93	0.0213
Polymethyl methacrylate (Acrylic, PMMA)	5.3%	0.0108	7.12	0.0766
Polystyrene (PS)	5.3%	0.0106	3.50	0.0371
Polyhydroxyalkanoates (PHA, PHB)	0.3%	0.0006	4.76	0.0028

Table 8.10. *Classification of materials on the basis of* I_{P2} *corrected by the amount of equivalent* CO_2 *per kg of raw material, found by an LCA and standardized*

8.6. Conclusion

The technique for choice of materials outlined here demonstrates the performance of polymers in comparison to that of other materials, in the case of a public-domain application: a bottle of water. The use of functional analysis enables us to extract all of the product's functions. In this chapter, we have presented the expression of these functions in terms of properties of the product and properties of the material.

When converting the functions into product properties, we are faced with the subjectivity of certain criteria (e.g. the esthetic criterion) or the difficulty in expressing other criteria (e.g. the criterion of watertightness). When we express the product properties in the form of material properties, the problem arises of the representativeness of the material properties available in the database (relating to color, say).

Certain properties, such as cost or density, are always included in a hierarchical classification. Other properties are selected by elimination on the basis of a limit value – such is often the case for temperature resistance. However, many limit values are difficult to determine. Indeed, these values would require the use of an often-complex model, limited to a specific use case. If we do not have a limit value, to avoid introducing an arbitrary limit value, we are forced to feed the property in question into a performance index. This technique enables us to take account of that property, but it is liable to favor materials whose performances may be excessive in relation to the function required of them. There are material properties, which are not subject to a quantitative physical measurement, or which are subject to a crude qualitative classification. In this case, the limit value can only be fixed arbitrarily by a classification relative to other materials whose performances are known.

There are not many processing properties available in the databases used to describe the required characteristics. In order to take account of the environmental impact, we need prior knowledge about the relative importance of the different stages in the lifecycle [ASH 04].

In the case of the bottle, the major impact is due to raw material production. This property enables us to use the data available in the database of some software packages. In other consumer sectors, it is other phases of the lifecycle, which have a predominant impact. The environmental impact of some of these phases can only be described by a lifecycle assessment, which is liable to yield a result that depends very closely on the perimeter of the study.

In addition, the taking into account of a single phase of the lifecycle or a single environmental criterion can lead to an incorrect choice if the performances of the chosen material are mediocre in relation to other criteria. This raises the question of determining an exchange value between the different environmental impact criteria [ASH 04].

The classification generated by our selection process demonstrates the good performances of polymers in the example of a water bottle. However, the only reason that metals are eliminated is a limited number of counter-performances (e.g. lack of transparency). Glass is still a promising candidate, but it is penalized because of the mass of the bottles manufactured. Of the various polymers, PET stands out from the rest, regardless of the criteria that are used. Technical polymers (PA and PC) are relegated to quite a low position in the classification. PVC and PP retain a good level of classification. For the biobased polymers, there is a very significant divide between PHA, which is relegated to the bottom of the classification, and PLA, which is in the fourth position. The classification of biopolymers needs to be considered in context, in relativity to the rapid evolution of the cost–performance ratio caused by the increasing interest being paid to these materials.

8.7. Appendix: modeling of cost index [ESA 03]

The cost C of processing a product can be modeled in the following manner:

$$C = \frac{mC_m}{(1-f)} + \frac{C_t}{n}\left\{Int\left|\frac{n}{n_t} + 0,51\right|\right\} + \frac{1}{\dot{n}}\left(\frac{C_0}{LT_{wo}} + \dot{C}_{oh}\right)$$

where:

m = mass of the product,

Cm = cost of material,

$1\text{-}f$ = fraction of initial material remaining in the finished product after waste extraction,

C_t = specific investment cost (tools),

n = number of products made,

n_t = number of products made by a tool before it is replaced,

$\dot{n}$ = production rate,

C_c = non-specific investment cost linked to production,

L = proportion of time for which the non-specific investment is devoted to the production of the product,

t_{wo} = time to return on the non-specific investment relating to production,

$\dot{C}_{oh}$ = level of indirect costs linked to production.

8.8. Bibliography

[ASH 00] ASHBY M.F., "Multi-objective optimization in material design and selection", *Acta Materiala*, vol. 48, pp. 359–369, 2000.

[ASH 04] ASHBY M.F., BRECHET Y.J.M., CEBON D., *et al.*, "Selection strategies for materials and processes", *Materials and Design*, vol. 25, pp. 51–67, 2004.

[DIE 97] DIETER G.E., "Overview of the materials selection process", *ASM Handbook*, 1997.

[ECO 10] ELIPSO, VALORPLAST, ECO E., Analyse du Cycle de Vie d'une bouteille PET, RDC Environment, June 2010.

[ESA 03] ESAWI A.M.K., ASHBY M.F., "Cost estimates to guide pre-selection of processes", *Materials & Design*, vol. 24, pp. 605–616, 2003.

[POT 98] POTHET J.P., "Emballage et environnement", *Techniques de l'Ingénieur*, A9730, 1998.

[REY 98] REYNE M., *Technologie des plastiques*, Hermès, Paris, 1998.

[VAL 11] http://www.valorplast.com/Front/docs/cms/synthese_acv_bouteille_pet-2010_20110110113759.pdf 2011.

9

Formulation and Development of Biodegradable and Bio-based Multiphase Materials: Plasticized Starch-based Materials

9.1. Introduction

Increasingly, it is accepted that the use of long-life polymers for application with a short duration (e.g. packaging, restoration, surgery, hygiene and others) is absolutely not appropriate. This is all the more true when the populace is increasingly concerned with the preservation of ecosystems and the environment. Indeed, at present, durable and non-renewable polymers are a major source of pollution, when they are "dispersed" into the natural environment. For example, plastic bags are well known for affecting submarine life, and constitute a burgeoning pollution source. In addition, conventional plastics affect overall waste management. In addition, collectives (municipal authorities, regional or national organizations) are increasingly becoming aware of the significant savings that can be made by using controlled and compostable waste. Indeed, conventional value-added processing of plastic waste creates certain problems. Energetic valorization generates certain toxic emissions (e.g. dioxin), when imperfectly controlled. Material valorization involves certain limitations relating to the difficulty in finding economically viable outlets. In addition, material valorization can create a negative ecological balance, which is detected on a lifecycle assessment (LCA) of the material given the necessity, in almost every case, of, say, washing plastic waste. This is associated with a significant energy consumption relating to the processes of grinding and transformation of the plastics.

Chapter written by Luc AVÉROUS.

Consequently, the potential of biodegradable polymers, and particularly those obtained from renewable resources, has long been exploited – especially for short-life applications [HAL 10]. However, at present, these polymers, although their use is growing significantly (by 10–20% each year), are used only in certain specifically targeted applications [STE 03] (e.g. packaging, agriculture or leisure).

For example, if we consider the set of biodegradable and/or bio-based polymers, a recent market study published by the organization *European Bioplastics* (http://en.european-bioplastics.org/) shows that world capacity for production of these polymers, in 2012, was only around 1.4 million tons, of which 0.6 million were accounted for by biodegradable polymers. Compare this with worldwide plastic consumption of 288 million tons, of which, 57 million tons were in Europe alone, according to Plastics-Europe (http://www.plasticseurope.org), in 2012. Recent projections – particularly that presented by *European Bioplastics* – show that these very rapidly growing polymers will, nevertheless, remain a niche market for the next 10 years. They will account for only a small percent of the world plastic market. Thus, it is not envisaged that these polymers can totally replace conventional plastics.

Biodegradable and bio-based polymers are, however, an interesting approach, particularly with a view to:

1) developing original macromolecular architectures, some of which would be difficult to obtain by conventional chemical pathways;

2) proposing a new end-of-life scenario for these materials. It should be noted that compostable materials can, for example, also yield added value through conventional recycling;

3) obtaining products which often have high added value, contributing to the viability and perennity of certain biorefineries;

4) dealing with the coming shortages of petrochemical resources – particularly in terms of certain fractions, which are becoming rare. Indeed, certain fossil resources (such as crude oil) which are in short supply could be partially replaced by more eco-friendly, renewable sources, arising from agricultural or marine resources, contributing to the reduction in CO_2 emissions.

In this overview, however, it should be noted that recent developments in the area of sources and production of shale gas (mainly in C1–C3) completely blur a vision of the future which was already fairly complex to sketch.

Without even considering the issue of management of water resources, agricultural resources could exhibit polemic (and potentially problematic) competition with food crops. However, it is also of note that there are no longer any industrial projects on bio-based materials which do not take account of this issue. In addition, there are no ready-made solutions to these issues, which relate more specifically to starch, protein and oil resources, for example. Only a long, in depth and dialectic study, conducted on a case-by-case basis, would yield a clear vision of this issue.

Note also that a certain number of terms used in this chapter (biopolymer, *biomacromolecule*, bio-based, etc.) were recently defined by International Union of Pure and Applied Chemistry (IUPAC) and presented in a reference article [VER 12].

9.2. Biodegradable polymers

9.2.1. *Concepts: biodegradability and renewal*

9.2.1.1. *Biodegradability and compostability*

There are a certain number of norms in existence, such as the American norm ASTM D-5488-94d, or the European harmonized standard EN 13432, giving definitions for terms such as *biodegradable*, *compostable* and others which are widely used – and sometimes incorrectly – as an argument in favor of the use of so-called "environmentally friendly" materials.

The norm EN 13432:2000 sets out the requirements for packaging recoverable through composting and biodegradation. EN 13432 was accepted by the decision of the European Commission (2001/524/CE) and published in the *Official Journal of the European Community*. It is a reference standard in the domain in Europe. It defines recovery through composting and biodegradation as follows (depending on whether it takes place in an aerobic or anaerobic environment): by the action of micro-organisms in the presence of oxygen, decomposition of an organic chemical compound into carbon dioxide, water and mineral salts (mineralization) with the creation of a new biomass; in the absence of oxygen, decomposition into carbon dioxide, methane and mineral salts, with the creation of a new biomass.

It is noteworthy that packaging materials and components of natural origin which have not been chemically modified, such as ligno-cellulose materials, starches and others are recognized as being biodegradable without needing to be subjected to the tests prescribed by the corresponding norm. However, they must be characterized in terms of their chemical composition (identification of the components, heavy metal content, organic carbon, dry solids, volatile solids, etc.) and conform to the criteria of disintegration and compost quality, particularly in terms of the ecotoxicity of the residual product.

The norm NF EN 13432 served as the basis for the elaboration of the norm NFU 52-001: "biodegradable materials for agriculture and horticulture", which came into force on 20 February 2005. Other norms are in place for biodegradable materials. Some of them are equivalents, such as EN 14046:2003 and ISO 14855:1999, both on evaluation of the biodegradability into compost. A series of norms define the conditions for testing biodegradability depending on the medium being used. These include, for example, the norms EN ISO 14851:2004 (aqueous medium, aerobic system, analysis of consumed oxygen), 14852:2004 (aqueous medium, aerobic system, analysis of released CO_2), 14853:2004 (aqueous medium, anaerobic system), 17556:2004 (soil) and 14855:2005 (compost). All these tests are performed by comparison against a control.

Biodegradable: in order to qualify a packaging material as biodegradable, EN 13432:2000 stipulates that the following tests be carried out:

– conditions:

- tests performed in a specifically defined medium (freshwater, saltwater, soil, etc.);

- test period lasting a maximum of six months;

– physical results and effects on the medium:

- the original mass of the material must be 90% degraded;

- therefore, the residual products must represent at most 10% of the original mass of the material being tested;

- the result of the biodegradation should not cause ecotoxic effects on the medium.

Compostable: in order to qualify a packaging material as compostable, EN 13432:2000 stipulates that the following tests be carried out:

– conditions:

- tests performed in an industrial composting facility (in a windrow or a heap);

- test period lasting 12 weeks maximum;

– physical results and effects on the compost:

- the residual products must represent at most 10% of the original mass of the material being tested;

- the particle size of the residues must be less than 2 mm (disintegration);

- there should be no negative effects on the composting process;

- the result of the composting should not cause ecotoxic effects on the compost.

It is difficult to compare the results obtained on the basis of different norms, because unfortunately, the test conditions (in terms of humidity, temperature cycle, etc.) vary from one norm to another.

A few general rules enable us to predict the evolution of a material's biodegradability. For example, an increase in factors such as hydrophobic nature, molar masses, crystallinity or size of crystalline regions harms a substance's biodegradability. On the other hand, the presence of polysaccharides, say (in the case of blends) favors biodegradation.

It should be noted that biodegradation/composting represents only one end-of-life scenario for these materials. As is true of most polymers, they are, of course, recyclable.

9.2.1.2. *Origin of carbon and sustainable development*

The "biodegradable" or "compostable" nature of a material refers to its end of life. The "bio-based" or "renewable" nature of a material, on the other hand, refers to its origin, and particularly that of the carbon (derived from biomass). This is illustrated in Figure 9.1 by the direct synthesis path (oblique arrow). In this diagram, the fossil resource (oil, etc.) can be seen as a carbon-sequestering intermediary product, which prolongs the carbon cycle.

Polymers which are both biodegradable and bio-based thus belong to a sustainable development approach extending throughout a sustainable lifecycle, represented by Figure 9.2. We speak of a *cradle-to-cradle* approach, but also of "carbon reincarnation". The composted materials can be used to enrich the soils, and thus create a new generation of biomass, which will later be used again in a biomass transformation unit (a bio-refinery), until the biodegradable polymer is obtained or extracted.

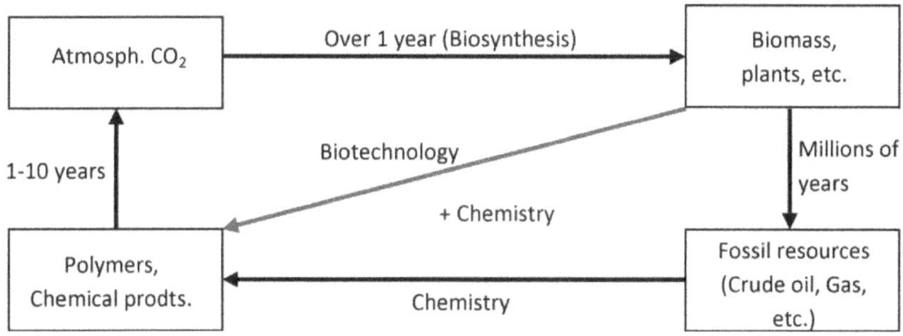

Figure 9.1. *Carbon cycle and means of polymer synthesis (direct and indirect) (adapted from Pr. Narayan)*

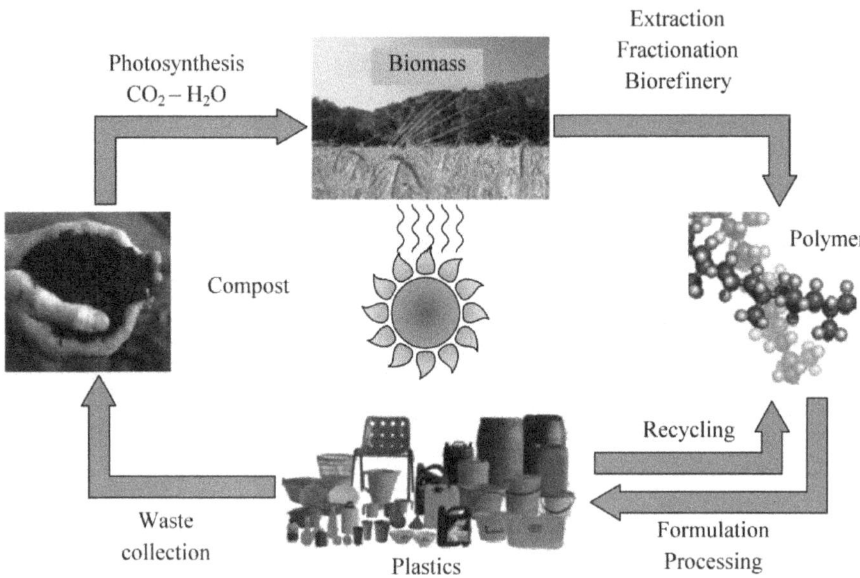

Figure 9.2. *Lifecycle of biodegradable and bio-based polymers (example: starch)*

9.2.2. *Classifications of biodegradable polymers*

Biodegradable polymers constitute an entire field in their own right – a field which is expanding [AVÉ 12, CHA 98, STE 03]. A large number of biodegradable polymers are synthesized or are formed in nature during the growth cycle of all organisms. Certain micro-organisms and enzymes capable of degrading them have been identified [CHA 98].

Depending on the synthesis process, different groups or types of biodegradable polymers have been classified. Figure 9.3 shows a proposal for a classification [AVÉ 04a] including four different categories. Only the first three categories (a–c) are obtained from renewable resources:

a) polymers derived directly from the fractionation of biomass, such as polymers from agroresources: agropolymers (e.g. starch, cellulose and proteins) and their derivatives;

b) polymers obtained by microbial production or fermentation (e.g. polyhydroxylalkanoates);

c) polymers synthesized conventionally, for which monomers are obtained from agro-resources and/or by biotechnology (e.g. poly(lactic acid – PLA));

d) polymers whose monomers and polymers are obtained conventionally, by chemical synthesis on the basis of fossil resources (e.g. polycaprolactone). However, at industrial level, there is now a trend toward replacing the source of these polymers, in favor of bio-based materials. Such was the case, relatively recently, of poly(butylene succinate) (PBS), which had previously been derived primarily from fossil resources.

We can classify the different biodegradable and bio-based polymers into two major families: agropolymers (category a) and biodegradable polyesters (categories b and c). To illustrate the latter, the next section focuses on the description of biodegradable polyesters, from synthesis to application.

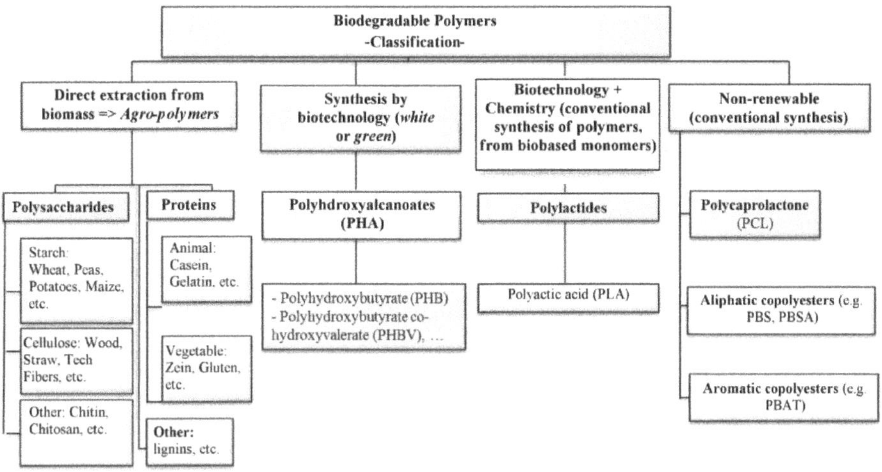

Figure 9.3. *Classification of biodegradable polymers [AVÉ 04a]*

9.2.3. *The case of biodegradable polyesters*

9.2.3.1. *Introduction*

Figure 9.4 illustrates the chemical structures of the different biodegradable polyesters. They can be classified into bio-based or non-renewable-resource-based polymers. Table 9.1 lists the main physico-chemical and mechanical properties of commercially used polyesters.

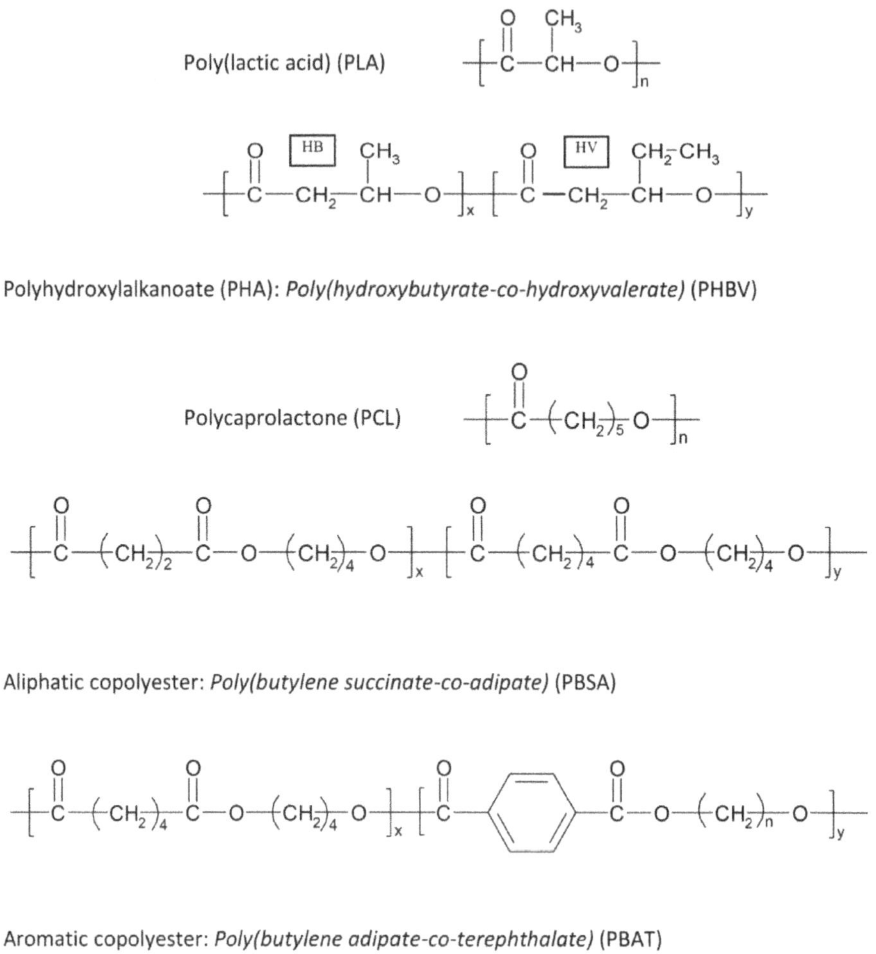

Figure 9.4. *Chemical structures of different biodegradable polyesters*

9.2.3.2. *Bio-based polyesters*

9.2.3.2.1. Polylactic acid (PLA)

Lactic acid can be produced chemically, or biologically, by fermentation of carbohydrate by lactobacillus, for example [SOD 02]. There are a number of ways to produce PLA [JAC 99], also written as "polylactic acid". The industrial and conventionally used means of obtaining high-molar-mass PLA is presented in Figure 9.5. This process is based on several steps:

1) the monomers (enantiomers D and L) are condensed and then de-polymerized and converted into lactides;

2) those lactides are then polymerized by “ring-opening polymerization” (ROP) to obtain a polymer with a high-molar mass [OKA 02]. Depending on the type of lactide used (see Figure 9.6) and therefore the intended L/D ratio, we can expect to obtain different PLAs.

	PLA	PHBV	PCL	PBSA	PBAT
	NatureWorks (94% L)	Monsanto (Biopol D400G -HV = 7 mol%)	Solvay (CAPA 680)	Showa (Bionolle 3 000)	Eastman (Eastar bio 14 766)
Density	1.25	1.25	1.11	1.23	1.21
Melting temperature – Tm (°C)[a]	152	153	65	114	110–115
Glass transition temperature – Tg (°C)[a]	58	5	–61	–45	–30
Crystallinity[b] (%)	0–1	51	67	41	20–35
Young’s modulus (MPa) (NFT 51-035)	2,050	900	190	249	52
Elongation at break (%) (NFT 51-035)	9	15	>500	>500	>500
Max. stress (MPa) (NFT 51-035)	–	–	14	19	9
Permeability to water vapor at 25°C (g/m^2/day)	172	21	177	330	550
Surface tension (γ) (mN/m)	50	–	51	56	53
γ_d (dispersive component)	37	–	41	43	43
γ_p (polar component)	13	–	11	14	11

a Determined by DSC.
b Determined on pellets, before use.

Table 9.1. *Physico-chemical data of various commercial polyesters [BOR 09b]*

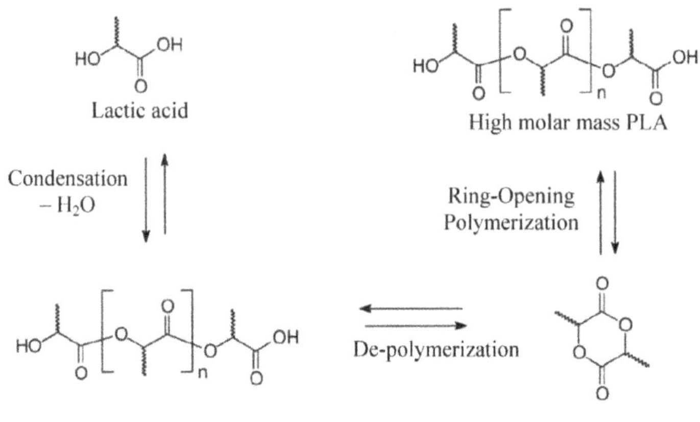

Figure 9.5. *Diagram of PLA synthesis*

In relation to other biodegradable polyesters, PLA is the biodegradable polymer which, at present, has one of the highest potentials, owing to its relatively low price (<€2/kg) and its availability on the market [AVÉ 12]. Estimated worldwide PLA production in 2012 was around 150 KT, according to *European Bioplastics*. A market leader in this field, Cargill-NatureWorks-PTT has developed a process based on maize and other raw materials to produce different grades of PLA as illustrated in Figure 9.6. Other companies such as Mitsui Chemicals (Japan), Mitsubishi (Japan), Shimadzu (Japan), Futerro (Total/Galactic – Belgium), Purac (the Netherlands), Teijin (Japan) and Zhejiang Hisun Biomaterials (China) produce different PLAs, with a broad range of L/D ratios. We can cite PLLAs (100% L) which have a high degree of crystallinity. There are also copolymers, poly(D,L)-lactic acid (PDLLA), which may be amorphous. Indeed, the D/L ratio determines many of the properties of PLA [AVÉ 13].

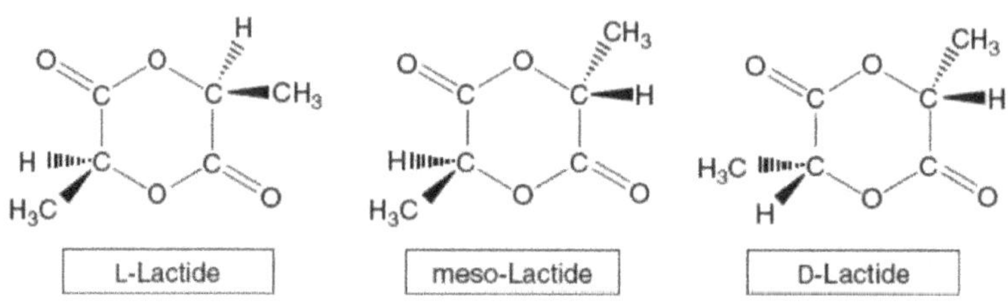

Figure 9.6. *The different types of lactide*

Depending on the preparation conditions, PLLA crystallizes in a variety of forms [AVÉ 13]. The α form exhibits a clearly defined diffraction diagram. This crystalline structure is more stable than the β form. It has a melting point (*Tf*) of 185°C, as compared with 175°C for the β form. The latter can be formed under significant strain at high temperature. Finally, the γ form is obtained by epitaxial crystallization. Furthermore, it has been observed that a (50/50) blend of PLLA and PDLA produces stereocomplexation. This stereocomplex exhibits better mechanical and thermal properties than those of pure PLLA or PDLA, and a high *Tf*, equal to 230°C.

PLA is a rigid polymer at ambient temperature (see Table 9.1). It crystallizes slowly, like PET. The crystallization kinetics of PLA has been extensively studied. Like PET, PLA can be oriented by thermomechanical treatment. Orienting the chains notably increases its mechanical strength.

The glass transition temperature (T_g) of PLA generally lies between 50 and 80°C. This T_g may change with plasticization. PLA can be plasticized using oligomers of lactic acid (O-LA) [MAR 01a], citrate [LAB 97] or poly(ethylene glycol)s (PEGs) with low-molar masses [MAR 01a], or even polyadipates [MAR 10]. The addition of a plasticizing agent increases the chain's mobility, favors organization of the PLA and accelerates its crystallization. Generally speaking, the properties of PLA can be significantly altered by its formulation [AVÉ 13, COU 11, COU 12].

The various properties of PLA and its availability mean that numerous projects are currently under way regarding its use in very different applications, such as packaging, biomedical applications, agriculture, health, leisure [DOM 11], and so on. The current trend in the use of PLA, however, is toward durable (i.e. long-lasting) materials in areas such as the automobile or construction sectors, where its biodegradability is not a desirable feature.

PLA biodegrades in soils more slowly than do other aliphatic polyesters. PLA is presumed to be biodegradable, although the respective and relative roles of hydrolytic de-polymerization and enzymatic degradation are still open to debate. It is a two-stage process, which begins with abiotic degradation (i.e. simple hydrolysis of the ester bonds without enzymes present) with reduction in the masses to the level of oligomers, followed by biotic degradation, with the participation of enzymes and micro-organisms. The two processes overlap partially. With regard to biodegradation into compost, the appropriate conditions can only be found in industrial composting facilities operating at a high temperature (50°C) and high, constant relative humidity to encourage mineralization of this polymer. The biodegradation of PLA causes no particular ecotoxicity [DOM 11].

9.2.3.2.2. Polyhydroxylalkanoates (PHAs)

PHAs are a family of intracellular biopolymers synthesized by numerous bacteria, as a means of energy and carbon storage [SUD 00]. PHAs are mainly produced by fermentation from renewable resources [POL 11]. A wide variety of bacteria accumulate PHA – it accounts for between 30 and 80% of their dry weight.

Biotechnology studies have shown that PHA is produced in controlled, balanced growing conditions, by feeding and then starving the micro-organisms of nutrients [ALB 11, ESC 11]. Depending on which hydrocarbon substrates and micro-organisms are used, different types of (co)polymers and macromolecular architectures can be obtained [POL 11].

PHAs are considered to be biodegradable. Thus, they are to be found in corresponding applications – e.g. short-term packaging or the agricultural sector. They are also held to be biocompatible for contact with living tissues, and can be used for biomedical applications (e.g. medicine capsules, tissue engineering and the like).

PHAs can be degraded by abiotic degradation. During the course of biodegradation, the enzymes degrade the residual products until they are completely mineralized.

PHAs are generally classified into "short chain length" (sCL) and "medium chain length" (mCL) molecules. For example, sCL-PHAs are based on 4 or 5 carbon atoms in their repeating patterns, while mCL-PHAs contain 6 or more carbon atoms in their repeating units.

The nomenclature of PHAs and their classification are still in the process of evolving, because new structures are continuing to be discovered. The main biopolymer in the PHA family is a homopolymer: polyhydroxybutyrate (PHB). The most common copolymer is poly(hydroxybutyrate-co-hydroxyvalerate) (PHBV), whose chemical structure is presented in Figure 9.4. However, there are others, such as poly(hydroxybutyrate-co-hydroxyhexanoate) (PHBHx), poly(hydroxybutyrate-co-hydroxyoctanoate) (PHBO) and poly(hydroxybutyrate-co-hydroxyoctadecanoate) (PHBOd), for example.

Figure 9.7 shows the generic formula for PHAs, where the value of x is 1 or more, and R can be either a hydrogen atom or an auxiliary group containing up to 16 carbons (denoted as C16). The main homopolymers of PHA are presented in Table 9.1.

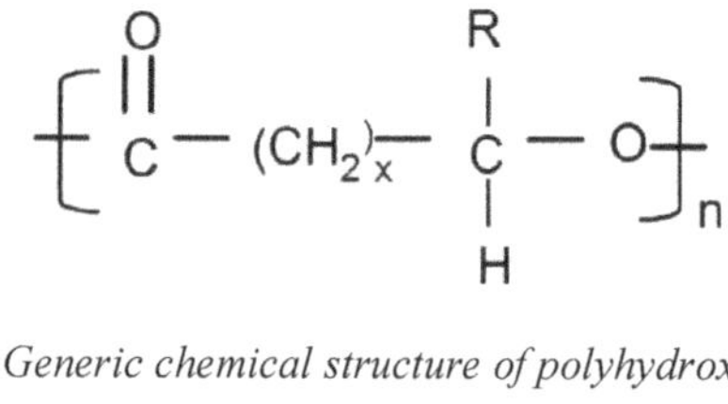

Figure 9.7. *Generic chemical structure of polyhydroxylalkanoates*

PHB or P(3HB) is a highly crystalline polyester (crystallinity greater than 50%) with a high melting temperature, *Tf* = 173–180°C, in relation to other biodegradable polyesters. Its T_g is around 5°C.

A wide range of homopolymers and copolymers of PHA have been produced – in most cases on a laboratory level. Some of these have attracted interest from industrial actors, and have been commercialized over the course of the past two decades. As Table 9.2 shows, PHA production is distributed between a large number of companies. Worldwide, over 24 companies are known for their involvement in PHA production.

Chemical name	Abbreviation	Value of *x*	R group
poly(3-hydroxypropionate)	P(3 HP)	1	Hydrogen
poly(3-hydroxybutyrate)	P(3HB)	1	Methyl
poly(3-hydroxyvalerate)	P(3HV)	1	Ethyl
poly(3-hydroxyhexanoate) or poly(3-hydroxycaproate)	P(3HHx) or P(3HC)	1	Propyl
poly(3-hydroxyhexanoate)	P(3HH)	1	Butyl
poly(3-hydroxyoctanoate)	P(3HO)	1	Pentyl
poly(3-hydroxynonanoate)	P(3HN)		Hexyl
poly(3-hydroxydecanoate)	P(3HD)	1	Heptyl
poly(3-hydroxyundecanoate)	P(3HUD) or P(3HUd)	1	Octyl
poly(3-hydroxydodecanoate)	P(3HDD) or P(3HDd)	1	Nonyl
poly(3-hydroxyoctadecanoate)	P(3HOD) or P(3HOd)	1	Pentadecanoyl
poly(4-hydroxybutyrate)	P(4HB)	2	Hydrogen
poly(5-hydroxybutyrate)	P(5HB)	2	Methyl
poly(5-hydroxyvalerate)	P(5HV)	3	Hydrogen

Table 9.2. *Main structures of homopolymers of PHA (see Figure 9.7)*

PHAs – even more than other polyesters – are sensitive to the conditions of thermomechanical treatment. With extrusion, we can obtain a rapid decrease in viscosity and molar mass, by increasing the degree of shearing, temperature and/or exposure time [POL 11]. Generally speaking, PHAs exhibit relatively high sensitivity to thermal and thermomechanical treatments [BOR 09a, HAB 08], which varies depending on the macromolecular architecture. During processing, the homopolymer presents a narrow window of useful thermomechanical input between melting and degradation. This window is wider for the copolymer PHBV – hence its development. For this substance, the properties vary depending on the HV content. An increase in the HV content causes an increase in shock resistance, and a lowering of the *Tf* and *Tg*, crystallinity, water permeability and traction strength [POL 11].

With regard to biodegradability, the kinetics of enzymatic degradation is variable depending on the crystallinity, the structure and the material's thermomechanical history. Bacterial copolyesters are more easily biodegradable than their equivalents obtained by chemical synthesis [AVÉ 12].

In comparison to PLA, world production of PHA is lower. Worldwide capacity for PHA production in 2012 was lower than 100 KT, according to *European Bioplastics*.

Company	Country	Commercial name	PHA	Pilot/industrial scale
Biomatera	Canada	Biomatera	PHBV	Pilot
Biomer	Germany	Biomer	PHB, PHBV	Pilot
Bio-On	Italy	Minerv PHA	PHB, PHBV	Pilot
Kaneka	Japan	Kaneka	PHBHx	Pilot, Industrial?
Meredian	USA	Meredian PHA	Copolymers	Pilot/ Industrial
Metabolix	USA	Mirel	Copolymers	Pilot, Industrial?
PHB Industrial/Copersucar	Brazil	Biocycle	PHB, PHBV	Pilot, Semi-industrial.
PolyFerm Canada	Canada	VersaMer PHA	PHBV and other copolymers	Pilot
Tianan	China	Enmat	PHBV	Industrial
Tianjin and DSM	China	GreenBio	3HB- and 4HB-based copolymers	Pilot
Tianzhu	China	Tianzhu	PHBHx	Pilot

Table 9.3. *Main commercialized PHAs and main producers in 2012*

For more than 10 years, new synthesis techniques have been under study at an industrial level. Metabolix (USA) has developed a means of PHA production using genetically modified cultures. In 2010, that company announced that it had performed conclusive tests on PHA production on modified tobacco plants, stemming from green biotechnologies.

PHA production may be destined to partially replace non-degradable synthetic polymers in various applications, such as: packaging, agriculture, leisure, rapid restoration, health and the biomedical domain [POL 11, SUD 00].

9.2.3.3. *Fossil resource-based polyesters*

Although the prevailing trend today is toward the development of 100% bio-based materials, a large number of biodegradable polyesters are still obtained from fossil resources (such as crude oil). They are chemically synthesized using non-renewable monomers [AVÉ 12]. It is possible to distinguish different (co)polyesters, either aliphatic or aromatic, depending on their chemical structures and their corresponding properties (see Table 9.1 and Figure 9.4). All these polyesters tend to be soft at ambient temperature.

9.2.3.3.1. Polycaprolactone (PCL)

Poly(ε-caprolactone) is produced by way of two methods: polycondensation of 6-hydroxyhexanoic acid (also known as 6-hydroxycaproic acid), or very often by ROP of ε-caprolactone, notably in the presence of aluminum isopropoxide, or tin ethylhexanoate [COU 06, OKA 02]. It is also possible to produce it by enzymatic catalysis using lipases [VAR O5]. PCL is widely used as a plasticizing agent for PVC or for applications in polyurethanes, such as polyol. Yet it also has applications based on its biodegradable nature in domains such as the biomedical (implants for controlled release of medicines, tissue engineering, etc.), compostable packaging, adhesives and so on. A variety of commercial grades used to be produced by Solvay (CAPA®), who sold that activity to Perstorp (Sweden). PCL is also produced by Dow Chemical (Tone®) and Daicel (Celgreen®).

Figure 9.4 shows the chemical structure of PCL. It is a highly crystalline material, and soft at ambient temperature, with very great strain at break (Table 9.1). PCL has a very low T_g (–61°C) and a low Tm (65°C), which can be a handicap for certain applications. Consequently, PCL is generally mixed [AVÉ 00b, BAS 98] or modified (e.g. by copolymerization or cross-linking [KOE 94]) to expand its field of application.

Tokiwa and Suzuki [TOK 77] examined hydrolysis of PCL and its fungal biodegradation. They demonstrated that PCL can easily be degraded by enzyme activity. According to Bastioli, biodegradability can clearly be claimed, but the rate of PCL's hydrolysis is very slow (slower than that of PLA) [BAS 98].

9.2.3.3.2. *Aliphatic copolyesters*

A large number of aliphatic copolyesters are biodegradable copolymers [AVÉ 12]. They are obtained by combination of diols such as 1, 2-ethanediol, 1, 3-propanediol or 1, 4-butanediol and dicarboxylic acids: adipic acid, sebacic or succinic acid. Until very recently, these (co)polymers were primarily produced from non-renewable resources. Now, increasingly, they are beginning to be bio-based. Indeed, in terms of bioproduction of succinic acid worldwide, there are a number of different projects to produce bio-based PBS. For example, there is a very recent project involving Faurecia, Bioamber (a succinic acid producer) and Mitsubishi to produce PBS, notably for the automobile market. The capacity for PBS production was 24 KTons in 2011, according to a recent study by Nova-Institut (Germany) (http://www.bio-based.eu); it is predicted to increase to 258 KTons by 2020.

Showa Highpolymer (Japan) developed a wide range of polybutylene succinate (PBS) by polycondensation of 1,4-butanediol and succinic acid. Polybutylene succinate–co-adipate (PBSA), shown in Figure 9.4, is obtained by the addition of adipic acid. These copolymers are commercialized under the brand name Bionolle® (Showa Denko K.K.). Industrial production of these polyesters from bio-based succinic acid was launched in 2012 by Showa Denko. Mitsubishi Chemical (Japan) produces and also commercializes partially bio-based PBS.

Skygreen® is a product of SK Chemicals (North Korea) and is obtained by polycondensation of 1,2-ethanediol and 1,4-butanediol with succinic and adipic acids. Nippon Shokubai (Japan), IreChem (North Korea) and Zhejiang Hangzhou Xinfu (China) also market an aliphatic copolyester under the respective names Lunare SE®, EnPol® and Biocosafe®. The properties of these (co)polyesters depend on their structures and macromolecular architectures.

These materials' biodegradability depends on their structures. For example, the addition of adipic acid, which decreases the crystallinity, tends to increase biodegradation into compost [FUJ 98].

9.2.3.3.3. *Aromatic copolyesters*

Aromatic and biodegradable copolyesters are, for the most part, based on terephthalic acid [AVÉ 12]. As with aliphatic copolyesters, aromatic polyesters are increasingly bio-based, notably with the future development of bio-based terephthalic acid production, alongside 100% bio-based PET projects.

Figure 9.4 and Table 9.1 respectively show the chemical structure and the properties of these materials. Most commonly used is poly(butylene adipate-co-terephtalate) (PBAT). BASF, DuPont and Zhejiang Hangzhou Xinfu (China) manufacture and market these aromatic polyesters, under the respective brands Ecoflex®, Biomax® and Biocosafe®. Biomax® has a high terephthalic acid content, which modifies some of its properties such as the melting temperature (200°C). According to Muller *et al.*, an increase in terephthalic acid content tends to decrease the rate of biodegradation into compost [MUL 97].

9.2.4. *Agro-polymers*

9.2.4.1. *Introduction*

Agropolymers are mainly extracted from plants. They are biopolymers, and for the most part they are biodegradable. There are different families such as polysaccharides, proteins, lignins and tannins. They share a relatively hydrophilic nature. Most of them can be processed directly, plasticized, used as fillers or chemically modified.

Polysaccharides are the most abundant biomacromolecules in the biosphere. These complex carbohydrate comprising glycosidic bonds are one of the main structural elements of plants and exoskeletal animals (cellulose, carrageenan, chitin, etc.).

The topic of cellulose and its derivatives has largely been discussed in many treatises. Cellulose is the most abundant agropolymer on Earth. It is a linear polymer made of units of D-glucose linked by β (1-4) bonds and synthesized by plants and bacteria. In order to obtain a thermoplastic material, cellulose is modified, e.g. by acetylation to obtain cellulose acetate.

Protein-based materials have also been abundantly discussed in other publications [ZHA 08]. The next section of this chapter is, therefore, primarily oriented toward one example of an agropolymer: starch. World production of starch-based materials for "bioplastic" applications is far from negligible. Put differently, it is greater than or equal to that of PLA.

9.2.4.2. *Native starch*

Starch is the main element used for storage by botanical resources (cereals, legumes and tubers). It is a raw material in abundant supply on the planet. For example, European starch production was 10 million tons in 2012, 38% of which was for non-foodstuff applications. Indeed, starch has many non-food-related industrial applications in varied domains, such as paper, textiles, construction and adhesives.

The main industrial starch productions are based primarily on four resources: maize (76%), cassava (12%), wheat (7%) and potatoes (4%). Other forms of starch represent less than 1%. The main production zones are North America (33%), China (33%), Europe (18%), South Asia (11%) and South America (5%). The following set: North America, China and Europe represent in total 85% of worldwide starch production. However, those three only correspond to around one-third of the world's population.

Starch is a polysaccharide made up of units of D-glucose. Starch is composed mainly of two different biomacromolecules: amylose and amylopectin. Amylose (Figure 9.8(a)) is linear, with little branching. It is formed mainly, though not solely, of α (1-4) bonds, and has a molar mass of 10^5–10^6. The chains are in the form of single or double helices. Amylopectin (Figure 9.8(b)) is a highly branched polymer with a high molar mass: 10^7–10^9. It is formed of α (1-4) bonds, but also of α (1-6) bonds constituting numerous branch points, which occur every 22–70 glucose units [XIE 12].

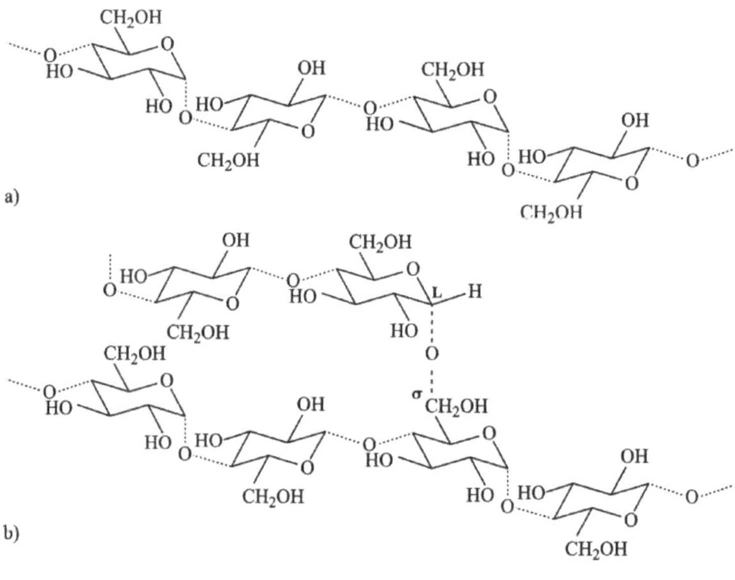

Figure 9.8. *Simplified chemical structures of amylose a) and amylopectin b)*

Depending on which botanical resource is used, the dimensions of the starch granules range from a micron to several hundred micrometers, with a very wide variety of shapes. Also depending on the botanical origin, the composition of starch is variable. Although amylose contents are typically between 20 and 30%, certain mutant plant species present specific compositions. We can find starch which is rich in amylose in the case of amylomaize (up to 80%) and starch very rich in amylopectin with waxy maize (>99%).

After the different industrial steps of fractionation and refining, the starch obtained exhibits some traces of lipids, gluten and phosphate. These co-products may affect the properties of the starch – e.g. by the formation of lipid complexes, or by Maillard reactions (with gluten) at the processing stage [XIE 12].

Co-crystallization phenomena with helicoidal structures of amylose have also been identified. Numerous types of crystallinity are observed in the granule.

9.3. Plasticized starch

9.3.1. *General points*

With the exception of applications as a filler for the production of reinforced plastics, native starch is generally modified and transformed [XIE 12]. Such is the case, for example, with de-structured or gelatinized starch. The de-structuring agent is usually water. Gelatinization of starch is achieved by a combination of water (a high content) and high temperature. Gelatinization causes the de-structuring of the starch, forming a viscous paste, with a reduction in the melting temperature (Tf) and the glass transition temperature (Tg). Depending on the level of de-structuring and the amount of water used, we can obtain different products for various applications, as illustrated in Figure 9.9. For example, we can obtain cellular structures with a fairly high water content. These foams have been developed to obtain packaging and shock conditioners, loose fills, isothermal materials and soundproofing [AVÉ 04a].

If the water content is decreased (to less than 20% weight), the melting temperature approaches the degradation temperature of starch. For example, dry starch has a Tf = 220–240°C, in comparison with 220°C, which is the temperature at which starch begins to degrade [AVÉ 04a].

To counter the loss of water during the process and a significant degradation of the material, to prevent, e.g., the formation of cellular structures output from an extruder, and to obtain a material with controlled properties, we use a non-volatile plasticizer such as glycerol or other polyols (sorbitol, xylitol, fructose, polyethylene glycol, etc.) [AVÉ 04a, XIE 12]. Blends of these different polyols are also used [CHI 10b]. Other, notably nitrogenous plasticizers (urea, ammonium derivatives, amines, etc.) can also be used. Plasticized starches are thus created. They are commonly known as "thermoplastic starch" [AVÉ 04a]. The earliest patents and articles in the domain were published in the late 1980s. Plasticized starch combines starch, a non-volatile plasticizing agent, with a high boiling point, and often water. Plasticized starch is generally transformed using conventional plastic processing machines (extruders, injection molding machines, etc.) [AVÉ 04a].

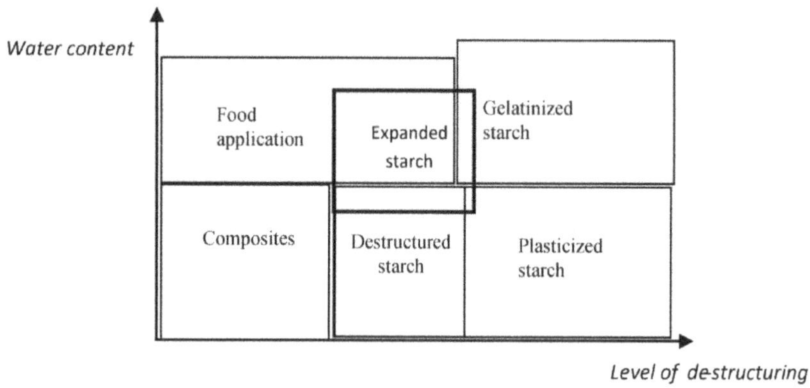

Figure 9.9. *Different applications of starch. Water content on the basis of the level of de-structuring*

9.3.2. *Implementation and rheology of plasticized starch*

The making of plasticized starch is based on the de-structuring of granular (native) starch with transformation of the initial semi-crystalline three-dimensional (3D) structure into a homogeneous and essentially amorphous material. This transformation is generally caused by the input of a significant amount of thermomechanical energy via the process. The combination of the thermal and mechanical energy inputs can be obtained, for example, by using a single-screw or twin-screw extrusion machine. The process for obtaining plasticized starch can be performed in one or two steps. The behaviors of the materials thus obtained are not the same. The exudation of the non-volatile plasticizing agent over time is significantly lesser in a two-step system.

In a one-step process, the extrusion machine, which generally has a double co-rotating screw, uses native starch as input. All along the cover, the different plasticizing agents (water, polyols, etc.) are introduced. Generally, before output into a value creation chain, a de-volatilization well is used to evacuate the excess water which helps to de-structure the native starch.

In a two-step process, the first step is the preparation of a "dry-blend", as happens in the case of plasticized PVC [AVÉ 04a]. In a high-speed powder mixer, the plasticizing agent is slowly added to the native starch until we obtain a homogeneous distribution. Then, the mixture is placed in a ventilated oven to diffuse the plasticizing agent through the grains. The plasticizing agent migrates totally into the starch grains, causing them to swell. After cooling, an appropriate amount of water is added to the mixture using a high-speed mixer. This dry-blend is then fed into an extrusion machine. Figure 9.10 diagrammatically represents the different mechanisms relating to extrusion. The starch granules are fragmented. Under the influence of temperature and shearing, the starch is de-structured, plasticized and melted. However, the thermomechanical treatment during extrusion also leads to a partial depolymerization of the polysaccharide.

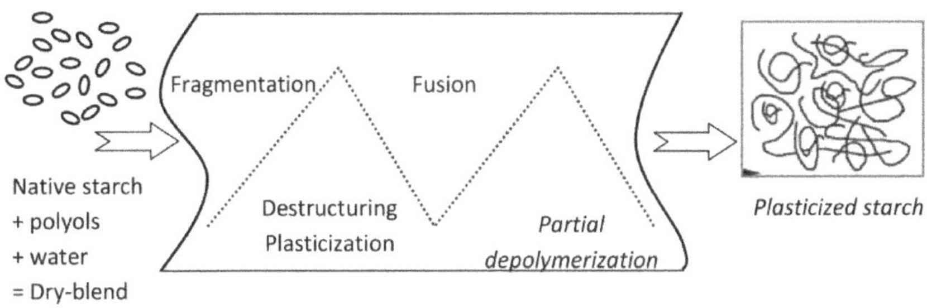

Figure 9.10. *Illustration of the process of manufacture of plasticized starch in two steps, with the mechanisms relating to extrusion*

After this treatment, we obtain a homogeneous melted system. The plasticized starch thus created can be taken up, in pellet form or otherwise, and used in different techniques such as injection, blowing and so on.

The rheological behavior in a melted system of plasticized starch is known to depend on the components of the mixture (water content, type and content of non-volatile plasticizing agent(s), type(s) of starch), but also on the temperature, and on the thermomechanical treatment applied [XIE 12]. The data on viscosity and from rheological models which take account of these different variables have been reported in the existing literature. A recent scientific review summarized the main rheological studies performed on plasticized starch, including the measuring tools developed and the model used [XIE 12]. One difficulty that is common in rheological studies on these specific materials is that a thermomechanical treatment is needed to obtain a melted material. Indeed, this material exhibits a particular behavior which is qualified as "thermomechano-plastic" [MAR 03, XIE 12]. At the same time, mechanical energy (specific mechanical energy – SME) and temperature are necessary to obtain a melted material, necessary for the corresponding rheological study. Consequently, the setups used in conventional rheology (plate–plate or cone–plate rheometers, capillary rheometers, etc.) are not fully appropriate for examining the rheological behavior of plasticized starch – particularly when the content of the plasticizing agent is low. Hence, tools specific for the task have been developed. In particular, these devices input the SME which is necessary to obtain a homogeneous melted material, which must be carefully controlled. These devices include a commercial tool: Rheoplast© [MAR 03, VER 87, XIE 12]. In addition, for this type of study, an extruder combined with tooled dies can conventionally be used, if we maintain control of the temperature and the SME [MAR 03].

9.3.3. *Behavior of plasticized starch in the solid state*

9.3.3.1. *Aging after processing*

Various authors [AVÉ 00b, AVÉ 04a, SOE 97b] have shown that after transformation, plasticized starch exhibits aging, with a significant change in its properties – particularly the Young's modulus, which increases over several weeks. The material presents two types of behavior, depending on the value of the T_g in relation to the storage temperature. Below its *Tg*, plasticized starch displays physical aging, with the material becoming denser. At a temperature higher than its *Tg*, plasticized starch exhibits a phenomenon of retrogradation, with a slow change in crystallinity and rearranging of the molecules of plasticizing agents within the material [SOE 97b], the longer it is stored. The kinetics of this retrogradation depends on the mobility of the (macro)molecules present, and on the type and amount of plasticizing agent.

9.3.3.2. *Crystallinity*

In comparison to native starch (crystallinity: 20–45%, depending on the plant resource used), plasticized starch displays a lower degree of crystallinity (only a few percentage). According to Van Soest and Vliegenthart [SOE 97a], two types of crystallinity are obtained after transformation of starch: residual crystallinity from the native starch (types A, B and C), and crystallinity induced by the processing (types E and V). The latter types of crystallinity are influenced by parameters such as the residence time in the extruder, the speed of the screw or the temperature. They are mainly caused by rapid recrystallization of amylose in a helicoidal structure. With regard to the aging of plasticized starch, Van Soest and Vliegenthart proposed the formation of a complex crystalline model, containing amylose and amylopectin in amorphous and crystalline forms, and probable co-crystallizations between both these polysaccharide types [SOE 97a].

9.3.3.3. *Effect of water and non-volatile plasticizing agent on the behavior of plasticized starch*

Certain authors have shown that at low concentrations of non-volatile plasticizing agent, an antiplasticization effect can occur [LOU 97a]. Because of the strong interactions between the plasticizing agent and the starch, a network of hydrogen bonds is formed, resulting in the hardening of the material. When the amount of plasticizing agent in the mixture increases, the interactions between the different molecules of the same plasticizing agent cause the material to swell and display classic plasticization behavior.

The change in the material's water content and the consequences of that change in terms of the evolution of the properties or transitions (such as the glass transition) is a key point in understanding the behavior of plasticized starch. Indeed, water is a volatile plasticizing agent, the amount of which will be balanced out, depending on the environment, by mechanisms of sorption/desorption. Godbillot *et al.* [GOD 06] put forward a phase diagram for a system comprising water, glycerol and starch which, for different relative humidity conditions, gives the water content of the material on the basis of the amount of non-volatile plasticizing agent (glycerol) present. These authors highlight the different interactions which take place in these complex multi-phase systems (Figure 9.11). Depending on the glycerol content and the relative humidity, the plasticizing agent may be more or less closely bonded to the polysaccharide chains. The plasticizing agent can occupy specific sorption sites which the water and non-volatile plasticizing agent content values are low, or become free when the sorption sites are occupied, in the case of high relative humidity and non-volatile plasticizing agent content. Between these two extreme cases, complex interactions between the different components could be established. Thus, in certain conditions, a phase separation occurs, generating multiphase structures with different areas that are rich and poor in non-volatile plasticizing agent.

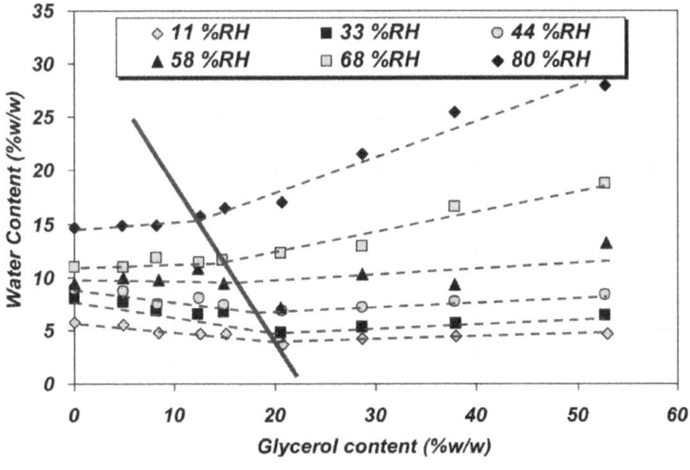

Figure 9.11. *Water content at equilibrium depending on the glycerol content for different degrees of relative humidity (%RH). Note: the oblique line corresponds to a change in the material's behavior in terms of phase segregation [GOD 06]*

9.3.3.4. *Physical properties of plasticized starch*

Depending on the ratio of plasticizing agent to starch, the plasticized starch evolves throughout a wide range of behaviors. The variations of Cp at the glass transition temperatures. Hence, in order to approximate the Tg values, dynamic mechanical thermal analysis (DMTA) is generally more appropriate than DSC. The evolutions of Tan δ on the basis of the temperature exhibit two very marked transitions: a significant and predominant relaxation (α transition), which may be associated with the glass transition of the plasticized starch or with the phases poor in plasticizing agent. The secondary relaxation (between –50 and –60°C) could be linked to the glass transition of the phases rich in non-volatile plasticizing agent [AVÉ 09]. The value of this latter relaxation could be an indicator of the level of interactions between plasticizing agents and polysaccharides [LOU 97b]. This relaxation temperature has a maximum for glycerol contents of around 15% with 50% RH. With higher content values, the decrease in sub-vitreous relaxation would be associated with the phenomena of phase separation.

A wide range of mechanical properties can be obtained by varying the plasticizing agent content, between a very soft material and a rigid, brittle material, with the Young's modulus varying from a few MPa to over a GPa, respectively. Also, like any plasticized system, starch exhibits different levels of permeability to water vapor and oxygen depending on its plasticizing agent content. Although the permeability to water may be high because of the polar nature of the matrix, its oxygen permeability is low in comparison to that of most biodegradable polyesters. The permeability considerably increases at the glass transition, and continues to increase to the rubbery plateau, in line with the plasticizing agent content. With a low content of plasticizing agent, plasticized starch is a good oxygen-resisting material.

9.3.4. *Issues and strategies*

As a material, plasticized starch offers interesting properties. It is a widely available material which is both biodegradable and renewable. Furthermore, in relation to synthetic thermoplastics, it is a relatively inexpensive material (around €0.50/kg). Also, in comparison to fossil resources, the prices of starch appear to be less unstable. Plasticized starch can easily be made and implemented with the conventional machines used in plastics manufacture. Depending on the level of plasticizing agent and the plant resource, it exhibits a broad range of properties and behaviors. Unfortunately, though, as a single material, plasticized starch presents a number of limitations, which leave it handicapped in terms of use for numerous applications. For example, plasticized starch is very sensitive to humidity, its mechanical properties are rather poor in comparison to conventional synthetic polymers, and its properties are liable to undergo significant change after processing. In order to overcome these weaknesses, a variety of strategies have been devised over the course of the past few decades.

Chemical modification of starch has been developed since the first half of the 20th Century, following on from the research on modified cellulose. In 1942, Mullen and Pacsu published a critical analysis on the method for preparing different starch esters [AVÉ 04a]. In 1943, the same authors presented an industrial use of these compounds. Since then, a very rich body of literature has been published on the subject. The esterification of starch (e.g. by acetylation) improves its water resistance [FRI 96], but can negatively impact its biodegradability beyond a certain degree of substitution (DS). We can control the DS between 0 and 3, to obtain a specific hydrophobic nature. In addition, these modified starches can be plasticized, with citrates, for example.

However, this strategy of chemical modification is greatly limited, in that the toxicity and diversity of the by-products obtained during the chemical reactions represent a very limiting factor. This has an impact on the cost of the synthesis procedures, because of the need for purification and elimination of the by-products formed. In addition, these chemical reactions often impact upon the length of the polysaccharide chains, which are partially degraded during the synthesis process. The respective mechanical properties suffer as a result [AVÉ 04a]. These modified starches do not fulfill all the requirements to be substituted for starch and its applications. However, new commercial products have very recently been released, and are managing to find a market. In particular, these new products use materials obtained by grafting onto starch. Such is the case, for example, with the recently launched Gaïalene® (Roquette, France), which is obtained by grafting with polyolefins by reactive extrusion. The material loses its biodegradability, and its recyclability is reduced.

For a few decades, another more promising strategy (one which is often friendlier to the environment) has been developed. It is a formulation-based approach, combining plasticized starch, and forming heterophase systems with other components such as biodegradable compounds (to preserve the overall biodegradability) and/or strengthening materials and fillers (to modify certain physical and mechanical properties). We can thus obtain different structures (composites, multilayers, blends, etc.) with modified properties which can be adapted for the intended applications. This approach raises new issues relating, for example, to the quality of the interfaces, which concerns interphase continuity and compatibility between the different phases introduced.

9.4. Biodegradable multiphase systems based on plasticized starch

9.4.1. *Structures of plasticized starch-based multiphase systems*

With a view to preserving the biodegradability of the final material, various compounds can be combined with plasticized starch:

– polymers such as biodegradable polyesters;

– agromaterials (lignins, cellulose, etc.) which also play the role of a filler or a strengthening agent;

– fillers or strengtheners in low proportions – particularly mineral-based compounds.

Different types of structures can be obtained by these associations, including:

– blends (with the particular case of the alveolar system);

– multilayers;

– composites;

– nanocomposites.

It should be noted that these different structures can be considered as the basic elements in the creation of new materials, with new use properties.

9.4.2. *Plasticized starch-based blends*

There are a certain number of advantages to blending different components. It is the easiest process to use in order to combine different polymers. It enables us to obtain materials with, often, improved or new properties, for an interesting cost. This approach is generally less expensive than developing new polymers with new macromolecular architectures. A key element in blend making is the compatibility between phases, which will determine the interface and interphase interaction of the final material.

Plasticized starch has widely been studied in a mixture with other polymers [AVÉ 04a, AVÉ 00a, AVÉ 00b, AVÉ 01a, MAR 01a, MAR 01b, SCH 04]. A large number of patents have been published on this subject [AVÉ 04a]. These research efforts have led to the commercialization of different biodegradable blends – some of them based on plasticized starch (Table 9.4).

Brand name	Company	Associated compounds	Note
Mater-Bi	Novamont (Italy)	PCL, PBAT	Current market leader
Bioplast	Bio-Sphere (France) and Biotec(Germany)	PBAT	Starch-based
Plantic	Plantic Technologies (Australia)	PVOH	Modified starch and high-amylose starch
Biolice	Limagrain (France)	Biodegradable polyesters	From maize flour
Vegemat	Vegeplast (France)	Plant proteins	Whole maize plant
Cereplast Compostables	Cereplast (USA)	PLA	
BIOPAR	BIOP Biopolymer Technologies (Germany)	Aliphatic copolyesters	Starch-based
Livan	Harbin Livan Biodegradable Product Co., Ltd. (China)	–	Biodegradable starch-based blend
Biograde B-F	Cardia Bioplastics (China) and Biograde Limited (Australia)	Aliphatic copolyesters	Plasticized starch-based
Terraloy	Cerestech (Canada) and Technor Apex (USA)	Biodegradable polyesters	Plasticized starch-based

Table 9.4. *Biodegradable blends based on starch, plasticized or otherwise, commercialized in 2012, and their corresponding suppliers*

These materials were developed for a variety of applications and economic sectors, such as:

– short-term packaging (e.g. supermarket bags, compost bags) and alveolar conditioning (shock absorption, loose fills, heat insulation);

– agriculture (e.g. pots, hooks, mulch film);

– health (e.g. cotton buds);

– leisure (e.g. building games, fireworks, picnic equipment);

– other domains (pens, bone prostheses for dogs, pharmaceutical gels, etc.).

Numerous non-biodegradable blends with synthetic polymers [ARV 99, STP 97], such as polyolefins, have been commercialized. As is currently the case for oxo-degradable polymers, these blends have, in the past, caused great controversy, because they were falsely presented as being biodegradable. Sometimes, the term "biofragmentable" was applied to these substances.

In order to preserve the final compostability, different blends of biodegradable materials have been developed. There is a vast body of literature available in this domain. We find certain associations with agropolymers such as proteins [ARV 99, FIS 00, OTA 99] or pectins [FIS 00], but most research focuses on blends of plasticized starches and biodegradable polyesters: PCL, PHA, PBSA, PBAT, etc. These polyesters, described previously, are produced industrially. They exhibit interesting properties such as a more hydrophobic nature, limited water permeability and improved mechanical properties, in comparison to polysaccharides. However, the cost of biodegradable polyesters is generally higher than that of starch.

Depending on the nature of the polyester used, we obtain different results. However, biodegradable starch–polyester blends are generally not compatible with phase separations and interfacial properties which tend to be poor. Co-continuous or granular-type structures can be obtained, depending on the starch/polyester ratio.

To improve compatibility between phases, the conventional compatibilization strategies have generally been applied, such as the addition of a compatibilizing agent, which can be performed by modification of at least one of the polymers initially present in the blend:

– functionalization of the polyester, with maleic anhydride [AVEL 00, MAN 01, MAN 99], with pyromellitic dianhydride [AVEL 00], or polyacrylic acid [KIM 01];

– functionalization of the starch with glycidyl methacrylate [WIL 98] or with formation of urethane functional groups by reaction with n-butyl-isocyanate [DUQ 01];

– cross-linking of starch into polyester with coupling agents such as peroxides [AVEL 02, or polyisocyanates [WAN 02];

– development of copolymers: starch grafted with polyesters. For example, to render starch and PCL compatible, various authors have developed starch macromolecules grafted with PCL [DUB 99, YDE 00]. The grafting is obtained by ROP of ε-caprolactone on the polysaccharide. The reaction can be catalyzed by tin octanoate, or aluminum alcoholates. We obtain a comb-shaped structure where the length of the grafts can be controlled [RUT 01, YDE 00]. The same approaches can be used with PLA [SCH 08].

The results, obtained by SEM, DSC or by exploiting mechanical properties, demonstrate better continuity between phases and lesser interfacial tension.

9.4.3. *Plasticized starch-based multilayers*

In comparison to blends, a multilayer structure may present additional advantages, particularly in terms of dealing with the problem of sensitivity to humidity, which is not completely resolved in a blend structure. In a multilayer material, a sandwich-type structure with, say, a polyester at the surface (this component is relatively hydrophobic and biodegradable) will help protect the (hydrophilic) plasticized starch at the core. On this subject, various patents have been filed. For example, Bastioli [BAS 98] put forward different processes (co-extrusion, molding, etc.) to protect starch-based materials, notably with wax [AVÉ 04a]. Examples of surface coatings are mentioned in the existing literature. The coating is laid down by pulverization, for example. For this purpose, various liquids such as dilute solutions of biodegradable polyesters have been used [AVÉ 04a].

Co-extrusion seems to be the best option, though, because it has the advantage of being a single-step process, continuous and polyvalent. Co-extrusion of multilayers has been widely used in recent decades to combine the properties of several polymers in a unique multilayer structure. However, certain problems inherent in the multiphase nature of the flow are liable to arise during the co-extrusion operations, such as non-uniform distributions of the layers, encapsulation of one polymer by another or interfacial instabilities. These issues directly affect the quality and functionality of multilayers. The phenomenon of encapsulation corresponds to the covering of the more viscous polymer by the less viscous [AVÉ 04a]. Figure 9.12 illustrates a phenomenon of interfacial instabilities with the formation of wavy interfaces and variations of the thicknesses of the different layers which interfere with the use properties of these materials. Very early on, in 1967, it was demonstrated that differences in viscosity can cause instabilities when exposed to the stratified flow [AVÉ 04a]. Certain authors have studied the factors responsible for the occurrence of instabilities, and suggested the existence of a critical value of shear stress, beyond which the occurrence of interfacial instabilities is likely. Maps of the stable zones can be drawn up. Certain pioneering work published from 1976 onward gave a detailed description of the factors responsible for the apparition of instabilities, such as the critical ratio of shear stresses at the interface, viscosity and elasticity of each layer, and the ratio of thicknesses. It has also been possible to demonstrate the effect of elasticity, and of viscosity on the interfacial stability, as a function of the geometry of the pipe and the ratio of the thicknesses of each layer. The elastic component of the polymers plays an important part in the mechanism of instability [AVÉ 04a].

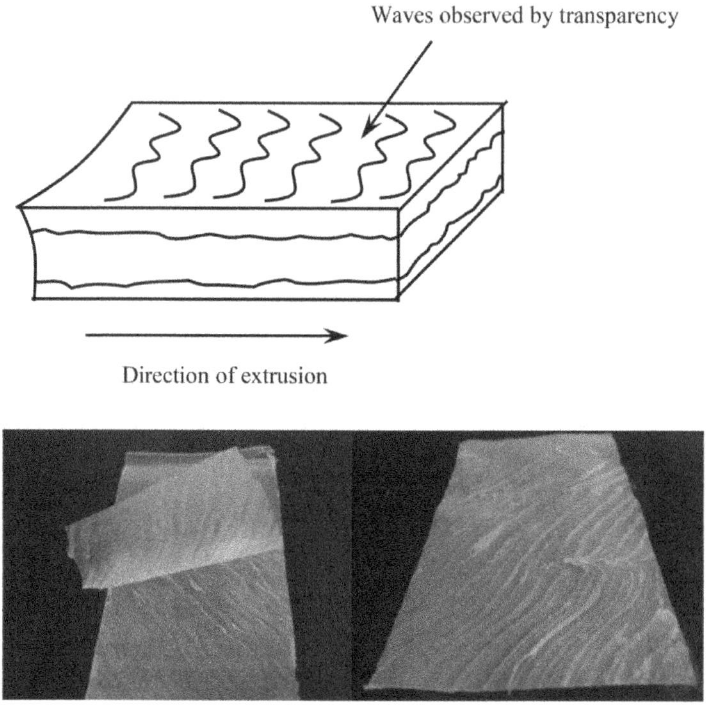

Figure 9.12. *Illustration of the phenomenon of interfacial instability with co-extrusion of plasticized starch/polyester systems [MAR 01b]*

In spite of the number and diversity of studies on multilayer flows and stability, only a few articles [MAR 01b, MAR 02, WAN 00] and patents [AVÉ 04a] have reported on the use of plasticized starch in co-extrusion. Various different stratified structures have been obtained by co-extrusion with different polyesters such as PCL, PBSA, PLA, PBAT or PHAs [AVÉ 04a, MAR 02, MAR 01b, WAN 00].

As a secondary point, there is a simple way to generate (pseudo)multilayer structures from polyester/plasticized starch-type blends. Indeed, the poor compatibility between the components causes very peculiar behavior. For example, during the injection process, we may witness preferential migration of the polyester with lower viscosity toward the surface of the mold [BEL 05, BEL 09]. After cooling, we obtain a material with a protective skin of polyester-based material and a core rich in starch. nuclear magnetic resonance (NMR) imaging has demonstrated this type of structure particularly with certain grades of PCL. This stratified structure exhibits a fairly good water resistance, owing to the protection given by the polyester surface [BEL 05, BEL 09].

In comparison to a conventional blend, the biodegradation of a multilayer is modified because of the stratification of its phases. The polyester-based skin first needs to be pierced for the micro-organisms to reach the core of the material, which will then biodegrade rapidly. The kinetics of biodegradation is therefore very specific, and depends, in this case, on the polyester which is used. Thus, it can also be modulated [GAT 02].

9.4.4. *Plasticized starch-based composites*

Plasticized starch-based composites have also been the subject of an abundant body of literature, particularly during the last decade, continuing on from the development of composites based on "natural" fibers.

9.4.4.1. *Biocomposites based on (ligno)cellulosic fibers*

Various different types of ligno-cellulosic fibers or microfibrils have been tested in combination with plasticized starch, such as potato microfibrils [DUF 00, DUF 98], bleached leafy fibers [AVÉ 04a, AVÉ 01b], or ligno-cellulosic fibers of jute and flax [WOL 98]. The authors have demonstrated excellent compatibility between the different types of polysaccharides introduced. For example, Avérous *et al.* noted a significant increase in *Tg* by adding 10% weight of cellulose fibers into a plasticized starch matrix [AVÉ 01b, AVÉ 04b]. This evolution is linked to the interactions between the fibers and the matrix, which decrease the mobility of the starch chains. As another example, Figure 9.13 shows an SEM image of a surface of cryogenic fracture; the fibers are fully embedded by the starchy matrix.

Figure 9.13. *SEM observation of the cryogenic fracture surface in a composite of plasticized starch reinforced with cellulose fibers (white line = 100 microns) [AVÉ 01b]*

The literature in the domain has indeed shown improvements in performance relating to the addition of this type of fibers. This evolution is connected to a strengthening of the matrix, because of excellent fiber–matrix interactions. This results in:

– a higher modulus [AVÉ 01b, AVÉ 04b];

– reduced water sensitivity because of fiber–matrix interactions which mobilize the OH functional groups, and because of the more hydrophobic nature of the cellulose, linked to its high degree of crystallinity [AVÉ 01b, AVÉ 04b, CUR 01];

– a higher heat resistance, due to the shift of the glass transition and to the increase in the level of the rubbery plateau;

– a reduction in aging after processing, because of the formation of a 3D structuring lattice between the different polysaccharides, formed of hydrogen bonds [AVÉ 09].

On this topic, a large number of patents have been published over the course of the past decade, with complex formulations, such as the introduction of proteins or of biodegradable polyesters.

9.4.4.2. *Other biocomposites*

Different types of fractionated or modified lignins (Kraft lignins, Acell© lignins, etc.) have been studied in association with plasticized starch. According to Baumberger *et al.*, lignins often act as fillers. However, the soluble fractions of lignin play the role of an interaction agent/plasticizing agent with the starch-based matrix [BAU 98].

Mineral microfillers have been tested in a plasticized starch matrix [CAR 01]. For example, micrometric particles of kaolin have been incorporated by extrusion. Due to a significant compatibility between the matrix and the filler, we note an increase in the glass transition temperature, a reduction in water absorption and an increase in the rigidity of the material. However, with the corresponding filler contents, these composites no longer satisfy the standards of biodegradation (at least 90% of the material has to be degraded).

There is a particular type of composites: nanobiocomposites. Because of the significant discussion surrounding this type of material, the next section is given over to this topic.

9.4.5. *The case of plasticized starch-based nanobiocomposites*

Nanobiocomposites are a specific class of nanocomposites [AVÉ 11, BOR 09a, CHI 09]. They are hybrid materials comprising fillers of nanometric size (nanofillers) incorporated into a matrix of biodegradable polymers. Depending on the nanofiller chosen for the purpose, nanocomposite materials exhibit a significant alteration in some of their properties: improved mechanical and barrier properties. We also note changes in heat and electrical conductivity [ALE 00]. These changes in properties are dependent on the geometry of the nanofiller, the interaction surface between the nanofiller and the matrix (e.g. 700 $m^2.g^{-1}$ for montmorillonite (MMT) when this lamellar clay is entirely exfoliated) and on the surface chemistry of the nanofiller [SIN 03].

According to the existing literature, two main families of nanoparticles have tended to be incorporated into plasticized starch to form nanobiocomposites:

1) nanofillers based on polysaccharides such as cellulose whiskers or starch nanocrystals;

2) nanoclays (MMT, sepiolite, etc.), either natural or organically modified [CHI 10 a, b, c].

Other nanofillers have been tested, such as carbon nanotubes, but in a less concerted manner.

9.4.5.1. *Nanobiocomposites with polysaccharide-based nanofillers*

9.4.5.1.1. Cellulose whiskers

Although chitin can, for example, be recovered by this method, whiskers are generally cellulose monocrystals. By virtue of a long pre-treatment, the whiskers can be isolated from the cellulosic biomass by acid hydrolysis with concentrated acids in conditions strictly controlled in terms of time and temperature [AZI 04, AZI 05]. Non-crystalline regions are consumed quickest by the acid hydrolysis of the cellulosic materials. Certain authors have used sea squirts (marine creatures) to produce whiskers [ANG 00, ANG 01, MAT 02, MAT 08]. They form thin stems, parallelepipedic in form, 500 nm to 1–2 µm in length, with a thickness of approximately 10 nm.

The interactions between the whiskers and the matrix are very significant. The high form ratio of the nanoparticles (50–200) and the high specific surface area (≈170 $m^2.g^{-1}$) enable us to obtain major phenomena at the interphase. Indeed, in relation to traditional biocomposites based on cellulose fibers or microfibrils, the overall behavior of whisker-based materials is primarily linked to the interface/interphase between the matrix and the nanofiller, which controls the properties and overall performances of the material (mechanical properties, permeability, etc.).

9.4.5.1.2. *Starch monocrystals [LEC 10]*

Nanocrystals of starch have been obtained by acid hydrolysis of native starch grain, strictly controlling the temperature of the process, the concentrations of acid and starch, the duration of hydrolysis and the rate of stirring. For example, nanocrystals of waxy maize starch are formed of platelets 5–7 nm in thickness, between 20 and 40 nm in length and 15–30 nm in width. Angellier *et al.* [ANG 06] demonstrated the strengthening and barrier effect of these nanocrystals in a plasticized starch matrix. The presence of starch nanocrystals slows down the degradation of the matrix as it ages in very humid environments. This is linked to the significant interactions between nanoparticles and matrix, which affect the phenomenon of crystallization [ANG 06].

9.4.5.2. *Nanoclay-based nanobiocomposites*

The most intensive research on nanobiocomposites relates to nanoclay-based materials [PAR 03, RAQ 07] – particularly using clays such as MMT because of its availability, polyvalence and low toxicity. It should be noted that other types of clay have successfully been tested, as happens in the case of sepiolite.

9.4.5.2.1. Montmorillonite-based nanobiocomposite structures

Depending on the conditions of the process or the matrix/clay affinity, different structures can be obtained [ALE 00 SIN 03, VAI 97]. Thus, we can distinguish:

– microcomposites: in this first type of structure, the polymer chains have not penetrated (or only to a very slight depth) into the interlamellar space between the clays. The size of the aggregates formed is micronic, and the corresponding behavior of the material is similar to that of traditional composites. In this case, the name "nanocomposite" cannot be correctly applied;

– intercalated nanocomposites: this structure exhibits alternating leaves of clay and intercalated polymer chains. The diffusion of the polymer and its intercalation into the space between the platelets leads to an increase in the interlamellar distance (d_{001});

– exfoliated nanocomposites: the platelets are individualized and totally dispersed throughout the matrix in a relatively homogeneous structure.

In order to reach a state of exfoliation, a variety of protocols and formulations have been tested. Nanofillers, organically modified by quaternary ammonium compounds, which tend to be hydrophobic, have been incorporated into plasticized starch (wheat potato [CHE 05 a, b, PAR 03, PAR 02], or maize [ZHA 07]) in concentrations ranging from 0 to 9% per weight. It has clearly been demonstrated that the incorporation of clays bearing quaternary ammonium compounds leads to the formation of microbiocomposites [CHE 05a, CHE 05b, CHI 05, PAR 02,

ZHA 07]. Higher values of d_{001} have been obtained with quaternary ammonium compounds carrying hydroxyl functional groups (e.g. Cloisite® 30B) and which therefore exhibit a more hydrophilic nature. In this case, a significant decrease in the intensity of the small-angle X-ray scattering [CHI 05, CHI 06, PAR 03], corresponding to a greater degree of dispersion of the filler within the matrix, has been noted.

Nanobiocomposites have been created with (natural) sodium montmorillonite (MMT-Na) because of the hydrophilic nature of starch and nanofiller [AVEL 05, CHE 05a, CHE 05b, CHI 05, CHI 06, CHI 07, CYR 08, HUA 04, PAN 05, PAR 02, PAR 03, ZHA 07]. These materials have been prepared using different starches, such as maize starch [HUA 04, PAN 05, ZHA 07], wheat starch [CHI 05, CHI 07] or potato starch [AVEL 05, CHE 05a, CHE 05b, CYR 08, PAR 03, PAR 02]. It has been shown that with a high glycerol content (greater than 10% per weight), such systems lead to the formation of an intercalated structure with a d_{001} of 18 Å – a value which is commonly reported in the literature and generally attributed to the intercalation of glycerol molecules [CHI 06, WIL 03].

Kampeerapappun *et al.* concentrated their attention on the use of chitosan as an ionic organo-modifier [KAM 07], but only a slight increase in the value of d_{001} was obtained. However, more recent results show that it is possible to achieve a state of exfoliation with such an approach. It has been successfully applied [CHI 08] with cationic starch (CS) as an organo-modifier of MMT (OMMT-CS). According to these authors, no diffraction peak is observed by small-angle X-ray scattering. This suggests an exfoliated morphology, which has been confirmed by MET analysis.

9.4.5.2.2. Properties of montmorillonite-based nanobiocomposites

The introduction of nanoparticles of clay into a plasticized starch-based matrix causes a significant improvement in the performances of the materials relating to the strengthening of the matrix, at an interface/interphase between nanofiller and matrix, and with a good state of dispersion [AVÉ 11].

Figure 9.14 shows that these nanocomposites display a substantial improvement in mechanical properties, relating to the clay content, for MMT-Na [CHI 08] or OMMT-CS [ZHA 07]. This evolution depends on the structure of nanobiocomposites. For example, Figure 9.14, illustrating the results of uniaxial traction testing, shows the differences between an exfoliated structure (with OMMT-CS) and a non-exfoliated structure (with MMT-Na). What is very clear is the evolution of the strain at break which increases slightly with increasing OMMT-CS content and decreases sharply in the case of a structure of microcomposites (classic behavior of composites) with MMT-Na. Similarly, the energy at break increases in the case of exfoliation and decreases in the case of a microcomposite structure.

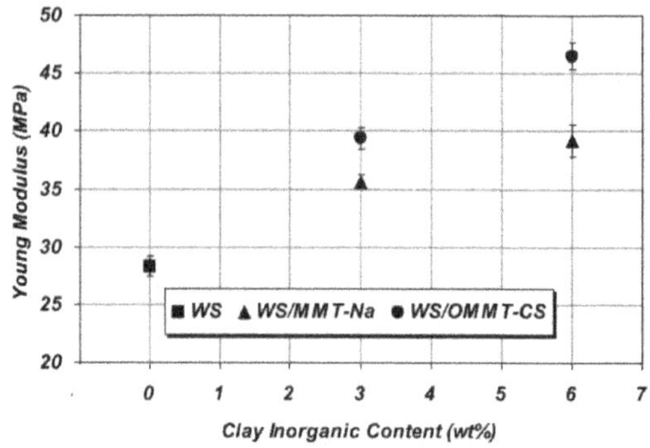

a)

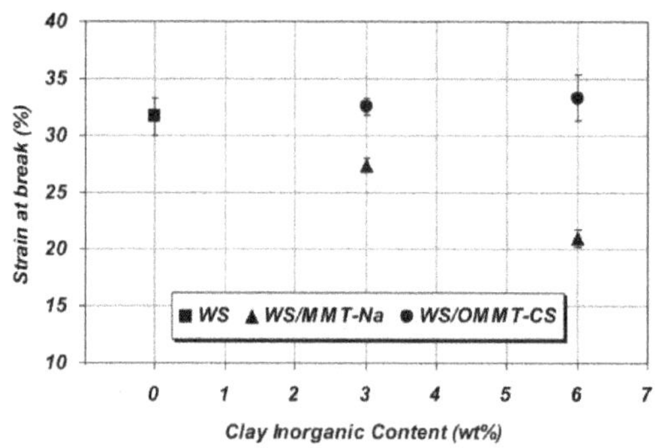

b)

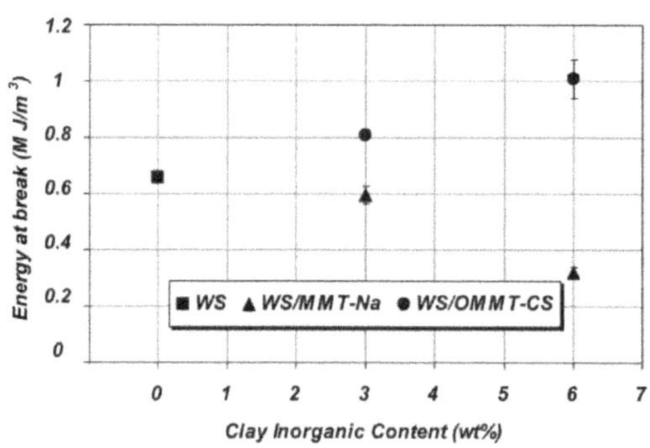

c)

Figure 9.14. *Evolutions, with differing nanoclay content, of a) the Young's modulus (MPa); b) the strain at break (%); and c) the energy at break ($MJ.m^{-3}$) c). Note: WS means plasticized [wheat] starch [CHI 08]*

We can show by DMTA and DSC that lamellar clays partially restrict the mobility of the starch chains [PAR 02]. Nanoclays are also known for having an impact on the permeability to water vapor of the corresponding nanocomposites, particularly when they are correctly oriented. Park *et al.* looked at the permeability of nanobiocomposites to water vapor with different types of clay [PAR 02]. According to those authors, all the films analyzed showed decreased permeability to water vapor in comparison to the matrix alone. Such results were not confirmed in the case of OMMT-CS, which has a very highly plasticized matrix [CHI 10a]. In this case, the high affinity of the clay platelets with starch-rich phases creates preferential diffusion pathways through the glycerol-rich phases, in which the platelets tend to be absent.

9.5. Acknowledgments

I would like to extend my heartfelt thanks to all the researchers (colleagues both from France and elsewhere, collaborators, postdocs, doctoral and MA students and others) who have accompanied me, since 1996, on my journey through this inspiring and dynamic field of research.

My particular gratitude goes to Dr. Patrice Dole and Dr. Eric Pollet, for their tireless and patient reading of this chapter and their valuable suggestions.

I also wish to thank all those who have funded the research I have carried out in the domain of plasticized starch, bio-based and/or biodegradable polymers – the various national and international enterprises, the regional agricultural council in Champagne-Ardenne, the local authorities (Champagne-Ardenne and Alsace), the universities (University of Reims Champagne-Ardenne and University of Strasbourg), INRA, the CNRS, ADEME or the former Agrice, Europe, Cofecub/Egide/FAPESP (collaborations with Brazil), ARC and the University of Queensland (collaborations with Australia).

9.6. Bibliography

[ALB 02a] ALBERTSSON A.C., VARMA I.K., "Aliphatic polyesters: synthesis, properties and applications", *Advances in Polymer Science*, vol. 157, pp. 1–40, 2002.

[ALB 11] ALBUQUERQUE M.G.E., MARTINO V., POLLET E., *et al.*, "Mixed culture polyhydroxyalkanoate (PHA) production from volatile fatty acid (VFA)-rich streams: effect of substrate composition and feeding regime on PHA productivity, composition and properties", *Journal of Biotechnology*, vol. 151, no. 1, pp. 66–76, 2011.

[ALE 00] ALEXANDRE M., DUBOIS P., "Polymer-layered silicate nanocomposites: preparation, properties and uses of a new class of materials", *Materials Science & Engineering R: Reports*, vol. 28, no. 1, pp. 1–63, 2000.

[ANG 00] ANGLES M.N., DUFRESNE A., "Plasticized starch/tunicin whiskers nanocomposites, 1. Structural analysis", *Macromolecules*, vol. 33, no. 22, pp. 8344–8353, 2000.

[ANG 01] ANGLES M.N., DUFRESNE A., "Plasticized starch/tunicin whiskers nanocomposite materials. 2. Mechanical behavior", *Macromolecules*, vol. 34, no. 9, pp. 2921–2931, 2001.

[ANG 06] ANGELLIER H., MOLINA-BOISSEAU S., DOLE P., *et al.*, "Thermoplastic starch-waxy maize starch nanocrystals nanocomposites", *Biomacromolecules*, vol. 7, no. 2, pp. 531–539, 2006.

[ARV 99] ARVANITOYANNIS I.S., "Totally and partially biodegradable polymer blends based on natural and synthetic macromolecules: preparation, physical properties, and potential as food packaging materials", *Journal of Macromolecular Science – Reviews in Macromolecular Chemistry & Physics*, vol. C39, no. 2, pp. 205–271, 1999.

[AVÉ 00a] AVÉROUS L., FAUCONNIER N., MORO L., *et al.*, "Blends of thermoplastic starch and polyesteramide: processing and properties", *Journal of Applied Polymer Science*, vol. 76, no. 7, pp. 1117–1128, 2000.

[AVÉ 00b] AVÉROUS L., MORO L., DOLE P., *et al.*, "Properties of thermoplastic blends: Starch-polycaprolactone", *Polymer*, vol. 41, no. 11, pp. 4157–4167, 2000.

[AVÉ 01a] AVÉROUS L., FRINGANT C., "Association between plasticized starch and polyesters: processing and performances of injected biodegradable systems", *Polymer Engineering and Science*, vol. 41, no. 5, pp. 727–734, 2001.

[AVÉ 01b] AVÉROUS L., FRINGANT C., MORO L., "Plasticized starch-cellulose interactions in polysaccharide composites", *Polymer*, vol. 42, no. 15, pp. 6565–6572, 2001.

[AVÉ 04a] AVÉROUS L., "Biodegradable multiphase systems based on plasticized starch: a review", *Journal of Macromolecular Science: Polymer Reviews*, vol. 44, no. 3, pp. 231–274, 2004.

[AVÉ 04b] AVÉROUS L., BOQUILLON N., "Biocomposites based on plasticized starch: thermal and mechanical behaviours", *Carbohydrate Polymers*, vol. 56, no. 2, pp. 111–122, 2004.

[AVÉ 09] AVÉROUS L., HALLEY P.J., "Biocomposites based on plasticized starch", *Biofuels, Bioproducts and Biorefining*, vol. 3, no. 3, pp. 329–343, 2009.

[AVÉ 11] AVÉROUS L., POLLET E., "Biorenewable nanocomposites", *MRS Bulletin*, vol. 36, no. 9, pp. 703–710, 2011.

[AVÉ 12] AVÉROUS L., POLLET E., "Biodegradable polymers", in AVÉROUS L., POLLET E. (eds.), *Environmental Silicate Nano-Biocomposites*, Springer, London, pp. 13–39, 2012.

[AVÉ 13] AVÉROUS L., BORREDON M.E., *et al.*, "9, Synthesis, properties, environmental and biomedical applications of polylactic acid", *Handbook of Biopolymers and Biodegradable Plastics*, William Andrew Publishing, Boston, MA, pp. 171–188, 2013.

[AVE 00] AVELLA M., ERRICO M.E., LAURIENZO P., *et al.*, "Preparation and characterisation of compatibilised polycaprolactone/starch composites", *Polymer*, 41, no. 10, pp. 3875–3881, 2000.

[AVE 02] AVELLA M., ERRICO M.E., RIMEDIO R., *et al.*, "Preparation of biodegradable polyester/high-amylose-starch composites by reactive blending and their characterization", *Journal of Applied Polymer Science*, vol. 83, no. 7, pp. 1432–1442, 2002.

[AVE 05] AVELLA M., DE VLIEGER J.J., ERRICO M.E., *et al.*, "Biodegradable starch/clay nanocomposite films for food packaging applications", *Food Chemistry*, vol. 93, no. 3, pp. 467–474, 2005.

[AZI 04] AZIZI SAMIR M.A.S., ALLOIN F., SANCHEZ J.Y., *et al.*, "Preparation of cellulose whiskers reinforced nanocomposites from an organic medium suspension", *Macromolecules*, vol. 37, no. 4, pp. 1386–1393, 2004.

[AZI 05] AZIZI SAMIR M.A.S., ALLOIN F., DUFRESNE A., "Review of recent research into cellulosic whiskers, their properties and their application in nanocomposite field", *Biomacromolecules*, vol. 6, no. 2, pp. 612–626, 2005.

[BAS 98] BASTIOLI C., "Biodegradable materials: present situation and future perspectives", *Macromolecular Symposia*, vol. 135, pp. 193–204, 1998.

[BAU 98] BAUMBERGER S., LAPIERRE C., MONTIES B., "Utilization of pine kraft lignin in starch composites: impact of structural heterogeneity", *Journal of Agricultural and Food Chemistry*, vol. 46, no. 6, pp. 2234–2240, 1998.

[BEL 05] BÉLARD L., DOLE P., AVÉROUS L., "Current progress on biodegradable materials based on plasticized starch", *Australian Journal of Chemistry*, vol. 58, no. 6, pp. 457–460, 2005.

[BEL 09] BÉLARD L., DOLE P., AVÉROUS L., "Study of pseudo-multilayer structures based on starch-polycaprolactone extruded blends", *Polymer Engineering & Science*, vol. 49, no. 6, pp. 1177–1186, 2009.

[BOR 09a] BORDES P., HABLOT E., POLLET E., *et al.*, "Effect of clay organomodifiers on degradation of polyhydroxyalkanoates", *Polymer Degradation and Stability*, vol. 94, no. 5, pp. 789–796, 2009.

[BOR 09b] BORDES P., POLLET E., AVÉROUS L., "Nanobiocomposites: biodegradable polyester/nanoclay systems", *Progress in Polymer Science*, vol. 34, no. 2, pp. 125–155, 2009.

[CAR 01] DE CARVALHO A.J.F., CURVELO A.A.S., AGNELLI J.A.M., "A first insight on composites of thermoplastic starch and kaolin", *Carbohydrate Polymers*, vol. 45, no. 2, pp. 189–194, 2001.

[CHA 98] CHANDRA R., RUSTGI R., "Biodegradable polymers", *Progress in Polymer Science*, vol. 23, no. 7, pp. 1273–1335, 1998.

[CHE 05a] CHEN B., EVANS J.R.G., "Thermoplastic starch-clay nanocomposites and their characteristics", *Carbohydrate Polymers*, vol. 61, no. 4, pp. 455–463, 2005.

[CHE 05b] CHEN M., CHEN B., EVANS J.R.G., "Novel thermoplastic starch-clay nanocomposite foams", *Nanotechnology*, vol. 16, no. 10, pp. 2334–2337, 2005.

[CHI 05] CHIOU B.S., YEE E., GLENN G.M., *et al.*, "Rheology of starch-clay nanocomposites", *Carbohydrate Polymers*, vol. 59, no. 4, pp. 467–475, 2005.

[CHI 06] CHIOU B.S., YEE E., WOOD D., *et al.*, "Effects of processing conditions on nanoclay dispersion in starch-clay nanocomposites", *Cereal Chemistry*, vol. 83, no. 3, pp. 300–305, 2006.

[CHI 07] CHIOU B.S., WOOD D., YEE E., *et al.*, "Extruded starch-nanoclay nanocomposites: effects of glycerol and nanoclay concentration", *Polymer Engineering & Science*, vol. 47, no. 11, pp. 1898–1904, 2007.

[CHI 08] CHIVRAC F., POLLET E., AVÉROUS L., "New approach to elaborate exfoliated starch-based nanobiocomposites", *Biomacromolecules*, vol. 9, no. 3, pp. 896–900, 2008.

[CHI 09] CHIVRAC F., POLLET E., AVÉROUS L., "Progress in nano-biocomposites based on polysaccharides and nanoclays", *Materials Science and Engineering R: Reports*, vol. 67, no. 1, pp. 1–17, 2009.

[CHI 10a] CHIVRAC F., ANGELLIER-COUSSY H., GUILLARD V., *et al.*, "How does water diffuse in starch/montmorillonite nano-biocomposite materials?", *Carbohydrate Polymers*, 82, no. 1, pp. 128–135, 2010.

[CHI 10 b] CHIVRAC F., POLLET E., DOLE P., *et al.*, "Starch-based nano-biocomposites: plasticizer impact on the montmorillonite exfoliation process", *Carbohydrate Polymers*, vol. 79, no. 4, pp. 941–947, 2010.

[CHI 10c] CHIVRAC F., POLLET E., SCHMUTZ M., *et al.*, "Starch nano-biocomposites based on needle-like sepiolite clays", *Carbohydrate Polymers*, vol. 80, no. 1, pp. 145–153, 2010.

[COU 06] COULEMBIER O., DEGEE P., HEDRICK J.L., *et al.*, "From controlled ring-opening polymerization to biodegradable aliphatic polyester: especially poly(beta-malic acid) derivatives", *Progress in Polymer Science*, vol. 31, no. 8, pp. 723–747, 2006.

[COU 11] COURGNEAU C., DOMENEK S., GUINAULT A., *et al.*, "Analysis of the structure-properties relationships of different multiphase systems based on plasticized poly(lactic acid)", *Journal of Polymers and the Environment*, vol. 19, no. 2, pp. 362–371, 2011.

[COU 12] COURGNEAU C., DOMENEK S., LEBOSSÉ R., *et al.*, "Ducruet V., effect of crystallization on barrier properties of formulated polylactide", *Polymer International*, vol. 61, no. 2, pp. 180–189, 2012.

[CUR 01] CURVELO A.A.S., DE CARVALHO A.J.F., AGNELLI J.A.M., "Thermoplastic starch-cellulosic fibers composites: preliminary results", *Carbohydrate Polymers*, vol. 45, no. 2, pp. 183–188, 2001.

[CYR 08] CYRAS V.P., MANFREDI L.B., TON-THAT M.T., *et al.*, "Physical and mechanical properties of thermoplastic starch/montmorillonite nanocomposite films", *Carbohydrate Polymers*, vol. 73, no. 1, pp. 55–63, 2008.

[DOM 11] DOMENEK S., COURGNEAU C., DUCRUET V., "Characteristics and applications of poly(lactide)", *Biopolymers*, 183–223, John Wiley & Sons, New York, 2011.

[DUB 99] DUBOIS P., KRISHNAN M., NARAYAN R., "Aliphatic polyester-grafted starch-like polysaccharides by ring-opening polymerization", *Polymer*, vol. 40, no. 11, pp. 3091–3100, 1999.

[DUF 98] DUFRESNE A., VIGNON M.R., "Improvement of starch film performances using cellulose microfibrils", *Macromolecules*, vol. 31, no. 8, pp. 2693–2696, 1998.

[DUF 00] DUFRESNE A., DUPEYRE D., VIGNON M.R., "Cellulose microfibrils from potato tuber cells: processing and characterization of starch-cellulose microfibril composites", *Journal of Applied Polymer Science*, vol. 76, no. 14, pp. 2080–2092, 2000.

[DUQ 01] DUQUESNE E., RUTOT D., DEGEE P., *et al.*, "Synthesis and characterization of compatibilized poly(epsilon-caprolactone)/granular starch composites", *Macromolecular Symposia*, vol. 175, pp. 33–43, 2001.

[ESC 11] ESCAPA I.F., MORALES V., MARTINO V.P., *et al.*, "Disruption of β-oxidation pathway in Pseudomonas putida KT2442 to produce new functionalized PHAs with thioester groups", *Applied Microbiology and Biotechnology*, vol. 89, no. 5, pp. 1583–1598, 2011.

[FIS 00] FISHMAN M.L., COFFIN D.R., KONSTANCE R.P., *et al.*, "Extrusion of pectin/starch blends plasticized with glycerol", *Carbohydrate Polymers*, vol. 41, no. 4, pp. 317–325, 2000.

[FRI 96] FRINGANT C., DESBRIÈRES J., RINAUDO M., "Physical properties of acetylated starch-based materials: relation with their molecular characteristics", *Polymer*, vol. 37, no. 13, pp. 2663–2673, 1996.

[FUJ 98] FUJIMAKI T., "Processability and properties of aliphatic polyesters, 'BIONOLLE', synthesized by polycondensation reaction", *Polymer Degradation and Stability*, vol. 59, nos. 1–3, pp. 209–214, 1998.

[GAT 02] GATTIN R., COPINET A., BERTRAND C., *et al.*, "Biodegradation study of a starch and poly(lactic acid) co-extruded material in liquid, composting and inert mineral media", *International Biodeterioration and Biodegradation*, vol. 50, no. 1, pp. 25–31, 2002.

[GOD 06] GODBILLOT L., DOLE P., JOLY C., *et al.*, "Analysis of water binding in starch plasticized films", *Food Chemistry*, vol. 96, no. 3, pp. 380–386, 2006.

[HAB 08] HABLOT E., BORDES P., POLLET E., *et al.*, "Thermal and thermo-mechanical degradation of poly(3-hydroxybutyrate)-based multiphase systems", *Polymer Degradation and Stability*, vol. 93, no. 2, pp. 413–421, 2008.

[HAL 10] HALARY J.L., *et al.*, "Polymer materials and sustainable development/Matériaux polymères et développement durable", *Actualité Chimique*, vol. 338–339, pp. 41–53, 2010.

[HUA 04] HUANG M.F., YU J.G., MA X.F., "Studies on the properties of Montmorillonite-reinforced thermoplastic starch composites", *Polymer*, vol. 45, no. 20, pp. 7017–7023, 2004.

[HUA 06] HUANG M.F., YU J.G., MA X.F., "High mechanical performance MMT-urea and formamide-plasticized thermoplastic cornstarch biodegradable nanocomposites", *Carbohydrate Polymers*, vol. 63, no. 3, pp. 393–399, 2006.

[JAC 99] JACOBSEN S., DEGEE P.H., FRITZ H.G., *et al.*, "Polylactide (PLA): a new way of production", *Polymer Engineering & Science*, vol. 39, no. 7, pp. 1311–1319, 1999.

[KAM 07] KAMPEERAPAPPUN P., AHT-ONG D., PENTRAKOON D., *et al.*, "Preparation of cassava starch/montmorillonite composite film", *Carbohydrate Polymers.*, vol. 67, no. 2, pp. 155–163, 2007.

[KIM 01] KIM C.H., CHO K.Y., PARK J.K., "Effect of poly(acrylic acid)-g-PCL microstructure on the mechanical properties of starch/PCL blend compatibilized with poly(acrylic acid)-g-PCL", *Polymer Engineering & Science*, vol. 41, no. 3, pp. 542–553, 2001.

[KOE 94] KOENIG M.F., HUANG S.J., "Evaluation of crosslinked poly(caprolactone) as a biodegradable, hydrophobic coating", *Polymer Degradation and Stability*, vol. 45, no. 1, pp. 139–144, 1994.

[LAB 97] LABRECQUE L.V., KUMAR R.A., DAVE V., *et al.*, "Citrate esters as plasticizers for poly(lactic acid)", *Journal of Applied Polymer Science*, vol. 66, no. 8, pp. 1507–1513, 1997.

[LEC 10] LE CORRE D., BRAS J., DUFRESNE A., "Starch nanoparticles: a review", *Biomacromolecules*, vol. 11, no. 5, pp. 1139–1153, 2010.

[LOU 97a] LOURDIN D., BIZOT H., COLONNA P., "Antiplasticization in starch-glycerol films", *Journal of Applied Polymer Science.*, vol. 63, no. 8, pp. 1047–1053, 1997.

[LOU 97b] LOURDIN D., BIZOT H., COLONNA P., "Correlation between static mechanical properties of starch-glycerol materials and low-temperature relaxation", *Macromolecular Symposia*, vol. 114, pp. 179–185, 1997.

[MAN 99] MANI R., BHATTACHARYA M., TANG J., "Functionalization of polyesters with maleic anhydride by reactive extrusion", *Journal of Polymer Science Part A: Polymer Chemistry*, vol. 37, no. 11, pp. 1693–1702, 1999.

[MAN 01] MANI R., BHATTACHARYA M., "Properties of injection moulded blends of starch and modified biodegradable polyesters", *European Polymer Journal*, vol. 37, no. 3, pp. 515–526, 2001.

[MAR 01a] MARTIN O., AVÉROUS L., "Poly(lactic acid): plasticization and properties of biodegradable multiphase systems", *Polymer*, vol. 42, no. 14, pp. 6209–6219, 2001.

[MAR 01b] MARTIN O., SCHWACH E., AVÉROUS L., *et al.*, "Properties of biodegradable multilayer films based on plasticized wheat starch", *Starch/Staerke*, vol. 53, no. 8, pp. 372–380, 2001.

[MAR 02] MARTIN O., AVÉROUS L., "Comprehensive experimental study of a starch/polyesteramide coextrusion", *Journal of Applied Polymer Science.*, vol. 86, no. 10, pp. 2586–2600, 2002.

[MAR 03] MARTIN O., AVÉROUS L., DELLA VALLE G., "In-line determination of plasticized wheat starch viscoelastic behavior: impact of processing", *Carbohydrate Polymers.*, vol. 53, no. 2, pp. 169–182, 2003.

[MAR 10] MARTINO V.P., RUSECKAITE R.A., JIMÉNEZ A., *et al.*, "Correlation between composition, structure and properties of poly(lactic acid)/polyadipate-based nano-biocomposites", *Macromolecular Materials and Engineering*, vol. 295, no. 6, pp. 551–558, 2010.

[MAT 02] MATHEW A.P., DUFRESNE A., "Morphological investigation of nanocomposites from sorbitol plasticized starch and tunicin whiskers", *Biomacromolecules*, vol. 3, no. 3, pp. 609–617, 2002.

[MAT 08] MATHEW A.P., THIELEMANS W., DUFRESNE A., "Mechanical properties of nano-composites from sorbitol plasticized starch and tunicin whiskers", *Journal of Applied Polymer Science*, vol. 109, no. 6, pp. 4065–4074, 2008.

[MUL 97] MULLER R.J., WITT U., DECKWER W.D., "Biological degradable polyester copolymers from petrochemical and renewable raw material", *Fett-Lipid*, vol. 99, no. 2, pp. 40–45, 1997.

[OKA 02] OKADA M., "Chemical syntheses of biodegradable polymers", *Progress in Polymer Science*, vol. 27, no. 1, pp. 87–133, 2002.

[OTA 99] OTAIGBE J.U., GOEL H., BABCOCK T., *et al.*, "Processability and properties of biodegradable plastics made from agricultural biopolymers", *Journal of Elastomers and Plastics*, vol. 31, no. 1, pp. 56–71, 1999.

[PAN 05] PANDEY J.K., SINGH R.P., "Green nanocomposites from renewable resources: effect of plasticizer on the structure and material properties of clay-filled starch", *Starch-Starke*, vol. 57, no. 1, pp. 8–15, 2005.

[PAR 02] PARK H.M., LI X., JIN C.Z., *et al.*, "Preparation and properties of biodegradable thermoplastic starch/clay hybrids", *Macromolecular Materials and Engineering*, 287, no. 8, pp. 553–558, 2002.

[PAR 03] PARK H.M., LEE W.K., PARK C.Y., *et al.*, "Environmentally friendly polymer hybrids Part I mechanical, thermal, and barrier properties of thermoplastic starch/clay nanocomposites", *Journal of Materials Science*, vol. 38, no. 5, pp. 909–915, 2003.

[POL 11] POLLET E., AVÉROUS L., "Production, chemistry and properties of polyhydroxy-alkanoates", in PLACKETT D. (ed.), *Biopolymers: New Materials for Sustainable Films and Coatings*, John Wiley & Sons, New York, pp. 65–86, 2011.

[RAQ 07] RAQUEZ J.M., NABAR Y., NARAYAN R., *et al.*, "New developments in biodegradable starch-based nanocomposites", *International Polymer Processing*, vol. 22, pp. 463–470, 2007.

[RUT 01] RUTOT D., DUQUESNE E., YDENS I., *et al.*, "Aliphatic polyester-based biodegradable materials: new amphiphilic graft copolymers", *Polymer Degradation and Stability*, vol. 73, no. 3, pp. 561–566, 2001.

[SCH 04] SCHWACH E., AVÉROUS L., "Starch-based biodegradable blends: morphology and interface properties", *Pol. Intern.*, vol. 53, no. 12, pp. 2115–2124, 2004.

[SCH 08] SCHWACH E., SIX J.L., AVÉROUS L., "Biodegradable blends based on starch and poly(lactic acid): comparison of different strategies and estimate of compatibilization", *Journal of Polymers and the Environment*, vol. 16, no. 4, pp. 286–297, 2008.

[SIN 03] SINHA RAY S., OKAMOTO M., "Polymer/layered silicate nanocomposites: a review from preparation to processing", *Progress in Polymer Science*, vol. 28, no. 11, pp. 1539–1641, 2003.

[SOD 02] SÖDERGÅRD A., STOLT M., "Properties of lactic acid based polymers and their correlation with composition", *Progress in Polymer Science (Oxford)*, vol. 27, no. 6, pp. 1123–1163, 2002.

[SOE 97a] VAN SOEST J.J.G., VLIEGENTHART J.F.G., "Crystallinity in starch plastics: consequences for material properties", *Trends in Biotechnology*, vol. 15, no. 6, pp. 208–213, 1997.

[SOE 97b] VAN SOEST J.J.G., KNOOREN N., "Influence of glycerol and water content on the structure and properties of extruded starch plastic sheets during aging", *Journal of Applied Polymer Science*, vol. 64, no. 7, pp. 1411–1422, 1997.

[STE 03] STEINBUCHEL A., *Biopolymers. General Aspects and Special Applications*, Wiley-VCH, Weinheim, 2003.

[STP 97] ST PIERRE N., FAVIS B.D., RAMSAY B.A., *et al.*, "Processing and characterization of thermoplastic starch/polyethylene blends", *Polymer*, vol. 38, no. 3, pp. 647–655, 1997.

[SUD 00] SUDESH K., ABE H., DOI Y., "Synthesis, structure and properties of polyhydroxyalkanoates: biological polyesters", *Progress in Polymer Science*, vol. 25, no. 10, pp. 1503–1555, 2000.

[TOK 77] TOKIWA Y., SUZUKI T., "Hydrolysis of polyesters by lipases", *Nature*, vol. 270, no. 5632, pp. 76–78, 1977.

[VAI 97] VAIA R.A., GIANNELIS E.P., "Polymer melt intercalation in organically-modified layered silicates: model predictions and experiment", *Macromolecules*, vol. 30, no. 25, pp. 8000–8009, 1997.

[VAR 05] VARMA I.K., ALBERTSSON A.C., RAJKHOWA R., *et al.*, "Enzyme catalyzed synthesis of polyesters", *Progress in Polymer Science*, vol. 30, no. 10, pp. 949–981, 2005.

[VER 87] VERGNES B., VILLEMAIRE J.P., "Rheological behaviour of low moisture molten maize starch", *Rheologica Acta*, vol. 26, no. 6, pp. 570–576, 1987.

[VER 12] VERT M., DOI Y., HELLWICH K.H., *et al.*, "Terminology for biorelated polymers and applications, IUPAC Recommendations 2012", *Pure and Applied Chemistry*, vol. 84, no. 2, pp. 377–408, 2012.

[WAN 00] WANG L., SHOGREN R.L., CARRIERE C., "Preparation and properties of thermoplastic starch-polyester laminate sheets by coextrusion", *Polymer Engineering & Science*, vol. 40, no. 2, pp. 499–506, 2000.

[WAN 01] WANG H., SUN X.Z., SEIB P., "Strengthening blends of poly(lactic acid) and starch with methylenediphenyl diisocyanate", *Journal of Applied Polymer Science*, vol. 82, no. 7, pp. 1761–1767, 2001.

[WAN 02] WANG H., SUN X.Z., SEIB P., "Mechanical properties of poly(lactic acid) and wheat starch blends with methylenediphenyl diisocyanate", *Journal of Applied Polymer Science*, vol. 84, no. 6, pp. 1257–1262, 2002.

[WIL 98] WILLETT J.L., KOTNIS M.A., O'BRIEN G.S., *et al.*, "Properties of starch-graft-poly(glycidyl methacrylate) PHBV composites", *Journal of Applied Polymer Science*, vol. 70, no. 6, pp. 1121–1127, 1998.

[WIL 03] WILHELM H.M., SIERAKOWSKI M.R., SOUZA G.P., "Wypych F., starch films reinforced with mineral clay", *Carbohydrate Polymers.*, vol. 52, no. 2, pp. 101–110, 2003.

[WOL 98] WOLLERDORFER M., BADER H., "Influence of natural fibres on the mechanical properties of biodegradable polymers", *Industrial Crops and Products*, vol. 8, no. 2, pp. 105–112, 1998.

[XIE 12] XIE F., HALLEY P.J., AVÉROUS L., "Rheology to understand and optimize processibility, structures and properties of starch polymeric materials", *Progress in Polymer Prog. Pol. Sci (Oxford)*, vol. 37, no. 4, pp. 595–623, 2012.

[YDE 00] YDENS I., RUTOT D., DEGEE P., *et al.*, "Controlled synthesis of poly(epsilon-caprolactone)-grafted dextran copolymers as potential environmentally friendly surfactants", *Macromolecules*, vol. 33, no. 18, pp. 6713–6721, 2000.

[ZHA 07] ZHANG Q.X., YU Z.Z., XIE X.L., *et al.*, "Preparation and crystalline morphology of biodegradable starch/clay nanocomposites", *Polymer*, vol. 48, no. 24, pp. 7193–7200, 2007.

[ZHA 08] ZHANG L., ZENG M., "Chapter 23: proteins as sources of materials", in NACEUR B.M., ALESSANDRO G., *Monomers, Polymers and Composites from Renewable Resources*, Elsevier, Amsterdam, pp. 479–493, 2008.

10

Different Strategies for Ecoplastics Development

10.1. Introduction

The phenomena of fashion aside, the aim of this chapter is to touch on the increasingly marked impact of plastics on our environment, but also and above all to put forward solutions to minimize the footprint of such materials. Many publications on the subject are available, to which readers can refer if they wish for further detail on these concepts [DUV 04, ROB 03]. After a few general points about the lifecycle of plastics, two concrete examples of ecoplastics are presented: one based on bio-based materials, and the other solely on recycled synthetic materials.

10.2. General points about the lifecycle of plastics

Without fully going into detail about the lifecycle of plastics, which would require a more in-depth analysis and the use of eco-design software suites such as SimaPro [PUY 14], it is possible to make a few general remarks to guide users of plastics toward the solutions which are least damaging for the environment. Looking at Figure 10.1 from the top left-hand corner to the bottom left-hand corner in a clockwise direction, we can see that it is possible to act at different levels in the life of plastics to reduce the trace they leave:

a) by choosing raw materials which do not impact on non-renewable resources, or whose use is as neutral as possible;

b) by minimizing the energy consumed during the extraction, purification, synthesis, transformation, use, transport, etc.;

Chapter written by Jean-François FELLER.

c) by minimizing emissions at every step in the life of the plastic.

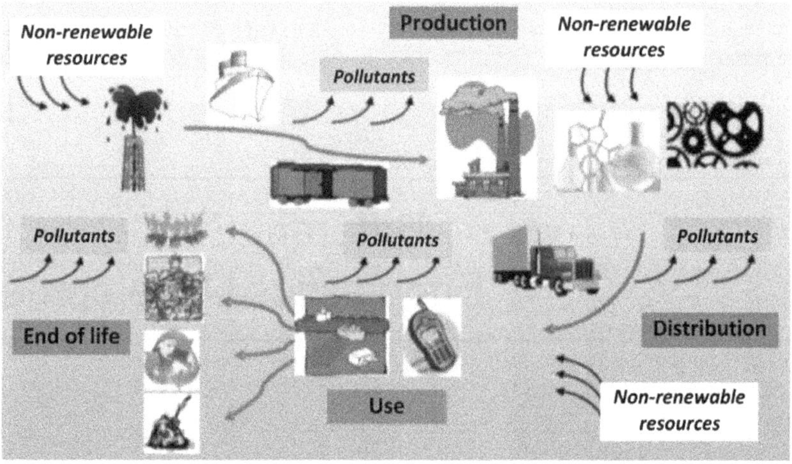

Figure 10.1. *Lifecycle of plastics*

10.3. Energy

A good way of reducing the energy impact of plastics during their lifecycle (at least up until the stage of their use) is to employ renewable energy sources such as biomass, wind power or photovoltaic (PV – solar) energy (Figure 10.2), which considerably improve their eco-balance.

Figure 10.2. *Renewable energies: wind power (left); solar/PV (right)*

This has already been understood by certain polymer producers such as Cargill Dow [VIN 03], who advocate the use of this type of energy in the production of PLA, poly(lactic acid). Indeed, if we compare the last two columns in the graph in Figure 10.3, relating to this bio-based polymer, we can see that the use of an energy scenario coupling biomass and wind power (Bio/WP) very significantly reduces fossil fuel consumption. In addition, the fact that PLA is derived from maize residues explains the absence of any fossil material consumption (hashed part in the graph). Figure 10.3 also shows that the poly(amide) family is almost twice as big an energy consumer as the poly(olefin) family.

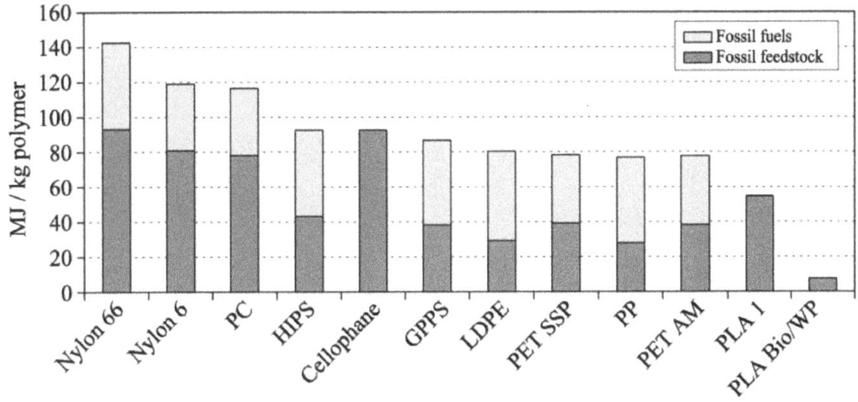

Figure 10.3. *Comparison of energy and fossil materials consumed in the synthesis of some commonly-used polymers [VIN 03]*

Furthermore, we can see in Figure 10.4 that while the production of bio-based polymers such as poly(hydroxybutyrate) PHBB, starch-poly(caprolactone) SPCL or poly(butylene succinate) PBSC often generates less carbonic gas, the energy consumption is comparable to that of oil-based polymers such as poly(styrene) PS, poly(ethylene terephthalate) PET or poly(ethylene) PE, including for a bacterial synthesis such as that used for poly(hydroxybutyrate) PHB (although it takes place at near-ambient temperature and pressure). The only energy gain we can expect at this level can be obtained by using waste as a raw material, because they do not require synthesis.

Finally, one more lever in reducing fossil energy consumption consists of limiting the various stages of transport of the material. As we might imagine, for example, an optimization of the logistical chain to favor local processing of plastic waste and local reuse of the material once regenerated would produce a notable reduction in fossil fuel consumption.

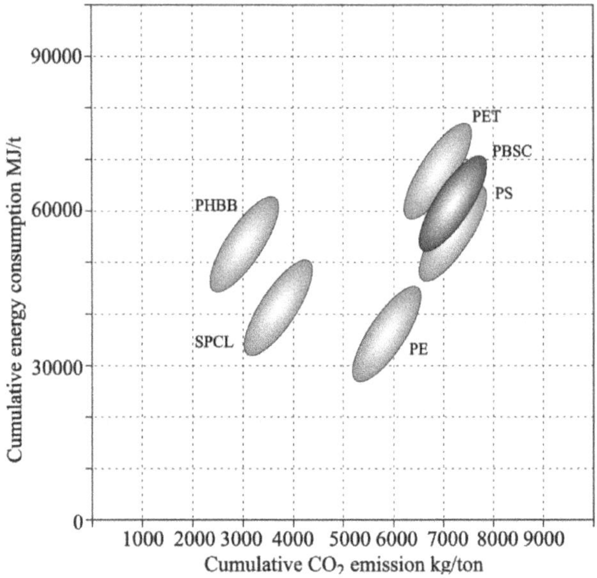

Figure 10.4. *Comparison of energy consumption and CO_2 emissions of the oil-based polymers (poly(ethyleneterephthalate) PET, poly(styrene) PS and poly(ethylene) PE), and the bio-based polymers (poly(hydroxybutyrate) PHBB, poly(butylene succinate) PBSC and starch-poly(caprolactone) SPCL) [PAT 05]*

10.4. Material

10.4.1. *Minimizing waste*

The ecological footprint of plastics can be reduced by decreasing the amount of plastic waste. It is possible to take action at several different levels, as Figure 10.5 shows. This pyramid representation shows that the earlier we intervene in the thought process about waste, the more effective that intervention can be – herein lies the crux of the advantage to the eco-design technique outlined in this book (see Chapter 7). It is desirable to take an end-of-life scenario into consideration from the very moment when we choose the material to make a plastic article.

Relatively little tends to be said about the reuse of plastic objects, but this is an interesting alternative in terms of prolonging their lifetimes. Imagine the energy gain made if we wash and reuse a plastic item 50 times rather than recycle it 50 times. This point has been demonstrated by the company *Solution Plastiques Industrie* [MAU 14], which suggested that producers should substitute cardboard egg boxes, which were for single use (to prevent bacterial proliferation), with washable plastic boxes.

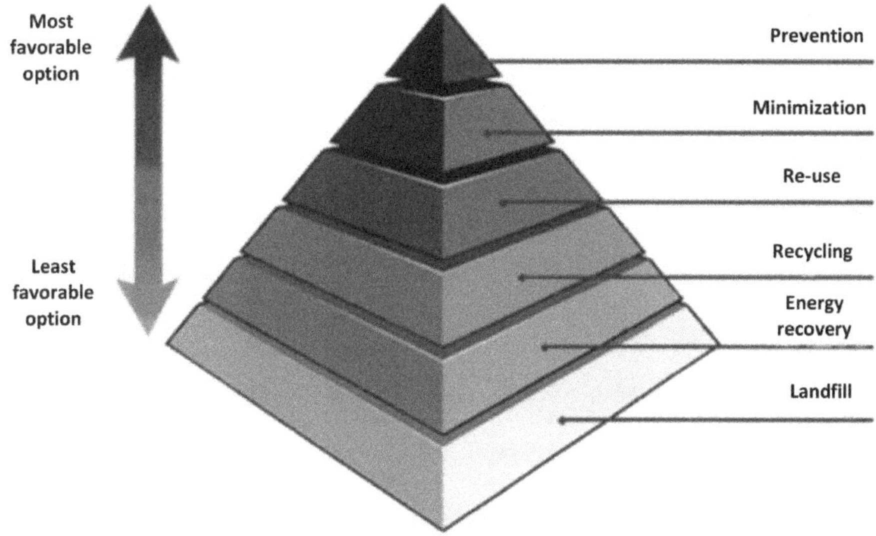

Figure 10.5. *Waste treatment pyramid*

Energy recovery may represent a good solution for certain multi-materials (composites, multi-layer packaging, etc.) whose components are difficult to separate, or those whose nature suggests a different end-of-life scenario, as is the case with poly(ethylene)/wood flour composites where the PE is recyclable and the wood compostable. In such conditions, it is interesting to make use of the good calorific capacity of plastics to assist the combustion of waste in incineration plans (which would otherwise use a fossil fuel at any rate). However, once the waste is burned, we have definitively lost the energy used to create them, so this solution should only be envisaged when all other options have been exhausted (see Chapter 2).

10.4.2. *Favoring sustainable and renewable*

If we stop at the level of "recycling" of the pyramid in Figure 10.5, it is possible to go a little further into detail about the type of recycling process and, again, distinguish the least environmentally damaging alternatives. Figure 10.6 represents the different possible avenues for plastic recycling, from re-use to the obtaining of monomers and reagents.

The most favorable scenario is, of course, re-use, followed by the combination of a certain percentage of production effluents with virgin material directly by transformation companies. This technique is now relatively well integrated into the industrial cycle for reasons of cost, but also under the influence of pressure from environmental standards. Most technical recycling solutions are at a stage of maturity, but the larger the loop, the greater the environmental impact.

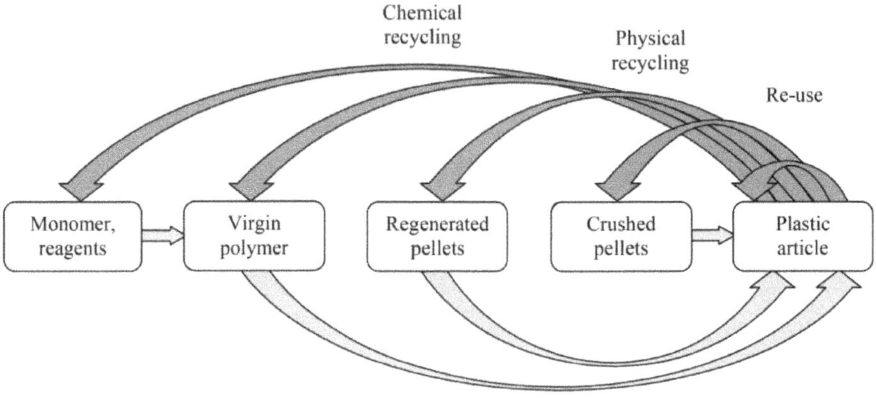

Figure 10.6. *Pathways for plastic recycling [DUV 04]*

Material recovery enables us to save on the virgin material and also on some of the energy associated with the manufacture, because the processes involved in the repeated rounds of recycling consume less than do synthesis processes. It appears that synthetic thermoplastics such as polyolefins are entirely appropriate for physical recycling, while biodegradable plastics can also be composted when their properties are no longer sufficient for any applications. Finally, thermosetplastic waste is best suited for thermal energy recovery, or possible chemical recycling or crushing, which explains why we seek to substitute thermoplastics as far as is possible [HOL 04].

These recycling techniques aside, it is also possible to extend the lifetime of plastics during their usage. It has been shown by certain studies that wood flour/poly(ethylene) composites used to build public benches could have a more favorable balance than wood itself, which is because of the non-biodegradability of PE [HOL 04]. Other authors have observed that the introduction of a certain quantity of wood flour into poly(propylene) helped increase its durability and improve its mechanical properties [LAM 08].

However, certain articles, by design, have very short lifetimes, ranging from only a few hours for certain forms of packaging to tens of months for mobile telephones, for example (see Figure 10.7).

Figure 10.7. *Scrapped mobile telephones*

It is tempting to think that for these short-term applications, biodegradable plastics offer a satisfactory compromise. It is probably true, for instance, with disposable covers or "green" bin liners for collecting compostable waste. However, as they decompose, these materials can release additives which are harmful for the ecosystem. Thus, we need to look carefully at these aspects [STA 06].

How are we to draw together all these notions into a summary? Given the complexity of the problem, only the use of eco-design software packages can bear fruit, as long as we correctly interpret the results and input sufficient quantities of relevant data (see Chapter7). However, as part of this approach, we need to take account of the political dimension, which lies in the choice and ranking of the impacts. Indeed, depending on whether we favor the impact on non-renewable material stocks or on the greenhouse effect or indeed on the ecosystem, we can expect rather different results. Also, the results may depend on the initial hypotheses and on the computation tools used, as we shall see in the following example. By contrasting the results of a comparison of the impacts of virgin PET, recycled PET and PLA (Figure 10.8), performed using SimaPro [KUM 10], with those shown in Figure 10.9, provided by the PLA fiber manufacturer [NAT], we can see a number of points of discord: if we disregard the scales, we can see that the classification is respected for non-renewable resources but inverted for climate change; that it is very favorable for PLA in terms of human health, but very unfavorable from the point of view of the quality of the ecosystem.

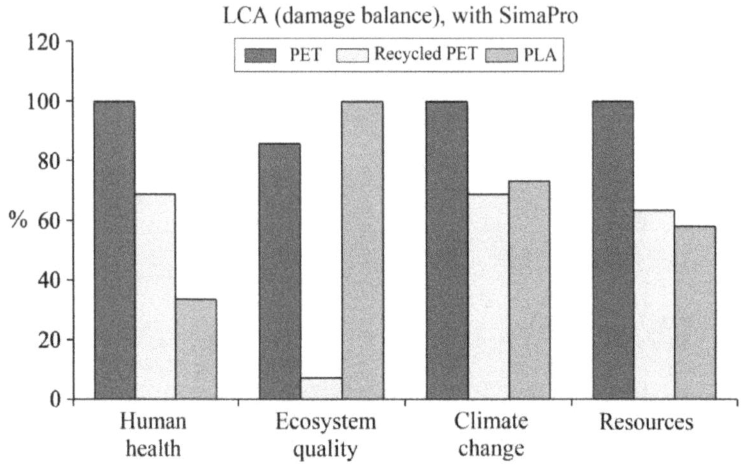

Figure 10.8. *Comparison of the impacts of virgin PET, recycled PET and PLA on human health, ecosystem quality, climate change and non-renewable resources [KUM 10]*

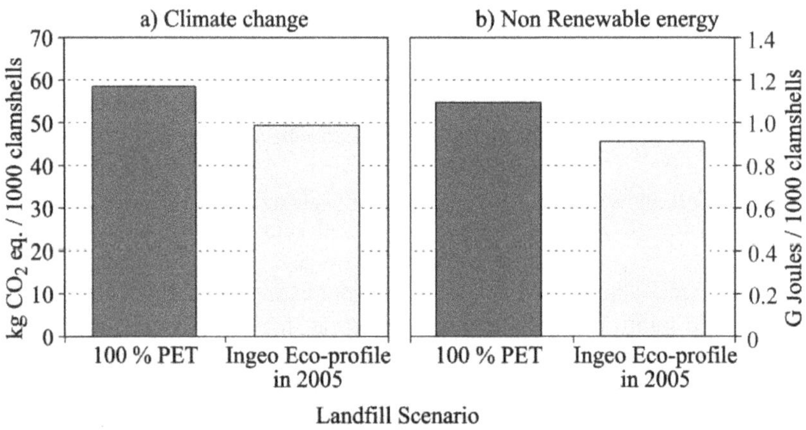

Figure 10.9. *Comparison of the impacts on climate change and non-renewable resources of Ingeo PLA fibers from Nature works and recycled PET [NAT]*

Hence, these impacts are not absolute measurements – they depend on the way in which they are calculated and on the person who evaluates them. Consequently, this type of assessment is increasingly of political and strategic importance for companies.

10.5. The solution of ecoplastics

In line with the lifecycle assessments presented earlier in this book, we can distinguish two classes of polymers which help to reduce a product's environmental impact. *Ecoplastics* include plastics derived from physical recycling and those derived from renewable resources (see Figure 10.10).

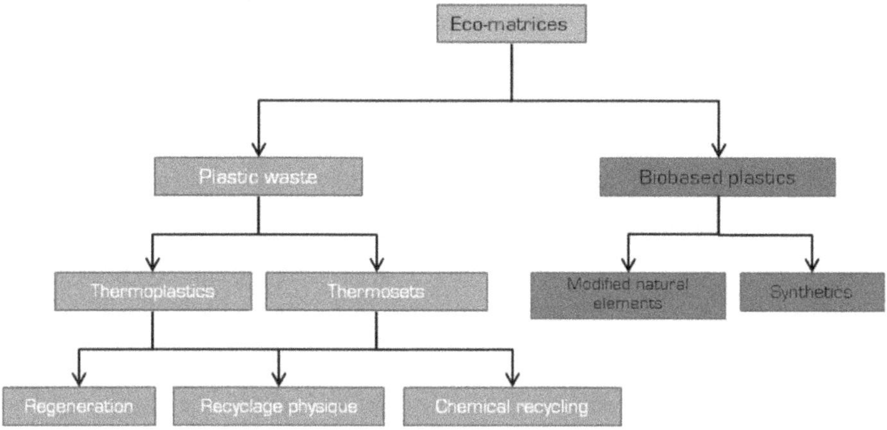

Figure 10.10. *Classification of different types of polymer biomatrices [FEL 09]*

Yet in practice, the boundary between the two classes is not quite so clear, because different combinations of these materials are possible. Indeed, we can very easily combine a cellulose biomatrix with biofibers of cellulose [SOY 09], biofillers with a recycled matrix [ADH 08], recycled fillers with a recycled matrix [ZRI 07], or two or more recycled matrices with one another [GRI 05]. All these combinations of ecoplastics may potentially be advantageous, but cross-bio/recycled solutions pose the problem of choosing their end-of-life scenario: recycling or composting, or finally incineration?

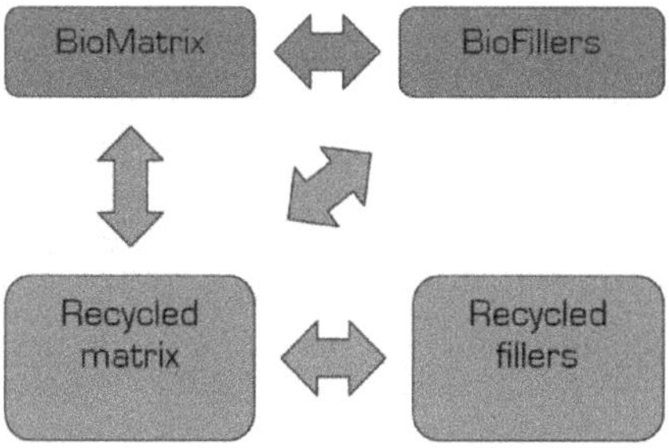

Figure 10.11. *Different possible combinations of matrices and fillers [FEL 09]*

In what follows, we are going to analyze, in turn, two scenarios for ecoplastic manufacture which are particularly attractive because of their low environmental impact.

10.6. Scenario with compostable ecoplastic

10.6.1. *Creation of a mixture of biodegradable polymers*

As we can see from the results in Figure 10.4, starch – which is a polysaccharide used for energy storage in plants – is an excellent candidate to use for making an ecoplastic; it is a totally biodegradable product, of which huge quantities are available (it is the second most abundant organic compound in nature, after cellulose) and very inexpensive (approx. €0.50 kg^{-1} in 2005). However, starch has poor mechanical properties, so it needs to be combined with another, more robust polymer. It also has a hydrophilic nature, so must be chemically modified if we combine it with a more hydrophobic polymer. Using an approach similar to that used to create Mater-Bi®, developed by Novamont, which won the Packaging Oscar (*Oscars de l'emballage*) in France in 2009 [MAT], we can closely combine macromolecules of modified starch and poly(caprolactone) to form a biodegradable and compostable plastic,conforming to the European norms UNI EN 13432 and UNI EN 14995.

10.6.1.1. *Structure of poly(ε-caprolactone)*

Poly(ε-caprolactone) PCL is a biodegradable-derived synthetic semi-crystalline polymer which exhibits the advantage of having good mechanical properties below its melting temperature of approx. 50°C. One positive consequence of this fairly low value is that its associated processing temperature is compatible with that of starch, and will not excessively degrade the starch during the transformation. The stages of PCL synthesis are illustrated in Figure 10.12, which demonstrates that this is a process of ring-opening polymerization and that PCL belongs to the polyester family.

10.6.1.2. *Structure of starch*

Starch appears as a powder, resembling ordinary flour. The grains which make it up are micronic in size, and therefore perfectly observable with an optical microscope, as we can see in Figure 10.13. Despite what we might think from their appearance, starch grains have a complex hierarchical structure, shown in detail in Figure 10.14. Among other things, we can see that the structure of the grains is semi-crystalline and that two types of macromolecular chains coexist: linear amylose chains and heavily branched amylopectin chains. At this scale, it is apparent that the chains can come together to form double helices in the native state.

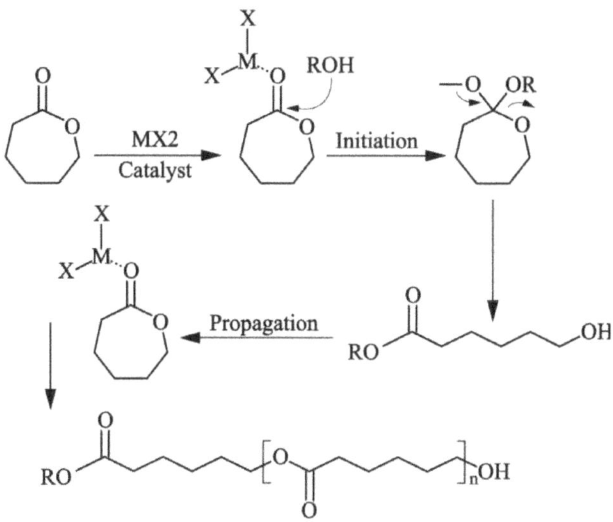

Figure 10.12. *Different stages in the synthesis of CAPA PCL from Solvay [SOL]*

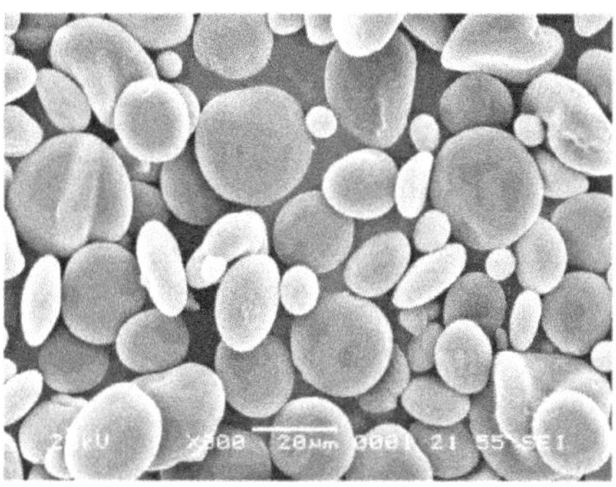

Figure 10.13. *Wheat starch observed with an optical microscope (OM) [DIV 04]*

Depending on which plant it comes from, the properties of starch change: in particular, the shape of the grains is different and the proportion of amylose and amylopectin varies. The starch used in this example was provided by the company *Roquette* (I59-113H10) and contained 25% amylose, 74.25% amylopectin, 0.1% lipids, 0.35% proteins and 0.3% minerals. The molar mass M_w was approx. 50×10^6 g.mol^{-1} and the degree of crystallinity X was between 35 and 40% [ROQ].

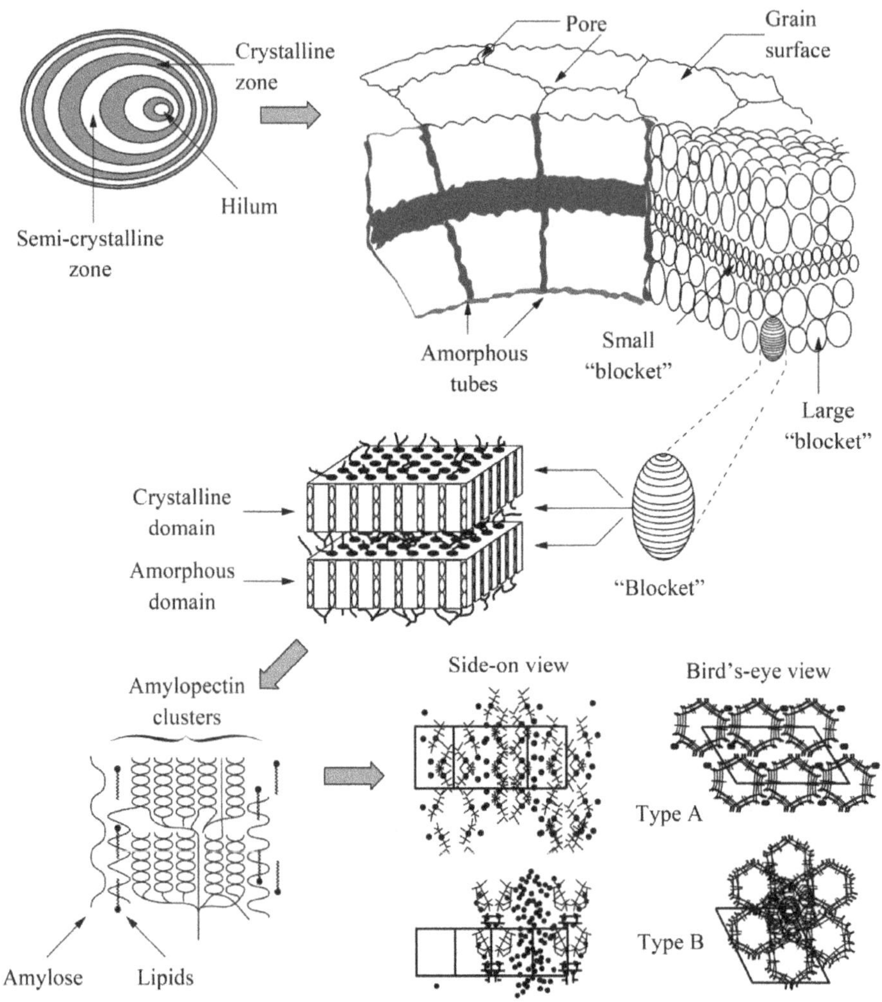

Figure 10.14. *Different levels of organization in starch grains [GAL 97]*

10.6.1.3. *Functionalization of starch by the action of formic acid*

As noted previously, it is difficult to closely combine starch and poly(ε-caprolactone), because there is little interaction between these two molecules to promote miscibility. For this reason, it may be advantageous to chemically modify the poly(saccharide) chains, e.g. by esterifying the hydroxyl functional groups of the α-D-glucose rings with formic acid (FA) as illustrated in Figure 10.15.

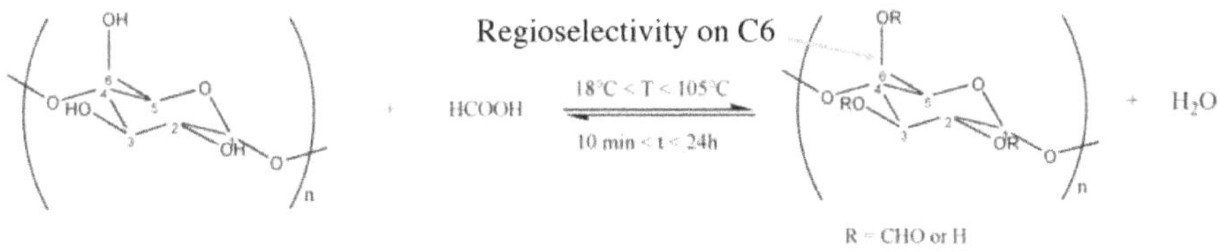

Figure 10.15. *Esterification of starch by formic acid [DIV 05]*

The maximum theoretical degree of substitution (DS) of these functional groups is three, but, in practice, it is fairly difficult to achieve a value of two without damaging the starch molecules too much. The acid has a two-fold effect: it functionalizes the macromolecules and de-structures the various molecular arrangements. This thermo-dependent process, similar to gelling (or gelatinization), may be seen under the OM by the disappearance of the grains (see Figure 10.16) or, more quantitatively, by rotational shear rheometry (RSR) in dynamic mode, by the change in the slope of G' (the modulus of conservation), as we can see in Figure 10.17.

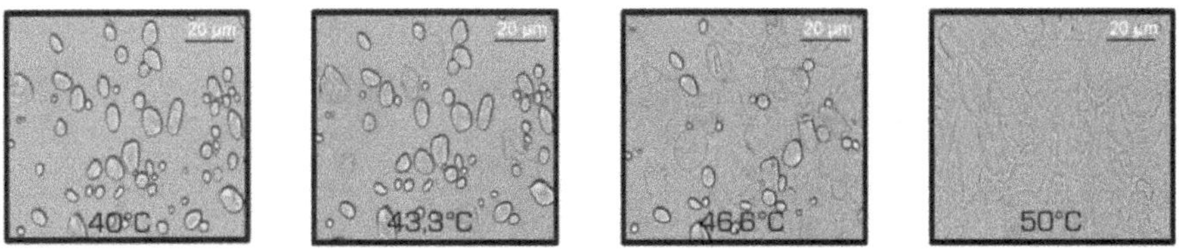

Figure 10.16. *De-structuring of starch grains in an aqueous solution containing 40% formic acid observed under an optical microscope (OM) [FEL 05]*

Using OM, we note that this transition occurs when the temperature exceeds 40°C, when the concentration of FA present is 40% v/v, whereas in rheometry, during frequency scanning, gelatinization is observed at around 32°C when the starch is in solution in pure FA. With this technique, it is also possible to monitor the degradation of the gel at higher temperatures. In this case, the slope of G', which is near to zero for starch in gelatinized form (characteristic of cooperative behavior) approaches 2, suggesting the behavior of dispersed chain fragments with little interaction, which is distinguishable from the behavior exhibited by the grains in suspension in the initial state (at ambient temperature).

In pure water, the temperature must be raised above 75°C to obtain gelatinization of the starch, as shown in Figure 10.18, which illustrates the influence both of the formic acid concentration and of the temperature on the gelatinization transition.

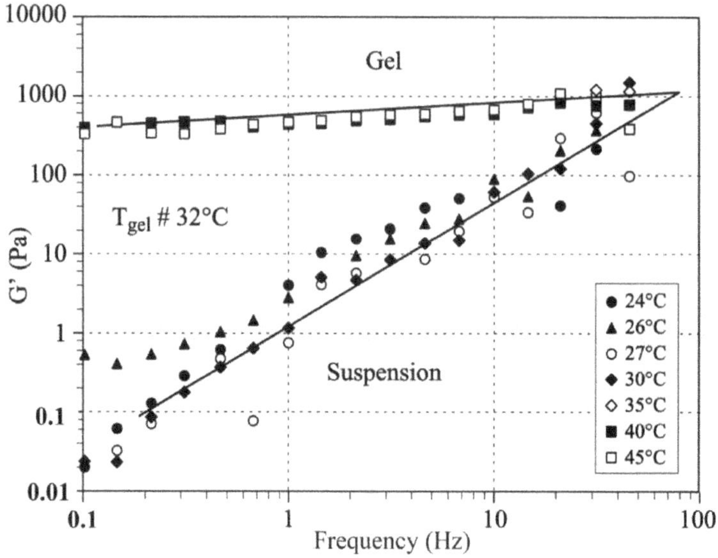

Figure 10.17. *Tracked gelatinization of starch [FEL 05]*

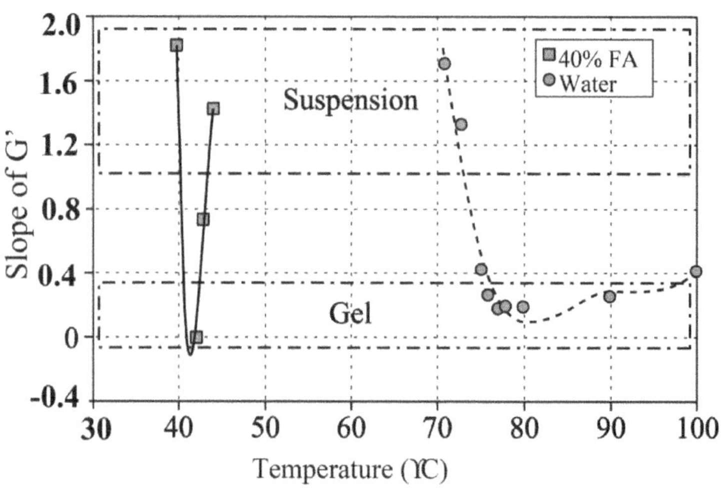

Figure 10.18. *Evolution of the slope of the complex elastic modulus with changing temperature and formic acid content in the solution, 0 and 40% v/v [FEL 09]*

Thus, it is possible to use these experimental parameters to adjust the level of de-structuring of the starch. In addition, we can supplement these characterizations by measuring the viscosity in solution, because if the reaction is not halted early enough, the formylation is accompanied by a reaction of hydrolysis of the macromolecular chains, which causes the molar mass to decrease. As molar mass is linked to viscosity by the Mark–Houwink equation (see equation [10.1]), we can characterize the scission of the chains by measuring the intrinsic viscosity:

$$[\eta] = K.M_v^a \quad [10.1]$$

According to the results given in Figure 10.19, we do indeed see a rapid drop in viscosity with increasing degree of substitution. It is therefore necessary to strike a balance between functionalization and decreasing of molar mass; the compromise should be around a DS of 1.

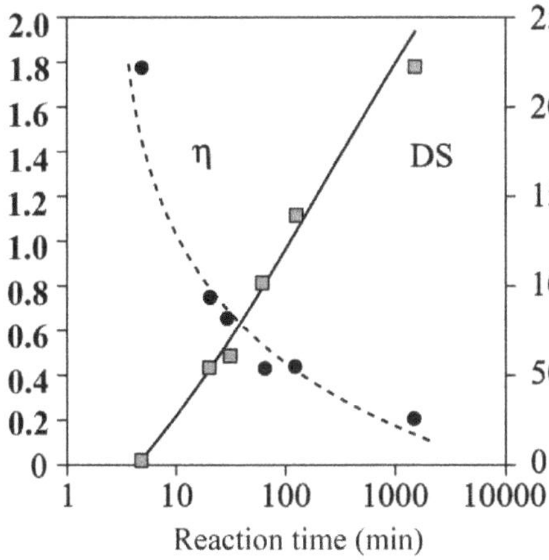

Figure 10.19. *Evolution of intrinsic viscosity and degree of substitution as a function of reaction time [FEL 05]*

In addition, as the objective of starch formylation is to improve its compatibility with PCL, we need to examine the effect of functionalization in terms of increasing the interactions between the two polymers. The compatibility between the two polymer phases is better when the Flory–Huggins interaction parameter χ_{12} is close to 0, i.e. when the solubility parameters δ_{pol} of the two polymers are similar, if we look at equation [10.2]:

$$\chi_{12} = \frac{V_0(\delta_{pol1} - \delta_{pol2})^2}{RT} \quad [10.2]$$

where:

χ_{12}: Flory–Huggins interaction parameter;

δ_{pol}: Hildebrand solubility parameter;

R: 8.314 $J.K^{-1}.mol^{-1}$: ideal gas constant;

T: absolute temperature in Kelvin;

V_0: molar volume of the monomer motifs.

By evaluating these parameters on the basis of data taken from the existing literature [BRA 99], we can see that a degree of substitution of around 2 would be optimal to promote miscibility of starch formate with PLC (Table 10.1). However, as we saw earlier, such a DS would require the chains to be heavily degraded by scission first.

Polymer	δ_t
Native starch	26.5
Formylated starch, DS = 1	23.1
Formylated starch, DS = 2	19.3
Formylated starch, DS = 3	16.5
PCL	18.7

Table 10.1. *Hildebrand solubility parameters [HIL 49]*

10.6.1.4. *Starch formate/PCL mixture*

One last point to take into account is the fact that even if we obtain good miscibility of the amorphous phases due to compatibilization by grafting, phase segregation inevitably takes place when the starch and the PCL crystallize separately. Indeed, if the de-structuring of the starch destroys the crystalline entities comprising branched amylopectin chains during biosynthesis, only a certain proportion of the linear amylose chains can crystallize during restructuring (retrogradation) – all the more so when the DS is low. Consequently, the mixture of the two polymers in solution may cause aggregates which are clearly visible in Figure 10.20, and give rise to a heterogeneous material as illustrated in Figure 10.21; it is therefore preferable to perform the formylation reaction in the molten state at the same time as mixing with the PCL [PIL 05].

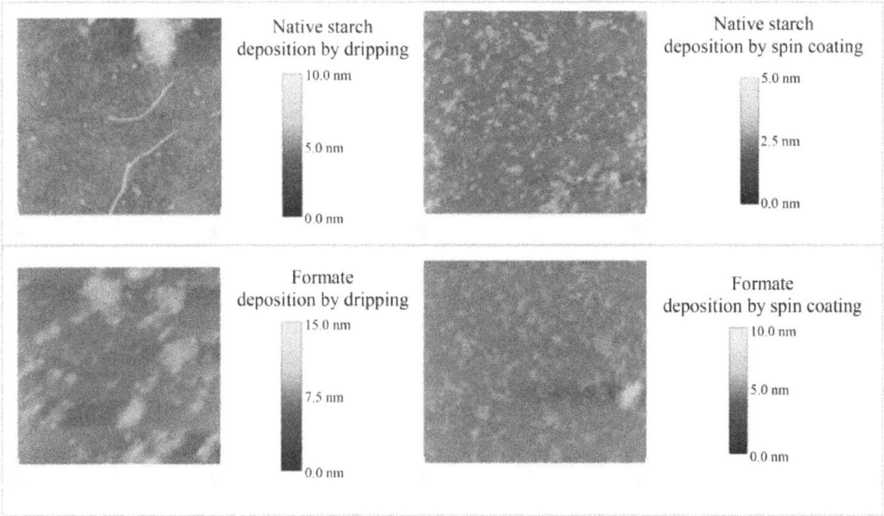

Figure 10.20. *Observation by atomic force microscopy (AFM) in tapping mode of the effect of formylation (DS = 1.25) on the restructuring of starch solutions in DMSO, deposited on cleaved mica, [Starch] = 10 mg.dm^{-3} [DIV 07]. (For a color version of this figure, see www.iste.co.uk/hamide/polymers.zip)*

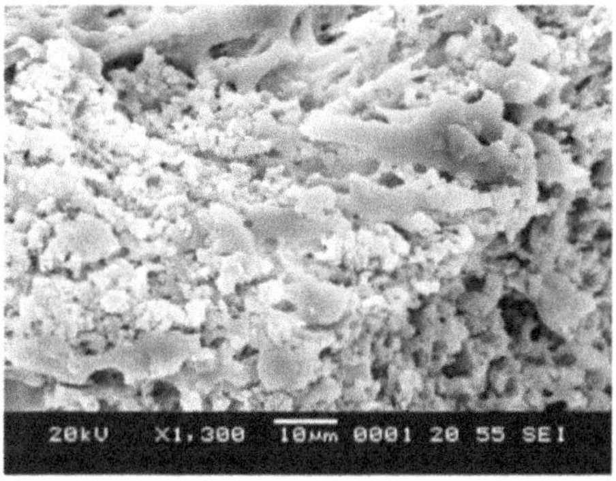

Figure 10.21. *Morphology of formate observed with scanning electron microscopy (SEM) after precipitation in methanol [PIL 05]*

Molten formulation also presents the advantage of preventing the use of solvents, which helps reduce the environmental impact, increase dispersion and reduce costs. The mixing and reactions can be performed by twin-screw extrusion, for example, with a rotation speed of 30 rpm and a temperature profile of 80, 85 and 90°C from the feed hopper to the screw.

10.6.2. *Characterization of starch/PCL mixtures*

10.6.2.1. *Mechanical properties*

A good way of assessing the compatibility of a mixture of polymers, and quite simply its effectiveness as a material, is to study its stretching properties. Figure 10.22 shows that we can fairly easily adjust the Young's modulus of the mixture by varying its composition. PCL lends the mixture a plastic nature (low stress, but significant stretching at break) while starch is more rigid, but also more fragile (see Table 10.2 – formulations F1 and F2). It is also worth noting that formylation increases the modulus of the mixture, which is consistent with the idea that better compatibility between the phases helps improve the mechanical characteristics.

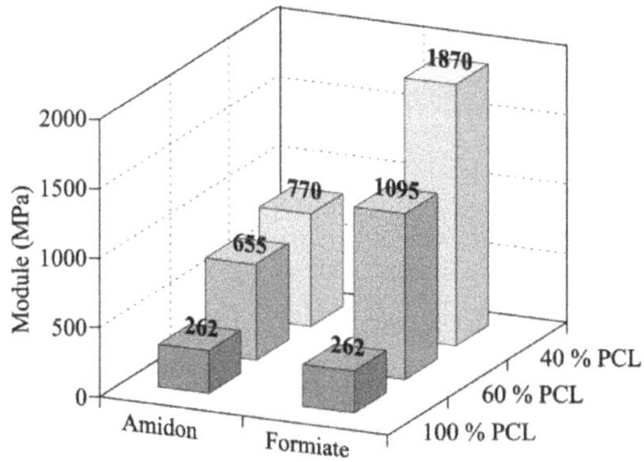

Figure 10.22. *Influence of the proportion of PCL on the elastic modulus of starch/PCL mixtures [FEL 05]*

Simple functionalization of the starch may prove insufficient to produce satisfactory properties for the different envisageable applications for biodegradable plastics. One of the solutions, therefore, is to use a coupling agent comprising an oligoester (OE) spacer and a cross-linking isocyanate (PPDI) to ensure greater cohesion to the material (see Table 10.3).

The formulations F5 and F7 in Table 10.2 are a good illustration of the compromises in terms of properties that can be achieved by combining PCL starch and oligoester starch, which respectively ensure significant rigidity and significant stretch at break. However, no formula is able to deliver a stress at break equivalent to pure PCL. On the other hand, this range of compositions is capable of providing solutions that can be adjusted to a given process (injection, extrusion, thermoforming, blow-molding, etc.) or application (disposable covers, shopping bags, plant pots, etc.).

Sample	Starch	Form	P1	P2	P3	P4	PCL	MVI ($cm^3.10\ min^{-1}$)	E (MPa)	σ_r (MPa)	ε_r (%)
Pure PCL							100	8.0	262	31.5	693.0
F1	40						60	4.7	655	8.9	6.2
F2		40					60	14.0	1095	10.1	1.4
F3	40		30				30	18.2	1002	2.8	3.6
F4	40			30			30	541	828	1.7	0.4
F5	40				30		30	22.8	832	7.8	11.8
F6	40					30	30	27.1	280	5.1	191
F7		40				30	30	38.0	266	4.0	247

Table 10.2. *Mechanical characteristics of formulations starch/PCL/OCL/PPDI [PIL 05]*

Oligomer	Nature	End chain	M_w ($g.mol^{-1}$)
P1: 105-42	1,6-hexane-diol adipate and phthalate	hydroxyl	2700
PS: 101-55	Glycol adipate and phthalate	hydroxyl	2000
P3: 105-15	1,6-hexane-diol adipate and phthalate	hydroxyl	7400
P4: 1063-35	Poly(caprolactone)	hydroxyl	2000

Table 10.3. *Linear oligomers of polyesters with hydroxyl (OE) chain end. (Source: Durez [PIL 05])*

The bioplastics created must, of course, have acceptable mechanical properties, but above all they must be able to be composted. It is therefore important to examine their biodegradability.

10.6.2.2. *Measuring the biodegradability of starch/PCL mixtures by respirometry*

The biodegradability of mixtures is evaluated by way of a respirometry. The rate of biodegradation is deduced from the aerobic degradation of the formulations in soil at 28°C. The principle of this test is volumetric dosing of the amount of CO_2 released by the respiration of the micro-organisms present in the soil, using the available carbon as a nutrient.

The CO_2 thus released is dissolved in a soda solution of known concentration, leading to the formation of sodium carbonate by the reaction shown in equation [10.3]:

$$CO_2 + 2NaOH \rightarrow Na_2CO_3 + H_2O \quad [10.3]$$

The sodium carbonate is then precipitated in solution with an excess of barium chloride by way of the reaction shown in equation [10.4]:

$$Na_2CO_3 + BaCl_2 \text{-> } 2\ NaCl + BaCO_3 \qquad [10.4]$$

The remaining soda is then dosed by an acid/base reaction with hydrochloric acid in the presence of thymolphthaleine as a color-based indicator.

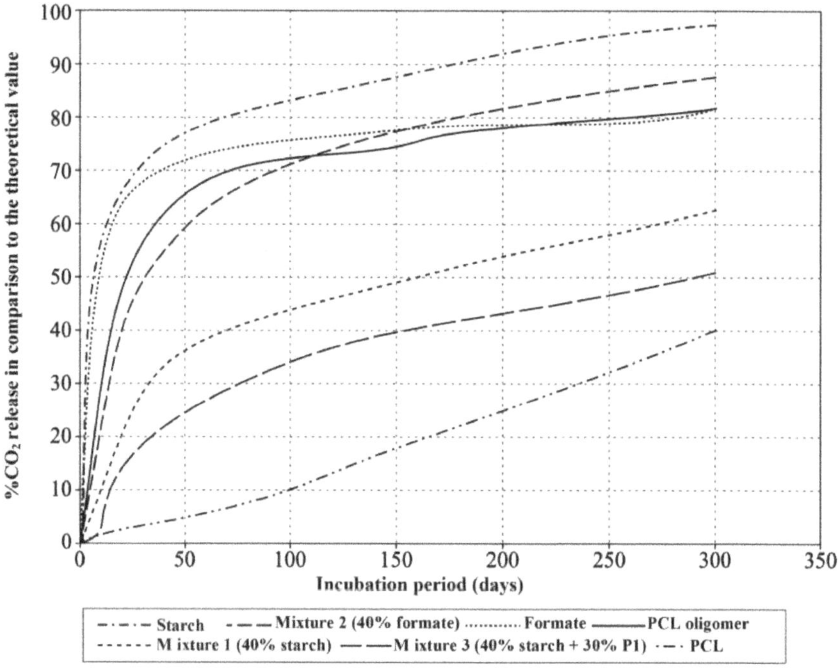

Figure 10.23. *Evolution of duration of biodegradation with changing composition [DIV 05]*

Thus, by applying these tests to the different formulations created, we can classify them in terms of their kinetics of biodegradation (reflected by the CO_2 release resulting from the bacterial activity). Figure 10.23 clearly shows that pure starch biodegrades to 80% within 50 days, while pure PCL is only 40% biodegraded after 300 days. We can clearly see the advantage presented by the various formulations in terms of being able to adjust the kinetics of biodegradation of bioplastics on the basis of the starch content and the presence or absence of oligoesters. Additionally, the values obtained are entirely in keeping with those presented by Novamont with regard to Materbi®.

10.7. Scenario with a recyclable ecoplastic

10.7.1. *Sources of plastic waste*

There is no shortage of sources of plastic waste on this planet (see Figure 10.24), but at present the potential resource they offer is only very partially exploited. Their use is not really a financially viable option, because while virgin plastics are inexpensive to manufacture, waste plastics need to be treated and transported, have poorer properties than the virgin material and the applications of recycled plastics have only little added value.

Figure 10.24. *Different sources of plastic waste [FEL 09]*

In addition, the production of plastics is continuing to increase regularly, as we can see in Figure 10.25, less quickly in Europe than on the world scale, but even then at an almost exponential rate. Thus, we can consider that the corresponding amount of waste will also increase ceaselessly – particularly if that waste is not re-absorbed as it is generated.

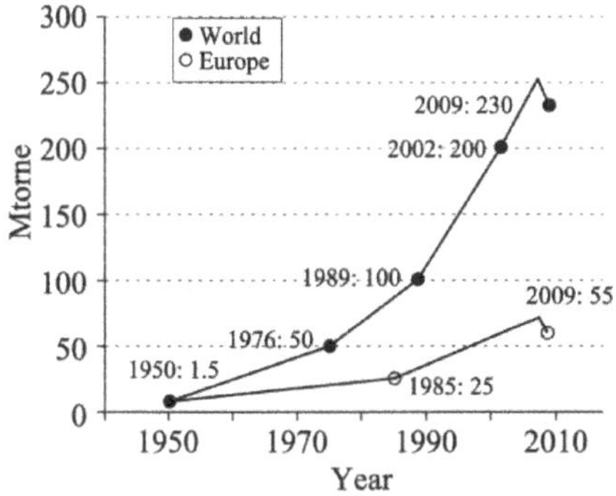

Figure 10.25. *World production of plastic in Gkg [PLA]*

If we analyze plastic consumption per sector, as we can see from Figure 10.26, for Europe in 2009, packaging was the most significant source of potential waste, followed by construction, the automobile industry and the electronics industry.

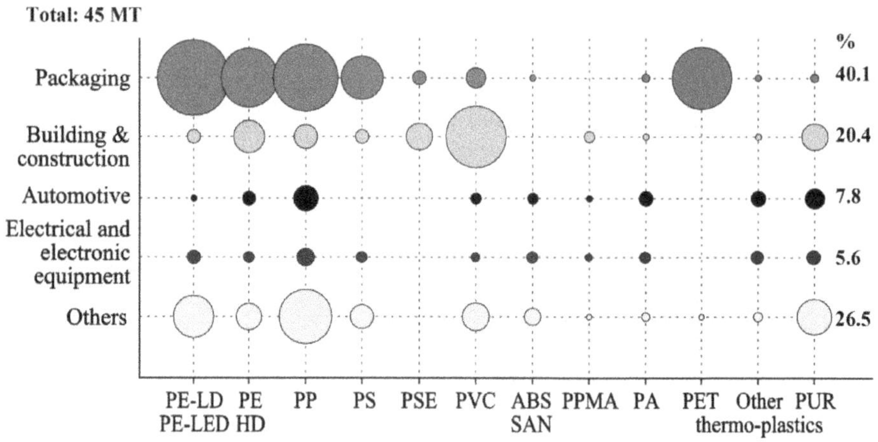

Figure 10.26. *Consumption of plastic products per activity sector in 2009 in Europe [PLA]*

Packing waste, unfortunately, represents a source which is difficult to process, because of its low mass-to-surface ratio, the variety of types of packaging waste and the fact that it is often soiled. However, the path of recycling poly(ethylene terephthalate) PET bottles into textile fibers to create fleece offers a good illustration of waste recovery for applications which are less demanding in terms of the properties of the material. Although the molar mass of PET falls during its second round of processing, beginning with a low grade (for extrusion, the viscosity in the molten state needs to be high) facilitates the material to be spun, which requires a higher degree of fluidity.

In contrast to this example, we can say that the use of recycled plastic waste is limited by the difficulty in transporting it as little as possible, sorting it for the lowest possible cost and regenerating the material with as few operations and additives as possible. Another limiting factor is that it is often difficult for a transformer to obtain recycled plastics of consistent and clearly known quality.

If we wish to circumvent these various problems, it is necessary to use clearly identified waste products, characterize their recyclability and find an application compatible with their lesser performances once recycled.

10.7.2. *Concept of recyclability*

Figure 10.27 illustrates the process whereby we can evaluate the recyclability of plastic waste. After initial crushing, injection-molded samples are analyzed simultaneously by different physico-chemical characterization techniques, such as dynamic mechanical analysis (DMA), differential scanning calorimetry (DSC) or RSR. These transformation/analysis cycles can repeat up to eight times, and we observe the evolution of the different characteristics at each stage.

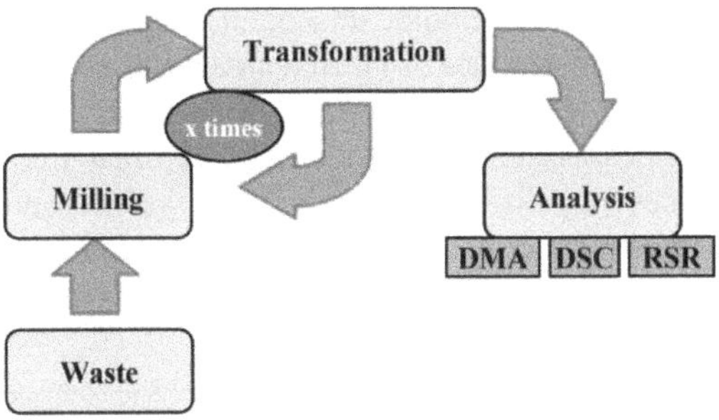

Figure 10.27. *Process for assessing recyclability*

10.7.3. *Recyclability of bisphenol A poly(carbonate) waste*

In this example, we have looked at the recyclability, by injection, of polycarbonate (PC) waste from cutting up temporary signposts based on extruded plates [SEL].

One advantage of these waste products is that they are clean, clearly identified and we can expect that, by beginning with a plastic such as PC, which is known for its good initial mechanical properties, it will be possible to preserve an acceptable standard of properties for common applications after recycling.

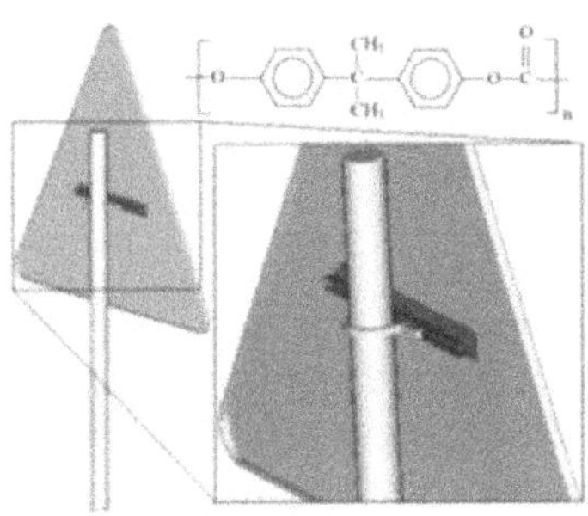

Figure 10.28. *PC signage panel [SEL]*

10.7.3.1. *Dynamic mechanical analysis (DMA) and differential scanning calorimetry (DSC)*

DMA presents the advantage not only of enabling us to monitor the evolution of the polymers' viscoelastic properties with changing temperature and frequency, but also of providing information about the molecular relaxations taking place within the material. Hence, this tool is entirely appropriate for revealing any degradation that has taken place in the wake of the recycling process. In this example, the samples are injected plates whose dimensions are stated in the insert in Figure 10.29. The samples are subjected to three-point sinusoidal flexural strain. The evolution of the conservation modulus of samples of PC, having undergone between one and eight rounds of recycling (PC1–8) during temperature scanning at 1 Hz, is shown in Figure 10.29. With linear scales, we can clearly see that the thermomechanical behavior of PC becomes very different from the third cycle onward (curve PC3). We can identify three temperature zones in which the influence of the recycling process is apparent.

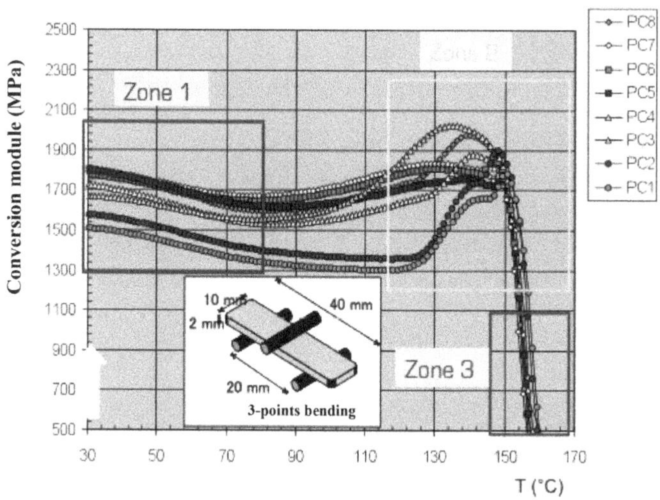

Figure 10.29. *Evolution of the conservation modulus E' (MPa) of PC at 1 Hz, as a function of temperature and the number of rounds of recycling [FEL 03b]. (For a color version of this figure, see www.iste.co.uk/hamide/polymers.zip)*

In zone 1, which corresponds to PC in the glassy solid state, the chains are not hugely mobile, and we can see that the more times the polymer is recycled, the more its conservation modulus *E'* increases, before stabilizing after the fifth round of recycling. This may seem surprising, as recycling is known for causing chain scissions, and the elastic modulus of a polymer generally increases as its molar mass increases. However, this has also been observed by Paul *et al.* [KAV 96]. This seemingly contradictory phenomenon can be explained, as we shall see shortly, by better structuring of the amorphous phase, rendered possible by the greater mobility of the chains.

Zone 2 in Figure 10.29 corresponds to a range of temperatures just below the glass transition point of PC, i.e. in which the chains with lower molar mass begin to become mobile. This capacity to move causes relaxation of the thermomechanical stresses which have accumulated during a previous use. Verdu [VER 82] reports that a similar phenomenon of structural relaxation can be observed by DSC after aging of the polymer. For both PC1 and PC2, we note the presence of two relaxation peaks which gradually shift toward lower temperatures with repeated rounds of recycling, and ultimately merge.

In zone 3, we can see the glass transition temperature of the PC slide toward lower temperatures with progressive rounds of recycling. This time, the evolution is consistent with a drop in molar mass. This tendency is confirmed by the DSC analyses shown in Figure 10.30. After expansion of the horizontal axis corresponding to zone 3, it is more apparent in Figure 10.31 that the T_g decreases from 157 to 153°C, and from 149.5 to 145.5°C, measured with DMA and DSC, respectively, which is a significant variation.

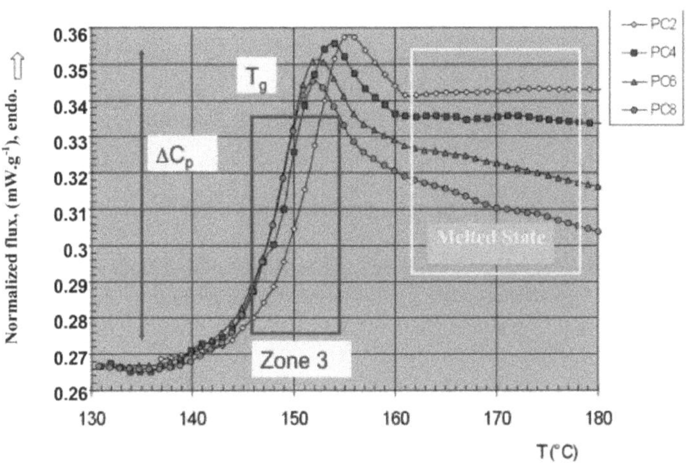

Figure 10.30. *Evolution of T_g of PC for 2, 4, 6 and 8 successive rounds of recycling, measured by DSC at 15°C.min^{-1}. (For a color version of this figure, see www.iste.co.uk/hamide/polymers.zip)*

The fact that this drop of around 4°C is observed at a different temperature is attributable to the different nature of the stresses to which the sample is subjected in DSC and DMA: simple application of thermal energy in the case of the former, and a combination of mechanical and thermal energies in the case of the latter. Consequently, in DMA, the transition temperatures depend on the frequency, and it is perfectly possibly to see an exact match with the DSC results when that frequency is less than 1 Hz. Figures 10.30 and 10.31 also show that the amplitude of ΔC_p decreases with successive recycling; yet Sperling [SPE 86] and Wünderlich [WÜN 97] have already pointed out that this type of behavior corresponded to the rigidification of the polymer, which is often associated with an increase in the degree of crystallinity. Given that PC does not have measurable crystallinity to speak of, we must suppose that, because of the decrease in their molar mass over the course of the successive rounds of recycling, the chains have a higher capacity to arrange and form structures in the amorphous phase. Thus, again, we see consistency with the aforementioned increase in modulus.

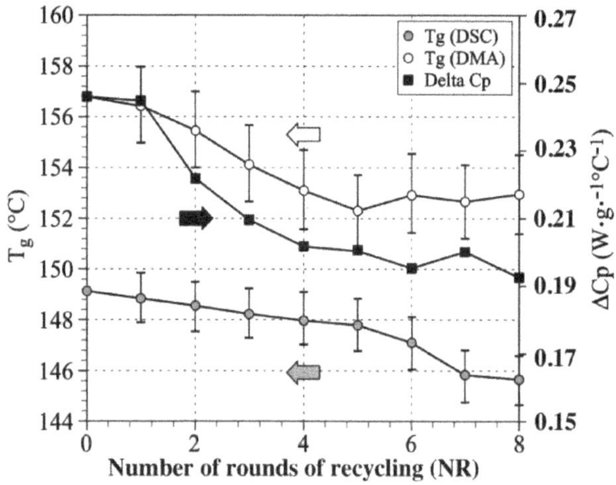

Figure 10.31. *Evolution of glass transition temperature T_g and variation in heat capacity C_p with increasing NR [STA 06]*

Finally, we can see that it is pertinent to characterize the effect of the successive rounds of recycling on the structure of PC, using DMA and DSC, but it gives us no direct information about the variations in molar mass. In order to proceed further in the interpretation of the above results, it is necessary to perform additional experiments by RSR.

10.7.3.2. *Rheological analysis*

Figure 10.32 illustrates a significant drop in the viscosity of PC with increasing shear (in view of the log scale) for successive rounds of recycling, measured with a plate–plate rheometer in steady-state mode at 240°C. This reinforces the idea that recycling causes a decrease in molar mass by thermo-oxidation phenomena, during the course of high-temperature, high-shear processing, by way of clearly identified mechanisms [RIV 02]. Indeed, beyond the critical mass of entanglement ($\overline{M}_c$), the viscosity of the molten polymers evolves, along with the molar mass, to the power of 3.4 [CAR 99], as shown by equation [10.5], derived from the Mark–Houwink equation for dilute polymer solutions, and modified by Bueche for molten polymers [SUN 04]:

$$\eta_0 = K_\eta \left[\overline{M}\right]^{3,4} \text{ si } \overline{\mathrm{M}} > \overline{\mathrm{M}}_\mathrm{c} = 2\overline{\mathrm{M}}_\mathrm{e} \qquad [10.5]$$

Thus, this strongly suggests that the drop in the viscosity of PC is due to a drop in its molar mass resulting from the successive rounds of recycling.

Furthermore, the behavior observed in Figure 10.32 is fairly typical – i.e. at low shear, the viscosity evolves little with the shear, and can therefore be qualified as Newtonian, whereas beyond a critical value γ_c of the shear rate, it decreases exponentially under the influence of the self-heating caused by friction between macromolecular chains, which characterizes pseudo-plastic (rheofluidizing) behavior. In Figure 10.32, we can see that γ_c increases as NR does, and therefore the shear domain wherein the polymer's behavior is Newtonian increases. This once again reinforces the hypothesis of a decrease in the size of the macromolecules.

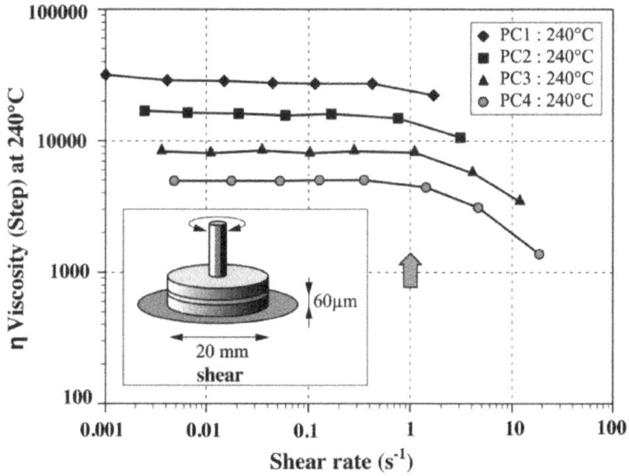

Figure 10.32. *Evolution of the viscosity of PC on the basis of the shear rate, measured by plate–plate rheometry in steady mode at 240°C, for successive rounds of recycling*

In addition to demonstrating the degradation of the polymer, these rheological data enable us to simulate processing by extrusion and injection, in light of the alterations in the viscosity of the recycled PC in comparison to the virgin polymer. Figure 10.33 shows examples of simulation of new parts for signs – typically a fixation rail (only one half of the item is shown in the drawing, but it is symmetrical so the other half is a mirror image of the one shown) and a handle, which can be made by extrusion and by injection-molding, respectively. *In fine*, it should be possible to use cut-up signage boards to create new parts to substitute the metal parts, eventually leading to an entirely plastic-based design of a sign.

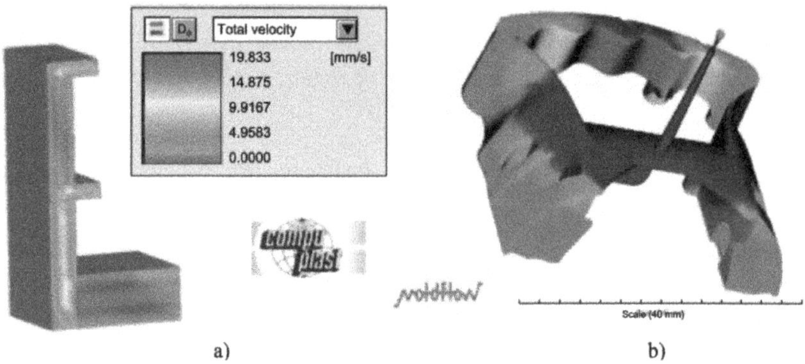

Figure 10.33. *Simulation of the flow of recycled PC in extrusion, using the software Compuplast a) and MoldFlow b) to create a rail and a handle for the sign. (For a color version of this figure, see www.iste.co.uk/hamide/polymers.zip)*

However, if these parts are to be used outdoors, it is necessary to check that the amount of stabilizers to counter ultraviolet rays is sufficient.

10.7.3.3. *Analysis of UV irradiation resistance*

We can assess the resistance to the Sun's UV radiation by exposing the recycled samples to cycles of artificially accelerated aging whereby, in 40 days in real time, a year's aging in external conditions is simulated. Then, we can monitor, e.g., the change in coloration of the surface of the samples with increasing length of exposure to UV. Poly(carbonate) is particularly vulnerable to sunlight if not properly stabilized against UV damage, and rapidly yellows in appearance. The intensity of this yellowing effect can be evaluated by colorimetry. In an "L, a, b" color space, as shown in Figure 10.34, the "*b*" component is particularly vulnerable to accelerated aging under UV. Indeed, even in the first few days of the aging process, the *b* component steadily increases, peaking at 20 days, both for virgin and recycled PC. In addition, the coloration is accompanied by an increase in the storage modulus *E'* in DMA, which indicates that the sample is being rigidified by the UV irradiation. Also, the fact that the change in color is much more pronounced on the recycled samples shows that a certain amount of the stabilizers must have been consumed during the recycling process.

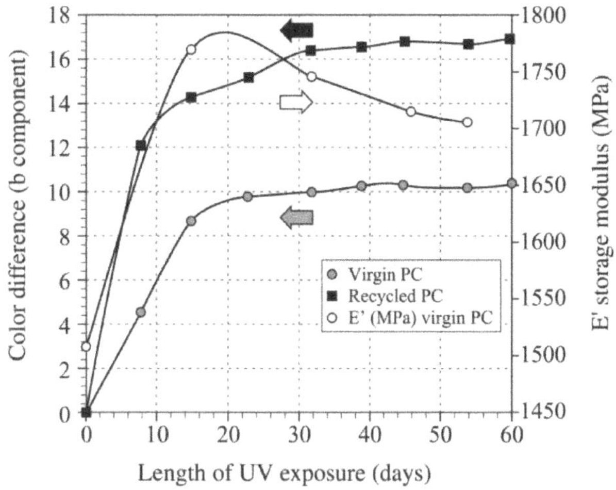

Figure 10.34. *Illustration of aging of PC from UV irradiation, measured bycolorimetry and DMA [STA 06]*

The phenomenon of rigidification can be interpreted both as the result of degradation of the PC and as a consequence of the loss of additives designed to plasticize the macromolecules. In both cases, any outdoor use of parts made of recycled PC must be preceded by a stage of reformulation to compensate for the loss of the stabilizing and plasticizing additives present in virgin PC.

As we have seen in this study, the recycling of poly(carbonate) tends to decrease its viscosity while increasing its rigidity, but also its fragility, which may be penalizing for the intended applications. One of the strategies to counter these phenomena is to increase the molar mass by chemical coupling the chains together. This can be done by using chain extenders such as pyromellitic dianhydride (PMDA), used by Incarnato *et al.* [INC 00] to increase the viscosity of poly(ethylene terephthalate) PET, or cross-linking agents such as methylene diphenyl diisocyanate as used by Pesetskii *et al.* [PES 09] with mixtures of PC and PET. However, these processes for mixing of reagents require good control of the reaction kinetics, and it is not always enough to properly control the structure of the chains, without preventing undesirable transfer and cross-linking reactions. If we wish to use recycled poly(carbonate) without performing chemical reactions to improve its mechanical performances, it is possible to employ physico-chemical processes using non-reactive mixtures.

10.7.4. *Bonding of rubber and poly(carbonate) waste*

Is it possible to reduce the fragility of recycled PC (rPC) by bonding it with a mixture of crumb rubber (CR), also recycled, by using a strategy similar to that use to increase the shock resistance of poly(styrene) by the introduction of a poly(butadiene) phase? In the case of this well-known couple of polymers, the poly(butadiene) PB is dispersed in the PS in the form of a multiple emulsion whose primary phases are of micronic size, and the inclusions are submicronic size. Evidently, it is not possible to obtain the same (salami-esque) type of morphology and the same positive synergy of properties by simply dispersing particles of solid elastomer through a molten PC matrix. However, it is possible to achieve a compromise in terms of mechanical properties which is acceptable for certain applications if we make an effort to improve the quality of the interface between the rPC and CR. This is the focus of the next section.

10.7.4.1. *Conditioning of materials and mixture*

The waste PC material used in this study is fairly clean, taken from cut-up signage panels as discussed in the previous section, and the rubber waste comes from used tires which have been crushed and milled, but have not been cleaned. In the interests of the financial viability of a plastic material entirely created from waste products, it is wise not to add too many stages of conditioning, the cost of which would be a penalizing factor for applications in opposition to virgin materials. For this reason, the processes envisaged here to make the waste products acceptable for use as raw materials are deliberately simple: degreasing of the surface by washing with a variety of commonplace solvents, surface activation by flaming with a torch, functionalization of the surface by sinking agents, but without a chemical reaction with peroxides, as was done by Mennig *et al.* [SCH 02], who created formulations of thermoplastic elastomers by dispersing microparticles (crumbs) of rubber in a poly(propylene) phase.

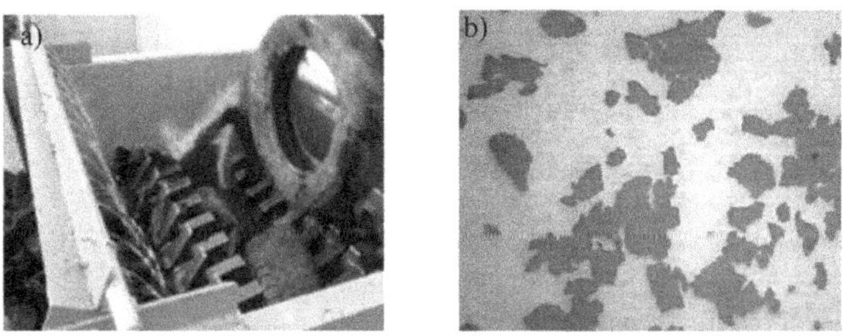

Figure 10.35. *Process of crushing of car tires a), to produce crumbs b) [ZRI 07]*

The crumb rubber (CR) provided by DeltaGom (France) came from crushed car tires (see Figure 10.35(a)) and came under the form of microparticles (see Figure 10.35(b)). The gravimetric composition of that CR is unknown, but the following approximation should be reasonably accurate: natural rubber NR (14%) + poly(styrene-co-butadiene) (27%) + carbon nanoparticles (CNPs) (28%). The images demonstrate, firstly, the soiled nature of the surface of the rubber, and secondly the complex shape of the crumbs once produced by crushing. Thus, post-treatment is necessary to eliminate the various substances such as oil, earth and brake dust present on the surface, and thus improve the adhesion between the CR and the PC matrix. In certain cases, however, it may be advantageous to use the residual oil as a plasticizing agent, as discussed by Lei and Wu [LEI 11] in a recent work regarding the recycling of oil barrels made of high-density poly(ethylene).

Although a preliminary study on the influence of the size of the CR showed the benefits of using CR with average particle size of <140 μm, so as not to damage the mechanical properties when the CR accounts for more than 4% of the content of the product (of particular note is the stretch at break) [ZRI 07], we have chosen to use CR with a diameter between 140 and 350 μm, the better to be able to see the influence of the surface treatment. Also, in order to obtain CR with dimensions less than 140 μm, the rubber must be frozen with liquid nitrogen before being crushed, which seems unrealistic in the context of industrial development. In addition, all the mixtures examined here have a mass composition of rPC80/CR20, which represents the maximum proportion of CR which is acceptable from the point of view of processing. Of course, however, the proportion of each component also influences the mechanical properties of the mixtures, as stressed by a previous study [ZRI 07].

In the present case, three types of simple treatments were envisaged: washing of the CR with dichloromethane, flaming with a propane torch for 1 min at around 800°C, at a distance of 20 cm, and the addition of two types of coupling agents (a three-block copolymer of ethylene-ethylacrylate-hydroxylmethacrylate (E-EA-MAH), referenced as Lotader AX4700, and a tri-block copolymer of ethylene-methyl acrylate-glycidyl methacrylate (E-MA-GMA), referenced as Lotader AX8920). The first treatment was to eliminate any surface impurities, and the second both to eliminate certain pollutants from the surface and to functionalize that surface (at the risk of degrading the macromolecules exposed to the flame). The final treatment serves to render the mixture compatible, by increasing the probability of coupling between the PC matrix and the CR.

The processing of rPC/CR mixtures is done with an internal mixer (Brabender 50 EHT with counter-rotating blades) after drying of the polymers at 90°C for 24 hours. Then, the combination of the CR and the PC is done at 240°C for 10 minutes at 40 rpm.The molten mixture is then pressed at 240°C, with 5 MPa of pressure, for 5 minutes to obtain plates measuring around $4 \times 250 \times 250$ mm^3, which are cooled to ambient temperature for 15 minutes, before being tooled with a digitally controlled machine to make the samples intended for the traction tests. After fracturing in liquid nitrogen, the samples typically present the morphology seen in Figure 10.36, where certain CR fragments are circled in white. It is worthy of note that the fracture surface of the sample suggests a more fragile behavior in the rPC phase than in the CR, and that the interface is not bad in spite of the cryofracturing undergone by the mixture to facilitate its observation.

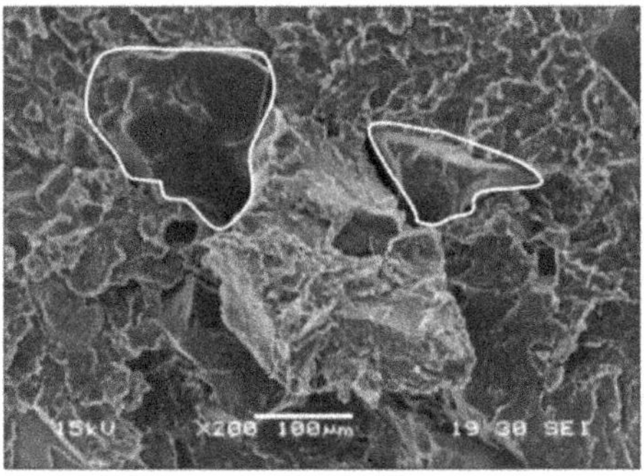

Figure 10.36. *Fractographic image of a mixture in proportions rPC80/CR20, washed in dichloromethane, observed in ×200 magnification under an SEM*

10.7.4.2. *Mechanical characterization*

The traction tests are performed on an MTS Synergie RT1000 machine, with samples conforming to the ISO527 standard, and a rate of traversing of 1 $mm.min^{-1}$. The experimental points correspond to an average value calculated on the basis of five samples.

Two categories of results are then obtained, depending on whether we are looking at the mechanical properties at the elastic limit (see Figure 10.37) or at break (see Figure 10.38). In the first case, it appears that the rate of increase in the stress and strain in the elastic zone (properties normalized in relation to the untreated mixture) is maximal for mixtures in which the CR has simply been washed in dichloromethane. Then come flame-treated mixtures, followed by those based on coupling agents, which remain competitive, particularly with regard to stretching, but far less so in terms of stress. It is worth noting that the modulus of all these mixtures in which the RC has been subjected to a surface treatment is lower than the reference modulus, which confirms their effectiveness. Indeed, as the modulus E of the CR determined by nanoindentation [BOU 07] (0.245, 0.489 and 0.431 GPa for untreated, washed and flamed CR, respectively) is much lower than that of the rPC (1.95 GPa), it is not surprising that good bonding between the fillers and matrix results in the lowering of the modulus of the mixtures.

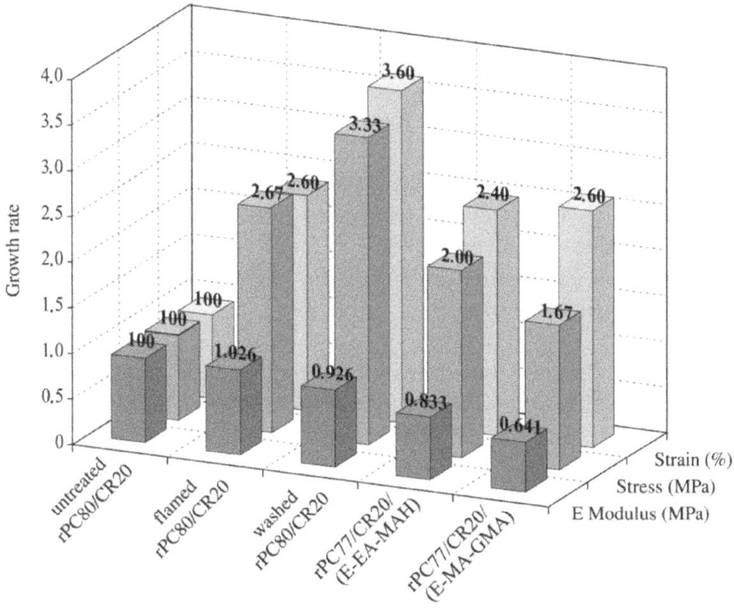

Figure 10.37. *Evolution of the mechanical characteristics of rPC/CR mixtures in the elastic domain [ZRI 07]*

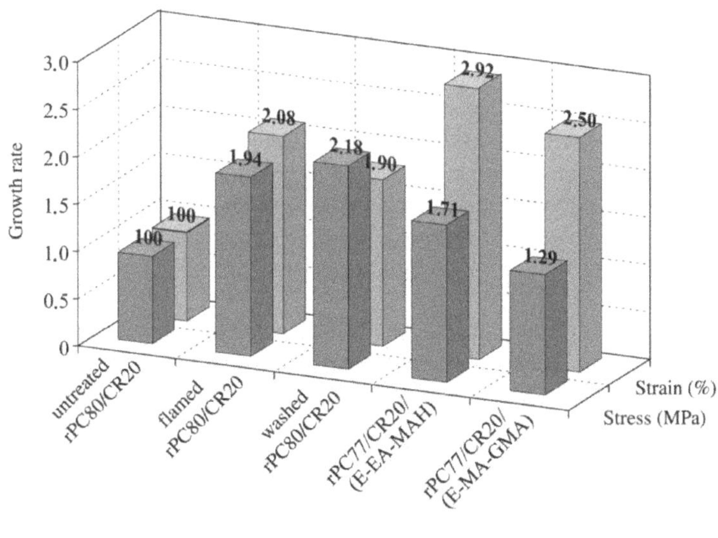

Figure 10.38. *Evolution of the mechanical characteristics of rPC/CR mixtures at break [ZRI 07]*

The second part of the results of the mechanical tests relates to the influence of the treatment undergone by the CR on the properties at break (see Figure 10.38). Overall, the same type of comment as for the results relating to the elastic zone can be made, with the difference being that this time, the treatment of the CR by coupling agents is more effective than the others in increasing the stretching at break of the materials. The presence of a ductile interphase is therefore beneficial if we wish to improve this characteristic. However, it is also worth noting that the maximum growth rate of the performances is only 2.92, as opposed to the 3.6 encountered in the previous case, so it is easier to improve the elastic properties than to improve the ultimate properties.

Finally, this study has shown that with only simple treatments (washing, flaming), the cost of which will not compromise the viability of the recycled material, it is possible to create rPC/CR mixtures without excessively compromising their mechanical properties, even in penalizing conditions (large size and high concentration of microparticles). However, we can also imagine that this type of mixture could be used for applications other than those which require good mechanical properties. In particular, the presence of around 30% CNPs in the CR should enable us to consider applications where good electrical conductivity is required, such as electromagnetic interference shielding [FEL 06], self-regulating ohmic heating [FEL 01], or indeed sensors [FEL 11]. It is this that we examine in the next section.

10.7.5. *Upgrading of waste for a higher-level application*

If, apart from satisfactory mechanical properties, a mixture of recycled polymers such as that which we have just examined (rPC/CR) can also exhibit interesting electrical properties, we will have succeeded in creating a multifunctional material simply from waste polymers. This can then be considered an "upgrading" of the material, because in this case it would be possible to target higher-level applications than those initially intended for the virgin materials.

Smart materials [FEL 11] such as conductive polymer composites (CPCs) could potentially offer such capabilities, and we shall now examine the capacities of rPC/CR mixtures in this domain in greater detail.

10.7.5.1. *Electrical conductivity of rPC/CR mixtures*

In order to render a plastic conductive although it is, by nature, an electrical and thermal insulator (with the exception of intrinsically conducting polymers, ICPs), we need to dope it with electrically conductive fillers such as steel microfibers (μFSs) [FEL 06], CNPs [FEL 01] or indeed carbon nanotubes [FEL 11]. By gradually varying the proportion of fillers in the polymer matrix, we see that its resistance goes through a transition from being an insulator to being a conductor – percolation – the value of which may be 5 decades or more. This exponential phenomenon is attributable to the fact that if the filler content is greater than the percolation threshold, then the interconnection of the fillers allows current to pass through, and we obtain a CPC. Plotting the percolation curves is generally very instructive about the state of dispersion of the fillers, their form factor and the intensity of their interactions with the macromolecules in the matrix. Figure 10.39 displays all of the percolation curves, showing the evolution of the resistivity depending on the amount of CR and CNP introduced for different rPC-based formulations. The curves can be adjusted with the power law shown in the insert in Figure 10.39, where ρ and ρ_0 are respectively the resistivity values for the composite and the filler, and where ϕ and ϕ^* are the volumetric fractions of filler at the point in question and at the percolation threshold, with t being the critical exponent which can be linked to the dimensionality of the fillers.

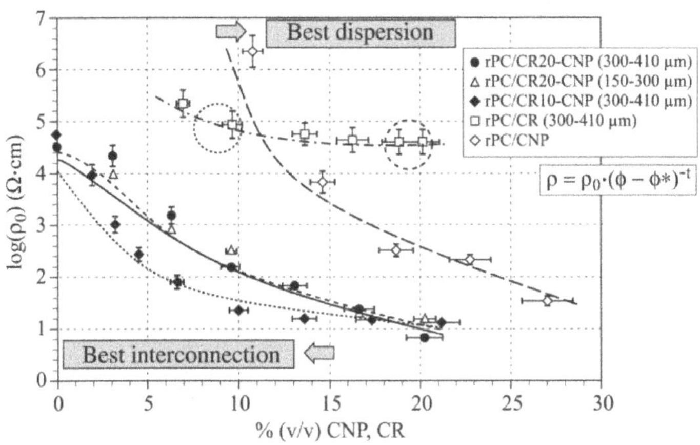

Figure 10.39. *Electrical percolation in rPC/CR mixtures depending on the CNP content [ZRI 06]*

By following the evolution of the percolation curve linking the white squares, we can see that the introduction of increasing fractions of CR from 7 to 20% v/v in the rPC matrix only decreases the composite's resistivity from 5.5×10^5 to 4.6×10^4 Ω.cm, which is not sufficient for the target applications (the CR is filled with 30% CNP, though, which should render it conductive). On the other hand, we can see that adding CNPs to rPC80/CR20 composites (curves connecting the black circles and the white triangles) singularly decreases their resistivity, which drops from 4.6×10^5 to 1×10^1 Ω.cm. However, comparing these two curves to one another also shows that decreasing the size of the CR grains from 300–410 μm to 150–300 μm does not yield a significant gain in conductivity. Yet if we decrease the fraction of CR from 20 to

10% v/v (curve linking the black diamonds), we can see that it is possible to reduce the resistivity of the composites by a decade for the same quantity of CNP.

These results can be attributed to the multi-scale structure of the conductor networks running through the rPC matrix. The different micro/nanofillers come together to form a double-percolation architecture, which is clearly illustrated in Figure 10.40. Thus, it is possible to create CPCs with only 10% v/v CNP and 10% v/v CR, whose resistivity is near to 1.5×10^1 Ω.cm, whereas in an equivalent rPC–CNP system (without CR – curve linking the white diamonds in Figure 10.39), the resistivity is 3×10^6 Ω.cm, i.e. nearly a million times greater. We can see that the CR has played a double role, increasing both the conductivity and the CNP exclusion volumes.

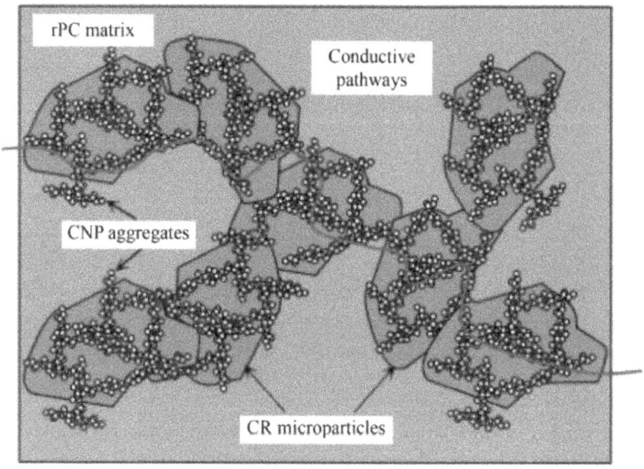

Figure 10.40. *Adsorption of the CNPs to the CR forms conductive pathways by double percolation [ZRI 06]. (For a color version of this figure, see www.iste.co.uk/hamide/polymers.zip)*

Now that a formulation of rPC/CR–CNP has been able to be optimized from the standpoint of electrical conductivity, it is possible to evaluate the "smart" properties of these materials, i.e. their capability, in particular, in self-regulating heaters, and for detecting vaporous volatile organic compounds (VOCs), which is discussed in the following sections.

10.7.5.2. *Thermoresistive effect of rPC/CR mixtures for self-regulating heating*

The ability of CPCs to facilitate ohmic heating with self-regulation of temperature comes from their heterogeneous structure and the nature of the components. It is not sufficient that the composite be conductive – a phase-change in the matrix must also cause the disconnection of the conductor network with a change in temperature, of significant amplitude. This is what is known as the positive temperature coefficient (PTC) effect, characterized by its amplitude I_{PTC} and a commutation temperature T_{com} beyond which the phenomenon is triggered. In Figure 10.41, we can clearly see that T_{com} is around 110°C, and that depending on the formulations, we can obtain a PTC effect whose amplitude lies between 1.5 and 4, which is perfectly acceptable for envisaging applications for self-regulating heating in the temperature range 120–140°C. It is also worth noting that the lower the CNP content, the higher the value of I_{PTC}, which is to be expected, because it is directly linked to an increase in the capacity for disconnection of the conductor network, as long as we are using fractions of fillers which are close to the percolation threshold.

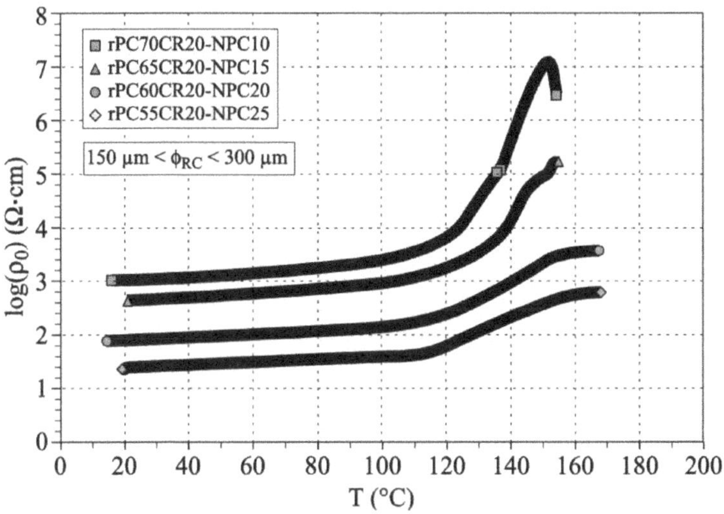

Figure 10.41. *Evolution of the amplitude of the PTC effect and the initial conductivity of rPC/CR–CNP mixtures depending on the CNP content for CR whose grain size is between 150 and 300 µm [ZRI 07]*

However, the most sensitive formulations (containing 10 and 15% CNP) are also those which give rise to a negative temperature coefficient effect above 160°C, which can lead to the destruction of the heating element (if it is used without an adjoining safety element). Conversely, the resistivity of the other two CPCs stabilizes at high temperature, up to 180°C. Regardless of the formulation chosen, it is necessary to make the heating element safe by surface cross-linking or by using a polymer phase with a higher phase-change temperature – these two strategies are commonly used in practice [FEL 01].

10.7.5.3. *Chemo-resistive effect of rPC/CR mixtures for detection of VOCs*

The constraints associated with the making of vapor sensors from CPCs are somewhat different from those which are required to make self-regulating heating elements. The voltages applied to the transducer materials are of a few volts only, whereas those which are used for energy dissipation by the Joule effect are several hundred volts. The physical phenomena associated therewith are therefore quite different, although the principle of variation of resistance caused by the disconnection of the conductor network remains the same.

Vapor sensors are based on the interactions of the VOCs being targeted for detection with the macromolecules of the CPC matrix. The partial dissolution which results from the diffusion of the VOCs disconnects the joints between conductive fillers, and results in an interpretable electric signal. This is exactly what happens when an rPC65/CR20-CNP15 transducer is exposed to toluene vapor, as we can see in Figure 10.42. This time, it is the formulations containing large-grain CR which are the most sensitive to the vapors, because their response is around 30% stronger. The signals are reproducible and stable. The response times are short (a few seconds) and the systems can operate at ambient temperature, which is a major advantage over their metal-oxide-based competitors, with require temperatures of over 200°C in order to function. We can just see a slight shift in the signal at the end of the second cycle, as it does not quite return to its initial level. This problem, which is indubitably connected to the accumulation of VOCs in the material, should be able to be solved by reducing the thickness of the transducers (a few mm in the case at hand).

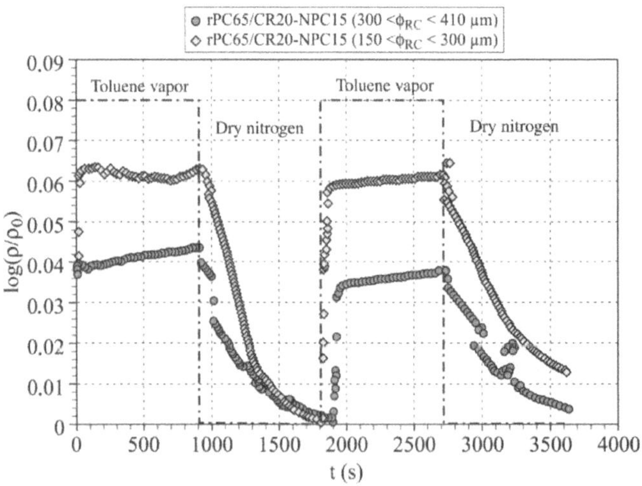

Figure 10.42. *Chemo-resistive behavior of rPC–CR–CNP mixtures in the presence of saturated toluene vapor at 25°C [ZRI 06]*

Finally, in view of the results presented above, we can say that materials formulated from polymer waste material PC and CR could very well be used as transducers for detection of VOCs for traditional molten processing of plastics, without any particular treatment of the waste products, which is indicative of a low cost.

10.8. Conclusion

In conclusion of this presentation on ecoplastics, let us recap the important points which have been made in the foregoing pages:

1) there is a genuine benefit of using ecoplastics because these new materials, while they cannot provide miraculous solutions to all problems, fit in with the logic of sustainable development which should help reduce the environmental damage done by the human race;

2) the quantity of plastic waste is constantly increasing, so there is an urgent need to find an acceptable solution to use that waste and avoid increasing the amount thereof; otherwise, there will inevitably be repercussions. Is it not a pity that the image of plastics should be so soured – plastics which have provided so many technological solutions and are continuing to contribute to our well-being – simply because we are not able to wisely manage their lifecycle and their environmental impact?

3) bioplastics provide a solution for an increasing number of applications. They are particularly appropriate for applications in packaging with short lifespan, but can also serve for making durable items, provided they are appropriately designed;

4) ecoplastics can be used to target functional applications with high added value. The examples chosen show that it is possible to create bioplastics with controlled biodegradability, and heating conductive composites sensitive to COVs, from plastic waste. In particular, CPCs offer genuine opportunities for upgrading of plastic waste because their requirements in terms of mechanical properties are not exacting.

10.9. Acknowledgments

For their technical contribution to this work, I thank Hervé Béllégou, Françoise Péresse, Dr. Thomas Divers, Dr. Karim Zribi, Dr. Isabelle Pillin, Dr. Alain Bourmaud, Dr. Eusebiu Grivei (Timcal), Jean-Charles Jégo (Self-Signal) and our financial partners, *Région Bretagne*, the Ministry for Research and Technology and the European Commission.

10.10. Bibliography

[ADH 08] ADHIKARY K.B., PANG S., STAIGER M.P., "Dimensional stability and mechanical behaviour of wood-plastic composites based on recycled and virgin high-density polyethylene (HDPE)", *Composites B*, vol. 39, pp. 807–815, 2008.

[BOU 07] BOURMAUD A., GROHENS Y., ZRIBI K., *et al.*, "Investigation of the polycarbonate/crushed rubber particle interphase by nanoindentation", *Journal of Applied Polymer Science*, vol. 103, pp. 2687–2694, 2007.

[BRA 99] BRANDRUP J., IMMERGUT E.H., GRULKE E.A., *Polymer Handbook*, 4th ed., John Wiley& Sons, New Jersey, 1999.

[CAR 99] CARROT C., GUILLET J., "Viscoélasticité linéaire des polymères fondus", *Techniques de l'Ingénieur*, vol. AM 3, no. 620, pp. 1–19,1999.

[DIV 04] DIVERS T., PILLIN I., FELLER J.F., *et al.*, "Starch controlled destructuration in diluted formic acid solutions followed by rheometry", *Starch*, vol. 56, no. 9, pp. 389–398, 2004.

[DIV 05] DIVERS T., Modification chimique de l'amidon par l'acide formique à l'usage d'additif pour polymères biodégradables, Doctoral Thesis, University of South Brittany, Lorient, 2005.

[DIV 07] DIVERS T., BALNOIS E., FELLER J.F., *et al.*, "The influence of O-Formylation on the scale of starch macromolecules association in DMSO and water", *Carbohydrates Polymers*, vol. 68, no. 1, pp. 136–145, 2007.

[DUV 04] DUVAL C., *Matières plastiques et environnement: recyclage, valorisation, biodégradabilité, écoconception*, Dunod, Paris, 2004.

[FEL 01] FELLER J.F., LINOSSIER I., LEVESQUE G., "Electrical properties measurement of conductive polymer composites (CPC) PBT/EEA-NC", *Entropie*, vol. 235, pp. 56–60, 2001.

[FEL 03a] FELLER J.F., GROHENS Y., "Bisphenol A poly(carbonate) recycling: rheological and calorimetric study, degradation", *Stabilization and Recycling of Polymers*, Prague, Czech Republic, 14–17 July 2003.

[FEL 03b] FELLER J.F., BOURMAUD A., "Following of rheological and calorimetric properties of poly(carbonate) after recycling", *Europe-Africa Meeting of the Polymer Processing Society*, Athens, 14–17 September 2003.

[FEL 03c] FELLER J.F., BOURMAUD A., GROHENS Y., "Bisphenol A poly(carbonate) recycling: rheological and calorimetric study", *Polymer Degradation and Stability*, vol. 82, no. 1, pp. 99–104, 2003.

[FEL 05] FELLER J.F., "Suivi de la déstructuration et la modification chimique de l'amidon par l'acide formique en vue de faciliter son association avec la poly(ε-caprolactone)", *Colloque MABIOLAC*, Roubaix, 16 November 2005.

[FEL 06] FELLER J.F., ROTH S., BOURMAUD A., "Conductive polymer composites: electrical, thermal, and rheological study of injected isotactic poly(propylene)/long stainless-steel fibers for electromagnetic interferences shielding", *Journal of Applied Polymer Science*, vol. 100, pp. 3280–3287, 2006.

[FEL 09] FELLER J.F., "Elaboration d'écoplastiques", *Stage pédagogique GFP. Formulation, mise en œuvre et cycle de vie des matériaux polymères*, Oyonnax, 2–5 June 2009.

[FEL 11] FELLER J.F., CASTRO M., KUMAR B., "Polymer CNT conductive nanocomposite for sensing", *Polymer Carbon Nanotube Composites: Preparation, Properties and Applications*, MCNALLY T. (ed.), Chapter 25, Woodhead Publishing, Cambridge, p. 848, 2011.

[GAL 97] GALLANT D.J., BOUCHET B., BALDWIN P.M., "Microscopy of starch: evidence of a new level of granule organization", *Carbohydrate Polymers*, vol. 32, pp. 177–191, 1997.

[GAU 98] GAUTRON P., "Valorisation et recyclage des déchets", *Techniques de l'Ingénieur*, A3830, 1998.

[GRI 05] GIRIJA B.G., SAILAJA R.R.N., MADRAS G., "Thermal degradation and mechanical properties of PET blends", *Polymer Degradation and Stability*, vol. 90, pp. 147–153, 2005.

[HAC 06] HACKWELL GROUP, Wood Plastics Composites in Europe, 4 235, 2006.

[HIL 49] HILDEBRAND D.J., SCOTT R., *Solubility of non-electrolytes*, 3rd ed., Reinhold, New York, 1949.

[HOL 04] HOLMGREN K., HENNING D., "Comparison between material and energy recovery of municipal waste from an energy perspective: a study of two Swedish municipalities", *Resources, Conservation and Recycling*, vol. 43, pp. 51–73, 2004.

[INC 00] INCARNATO L., SCARFATO P., DI MAIO L., *et al.*, "Structure and rheology of recycled PET modified by reactive extrusion", *Polymer*, vol. 41, no. 8, pp. 6825–6831, 2000.

[KAV 96] KAVANO Y., KESKKULA H., PAUL D.R., "Effect of polycarbonate molecular weight and processing conditions on mechanical behaviour of blends with a core-shell impact modifier", *Polymer*, vol. 37, no. 20, pp. 4505–4518, 1996.

[KUM 10] KUMAR B., Smart multi-reactive textiles obtained from eco-conductive polymer nanocomposite, Doctoral Thesis, Université de Bretagne Sud, Lorient, 2010.

[LAM 02] LA MANTIA F.P., *Handbook of Plastics Recycling*, Rapra Technology, Shawbury, 2002.

[LAM 08] LA MANTIA F.P., MORREALE M., "Accelerated weathering of polypropylene/wood flour composites", *Polymer Degradation and Stability*, vol. 93, pp. 1252–1258, 2008.

[LEI 11] LEI Y., WU Q., "Recycling engine oil containers to prepare wood-plastic composites", *Journal of Applied Polymer Science*, vol. 122, no. 2, pp. 964–972, 2011.

[MAT] MATERBI, http://www.materbi.com/.

[MAU 14] MAUPIN P., SPI, Lannion, http://lannion.cylex-france.fr/entreprises_11476292/solutions-plastiques-industrie.html, 2014.

[NAT] NATUREWORK, http://www.natureworksllc.com/.

[PAT 05] PATEL M., NARAYAN R.,"How sustainable are biopolymers and biobased ... reality", in MOHANTY A.K., MISRA M. DRZAL L.T. (eds.), *Natural Fibers, Biopolymers, and Biocomposites*, CRC Press, Boca Raton, vol. 27, pp. 834–851, 2005.

[PES 09] PESETSKII S.S., FILIMONOV O.V., KOVAL V.N., *et al.*, "Structural features and relaxation properties of PET/PC blends containing impact strength modifier and chain extender", *eXPRESS Polymer Letters*, vol. 3, no. 10, pp. 606–614, 2009.

[PIL 05] PILLIN I., DIVERS T., FELLER J.F., *et al.*, "Elaboration and characterization of starch/poly(caprolactone) blends", *Macromolecular Symposia*, vol. 222, pp. 233–238, 2005.

[PLA] PLASTICS EUROPE, http://www.plasticseurope.org/.

[PUY 14] PUYOU J.B., *SimaPro LCA software* was developed in the Netherlands by *PRé Consultants*, http://www.pre.nl/simapro/, and is distributed in France by *EVEA conseil*, http://www.evea-conseil.com/, 2014.

[RIV 02] RIVATON A., MAILHOT B., SOULESTIN J., *et al.*, "Comparison of the photochemical and thermal degradation of bisphenol-A polycarbonate and trimethylcyclohexane-polycarbonate", *Polymer Degradation and Stability*, vol. 75, no. 1, pp. 17–33, 2002.

[ROB 03] ROBIN J.J., "Recyclage des thermoplastiques", *Techniques de l'Ingénieur*, vol. AM 3, p. 832, 2003.

[ROQ] ROQUETTE, www.roquette.fr/.

[SCH 02] SCHOLZ H., PÖTSCHKE P., MICHAEL H., *et al.*, "Morphology and mechanical properties of elastomeric alloys from rubber crumb and thermoplastics", *Kautschuk Gummi Kunststoffe*, vol. 55, no. 11, pp. 584–589, 2002.

[SEL] SELF SIGNAL SIGNALISATION, http://www.selfsignal.fr.

[SOL] SOLVAYCAPROLACTONES, www.solvaycaprolactones.com/.

[SOY 09] SOYKEABKAEW N., NISHINO T., PEIJS T., "All-cellulose composites of regenerated cellulose fibres by surface selective dissolution", *Composites A*, vol. 40, pp. 321–328, 2009.

[SPE 86] SPERLING L.H., *Introduction to Physical Polymer Science*, Wiley & Sons, London, 1986.

[STA 06] STASSIN F., "Resources, conservation and recycling, bioplastics in the millenium: a look at technological and commercial developments", *BPRI and SRC Green Polymers*, Louvain-la-Neuve, 11 October 2006.

[SUN 04] SUN S.F., *Physical Chemistry of Macromolecules: Basic Principles and Issues*, 2nd ed., Wiley & Sons, London, 2004.

[VER 82] VERDU J., "Analyse physico-chimique des polymères", *Techniques de l'Ingénieur*, PE 3 760, 1982.

[VIN 03] VINKA E.T.H., RABAGO K.R., GLASSNER D.A., *et al.*, "Applications of lifecycle assessment to NatureWorks polylactide (PLA) production", *Polymer Degradation and Stability*, vol. 80, pp. 403–419, 2003.

[WÜN 97] WÜNDERLICH B., "Thermal characterization of polymeric materials: the basis of thermal analysis", in TURI E.A. (ed.), *Thermal Characterization of Polymeric Materials*, 2nd ed., Academic Press, New York, p. 385, 1997.

[ZRI 06] ZRIBI K., FELLER J.F., ELLEUCH K., *et al.*, "Conductive polymer composites obtained from recycled poly(carbonate) and rubber blends for heating and sensing applications", *Polymers for Advanced Technologies*, vol. 17, pp. 727–731, 2006.

[ZRI 07] ZRIBI K., Eco-composites polymères. Conducteurs stimulables obtenus à partir de matériaux recyclés : matrice poly(carbonate)/microparticules de caoutchouc, Doctoral Thesis, University of South Brittany, Lorient, France and ENIS, Sfax, Tunisia, 2007.

[ZRI 07] ZRIBI K., ELLEUCH K., FELLER J.F., *et al.*, "Eco-plastics: morphological and mechanical properties of recycled poly(carbonate)-crushed rubber (RPC-CR) blends", *Polymer Engineering and Science*, vol. 47, no. 11, pp. 1768–1776, 2007.

11

Thoughts about Plastic Recycling. Presentation of a Concrete Example: End-of-Life Polypropylene

11.1. Why do we use plastics?

Because of its intrinsic qualities – lightweight, sturdiness, etc. – plastic is becoming an ever-increasing presence in our daily lives, and gradually replacing many traditional materials, such as wood, metals, pottery, glass and so on.

Nowadays, plastics have very numerous and very diverse applications. The use of plastics on a massive scale is a relatively recent phenomenon (less than a century). Yet even now, there is a long list of new functions which can only be served by plastics.

Some such functions are common and day to day; others are high-tech. Packaging, by far, represents the majority of plastic use. Other applications for plastics notably include functions in clothing, automobiles, bicycles, household goods, construction, paving, toys, mobile telephones, computers and medical accessories: the uses of plastic extend, quite literally, from a small plastic bag to gargantuan airliners.

A multitude of different types of plastics may be used, depending on their characteristics (rigidity, temperature resistance, fluid-, UV- or shock resistance, etc.). The main polymers used are polyolefins (around 40% market share in all industrial sectors), followed by styrenics, polyamides, PET, PVC, etc.

Chapter written by Frédéric VIOT.

The automobile sector has experienced that same evolution in terms of the materials used in composition of the products: while in the 1970s, plastics represented only 3–5% of the weight of a typical vehicle in Europe, today they represent around 15%, with a regular growth curve projected to continue until 2025.

The distribution is similar, with around 50% polyolefins, styrenics, polyurethanes as well as a small proportion of thermo-hardened plastics (such as polyesters with fiberglass fillers), which is predicted to increasingly replace structural parts or bodywork panels which are currently made of steel. This will reduce the weight of the vehicle and, consequently, it is hoped, its CO_2 emissions.

11.2. What are the regulations governing the "end of life" of plastics?

A variety of European legislation documents impose recycling requirements for certain percentages of end-of-life (EOL) products in various industrial sectors. The results in France (which we shall use as a case study) are not hugely satisfactory, as shown in Table 11.1.

Avenue/source	Recycling rate	Objectives
Household waste and similar	Material and organic recycling rate: 37% in 2013	45% in 2015 60% in 2020
Household packaging	67% in 2013	75% in 2012
Industrial waste (apart from building and civil engineering, agro-food industry, agriculture and specific activities)	70% in 2012	not really defined (75%?)
End-of-life vehicles (ELVs)	In 2012 Reuse and recycling: 79.8% Reuse and recovery: 81.5%	85% of material from ELVs must be recycled or reused and 95% recovered in 2015

Table 11.1. *End-of-life scenarios for some common plastics*

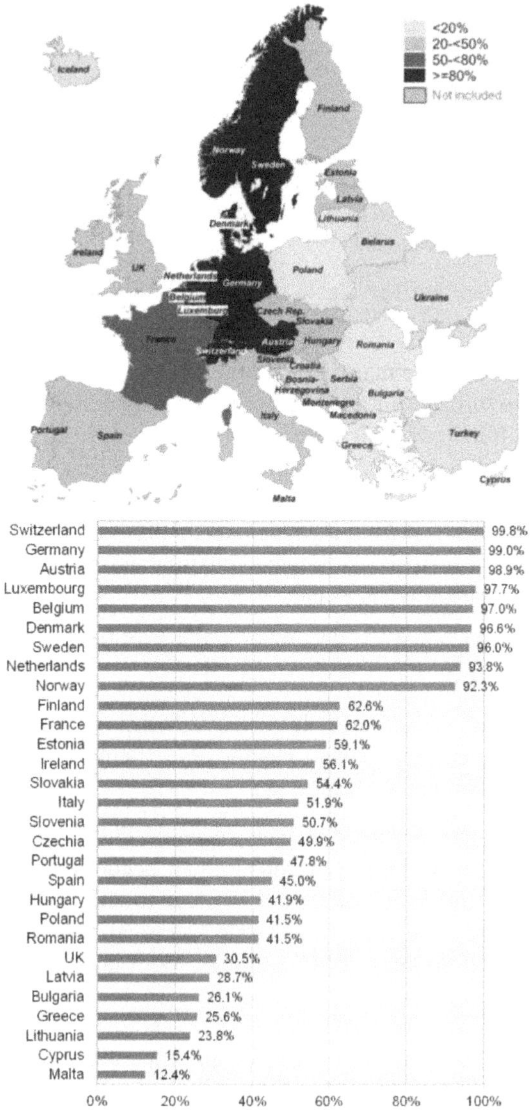

Figure 11.1. *Overview of plastic recycling in Europe (Plastic Europ 2012)*

Generally speaking, France ranks very poorly in European statistics on waste recovery rates – even in comparison to countries which are economically, socially and demographically similar to it, such as Germany, Belgium, Spain, Italy and the

Netherlands. In relation to these countries, France, which was in third position in 1997, slipped to last position (between 2008 and 2010), and remained constantly in the last place in terms of plastic recycling between 1997 and 2010, as shown in Figure 11.1 (Plastic Europ 2010).

As we can see from these graphs, even the best performers only manage to recycle 40% of material (and even then, barely), with the rest being used for energy recovery, for a variety of reasons:

1) plastics which are not recyclable or are difficult to recycle (thermo-hardened plastic, foam, etc.);

2) plastics which undergo major degradation during their use (polyamides which prolonged exposure to heat, oil, etc.). When these polymers reach end of life, they have all of these differences, which are accentuated by their different lifecycles (the plastics used in a washing machine do not have the same lifecycle as those used in a domestic iron or a dashboard or a yoghurt pot); they are not subjected to the same aging elements (UV, heat, fouling, liquid, chemical products, etc.). Exposure to such elements leads to the scission of polymer chains, and the apparition of various molecules due to degradation and usage;

3) plastics containing additives which are now banned (e.g. styrenics with certain bromine-based additives);

4) plastics in small quantities which are difficult to identify, multi-layered plastics or plastics bonded together, making it difficult to retrieve the original polymers, foam- or film-coated plastics, etc.

There is no shortage of examples that we could cite; sorting technologics do exist, but they need to be improved and used in accordance with the particular source at hand. It is fruitless to put sensors in place to eliminate bromide plastics if we have NO polypropylene (PP) from waste electrical and electronic equipment (WEEE or W3E) in the sources of EOL plastics. W3E does often contain bromine, unlike other sources.

All these difficulties account, in part, for the low rates of plastic recycling, but they are not the only reasons; there are also numerous barriers:

– *the sources of EOL material are not constant and reproducible, in terms of both quality and quantity.* Overall collection is not well organized, and, at present, EOL polymer materials are only made available for recovery on the initiative of a few organizations driven by regulations or legislation, or a few private companies;

– *the batches of materials collected come in the form of heterogeneous mixtures, which may contain impurities* that can lead to a degradation of the properties, or indeed the prohibition of the material (as is the case with certain additives containing lead or other heavy metals). Plastics are, in general, not very stable, so in order to limit degradation of the material, at the stage of manufacture, transformation or use, additives are mixed in (anti-oxidants, fluidifiers, nucleating agents, anti-UV products, anti-leaching agents, fireproofing, modulus modifiers, etc.) which are specific to the intended final use of the product: we will not find the same mixture of additives in external or internal car body parts, and the difference is more significant still between the automobile and electrical industries, or – within the same industrial sector – between major appliances (white goods) and small appliances (brown goods), not to mention household waste (yoghurt pots, fruit punnets, etc.);

– *the industrial sector of EOL general equipment* (electrical goods, construction, automotive industry, furniture, etc.) generally comprises two main types of actors:

- large corporations, capable of processing huge volumes of material, with significant means of investment, but also with profit margins to consider, which orientates the sorting toward materials present in larger volumes and with higher economic value (plastics generally do not fall into this category). In such cases, the time taken to change tack with regard to sorting is inevitably longer;

- smaller companies – often family-run SMEs – which are more reactive and receptive to change, but have lesser investment capabilities.

Ultimately, both these groups tend to specialize in ferrous and non-ferrous metals, which are at the heart of their profession (they are capable of recognizing the quality of a given type of steel almost at a glance, but would sort plastics by color, for example).

The final barriers to the development of recycled polymers (not to mention social acceptance or traceability) lie with the compounders and the end users.

The principle is simple: if we wish to expand the market for recycled plastic, we have to win back a market share from virgin plastics, which have very good technical performances. It is *obviously* necessary to supply industries which require "low-grade" plastics, but if we wish to develop the avenue of recycled plastics, it is essential to make recycled polymers whose performances are similar to those of virgin materials, thereby helping to increase volumes and margins.

11.2.1. *Compounding of collected plastic*

Now we need compounders capable of modifying the batches of materials arising from the collection and sorting operations, whose performances are not sufficient to satisfy all that is required of plastic products, in terms of both the content of the technical specifications of the finished products (mechanical strength, heat resistance, fire- or aging-resistance, etc.) and the esthetic level. These modifications are made by mixing in additives, fillers, by reactive extrusion or new processes. This requires time, and also skills which not all compounders in Europe possess ... if only because the virgin plastics used in 2013 are not the same as those used in 1990 (the average age of an ELV in Europe is 13 years); for example, in terms of fluidity, modern materials are between 5 and 10 times more fluid, for reasons of cycle time or machine power.

If no chemical reaction takes place or if additives are not mixed in at the time of granulation, it is not possible to mold a part using recycled material. Therefore, it is clearly important for recycling companies to acquire skills in chemistry in order to modify the recycled polymers and lend them characteristics which nearly match those of the virgin polymers.

In addition, the quality standards and the end users' expectations require the recycled materials to have similar qualities in terms of reproducibility as do the virgin materials.

11.2.2. *Tools for working with recycled plastics*

The available tools are not always helpful for transforming recycled material.

The design of plastic products and of the manufacturing tools associated therewith is closely linked to the performances of the material used.

Today, professionals in the field do not have enough perspective in hindsight or enough data about the specific rules of design relating to recycled materials arising from EOL processing. What they tend to do is design products with a view to using virgin material, and the recycled material needs to adapt to the constraints of the existing process and tools, whereas it should be the other way around in order to increase the use of recycled materials.

The problem is the same for the other stages of processing which take place after the process of transformation (injection, compression, extrusion, assembly, stamping, bonding, painting, etc.). It is crucial for transformers/machinists to acquire skills in the transformation and usage of recycled materials.

11.3. Armed with these observations, how did we proceed?

We attempted to apply all of these good practices for ourselves.

One of the first things we noted was that it was necessary to secure a steady supply of materials (i.e. of EOL polymers). This was done by setting up *Plastic Recycling* (a 50/50 joint venture between *Plastic Omnium* and *Derichebourg*). As Derichebourg is the European market leader in the crushing of EOL products (ELVs, W3E, furniture, etc.), in theory it is the largest "owner" of EOL plastics.

As, in the final analysis, plastics-sorting is not central in Derichebourg's activities or interests, we began by developing skills in the characterization of sources and sorting, with the help of university-based partners in multi-player projects.

The plastics of interest are mainly PP. In theory, these polymers are easier to sort because they are the only ones which float alongside polyethylene (PE) and foams, wood and other pollutants. Fairly rapidly, we devised simple solutions to eliminate a large proportion of the foam and wood. A simple sorting process to yield only PP without PE seemed far more complicated industrially. Based on the same principle as is used for vehicle fenders with PP/PE copolymers, we envisaged retaining this mix of PP and PE, but rendering them compatible to obtain a PP/PE compound similar to a copolymer in terms of shock resistance, fluidity and injectability.

In conjunction with the *Laboratoire ingénierie des materiaux polymères* (Polymer Engineering Lab) in Lyon (IMP-INSA/Lyon1), we worked and found a compatibilizing agent which yielded a compound similar to a copolymer obtained by reaction during polymerization. We believe that the compatibilizing agent creates a bond between EPR and PE and EPR and PP (see Figure 11.2).

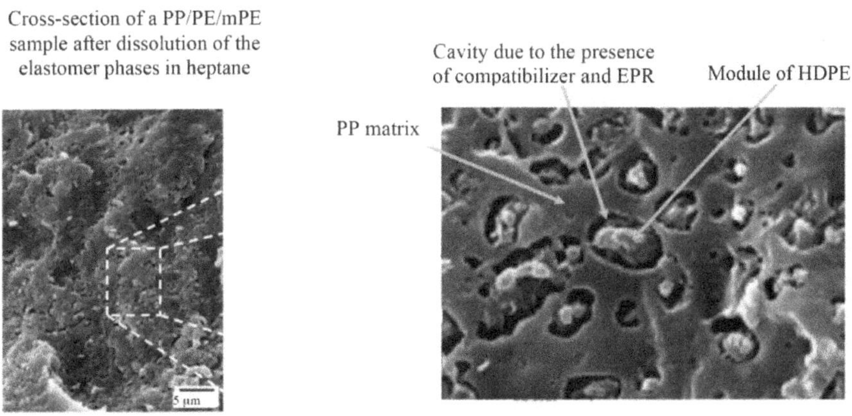

Figure 11.2. *Cross-section of a sample of recycled PP and PR, and compatibility*

In the wake of this initial result, project *Greenlene* was launched at the start of 2009. Thus, we "still" had in mind to:

– industrialize sorting;

– industrialize the extrusion process;

– sell the idea to a customer and manufacture acceptable parts.

We already had the ability to make molds, assemble parts, paint pieces, etc., all to a level of quality compatible with the requirements of automobile manufacturers.

The idea was very quickly sold to a manufacturer and – to some extent through over-confidence and to a large extent through lack of time – we neglected the necessary points evoked earlier:

– indeed, our partner was not ready to undertake industrial sorting of polyolefins, and even less so with the reproducibility necessary for a latter-stage industrial process;

– plastic recycling was beginning to acquire the skills needed for reactive extrusion;

– Plastic Omnium believed that, as with the virgin material, Greenlene could be easily adapted to the needs of the project and above all we had not anticipated the technical difficulties in satisfying the quality requirements both of the PO factory and of the final customer – Peugeot Ltd.

We cannot easily give a technical explanation of the various solutions found, because many of them are protected as intellectual property. However, we can give the following examples of the problems encountered during this development which, let it not be forgotten, yielded a worldwide start: "industrial creation of a painted car body part from 100% recycled material".

Figure 11.3. *Rear fender of the Peugeot 208*

11.3.1. *List of problems encountered*

To begin with, we had not given sufficient training to the "upstream" actors (those collecting, sorting and extruding the materials) about the difficulty of their task and the influence of their results on events further along the chain (injection, processing, appearance after painting, shock resistance, paint retention, etc.), and it is very important for each of the actors to be aware of their own role in this cascade-type process.

Similarly, we had not made the "downstream" actor sufficiently aware of the drastic impacts that they would cause by changing the process or the characteristics (to double the fluidity of a polymer or to decrease its shrinkage has significant impacts on the extrusion process).

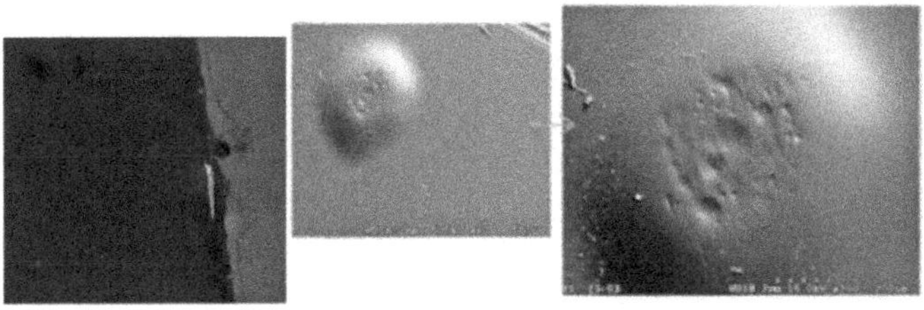

Figure 11.4. *Example of faults in appearance after painting*

We had the good fortune to be completely integral (from the sorting to the delivery of the parts) and to have the scientific support of the IMP lab to anticipate these problems, but overtaken by the progress of a project that had already been sold to the customer, to manufacture the rear fender for the Peugeot 208, we had to forego certain necessary steps and resolve the problems during the development stage (design, mold, material, injection, surface treatment and painting to mention only those of greatest influence).

Having overcome these difficulties in terms of training and skills required, the technical points resolved were as follows. There is often significant interplay between these points:

– composition of the input materials – a concentrate of polyolefins: creating a mix of various sources with similar and reproducible composition, with detailed characterization of the components and establishing of ranges;

– modification of fluidity: controlled attacking of the PP by peroxide and rapid stabilization to block the fluidity and avoid the formation of degradation products;

– controlled shrinkage: addition of a specific talc into the twin-screw extrusion mechanism with a suitable screw profile;

– reproducibility of batches: putting in place a system for continuous control of the twin-screw machine + quality control and traceability assurance;

– molding: modification of the ejection system, reinforcement against stripping, specific injection system;

– injection: control of the injection process by a control system using opening/closing of nozzles, pressure, etc.;

– shock-resistance: inclusion of additives which do not alter the paint retention or fluidity;

– appearance after painting: specific process of washing and extrusion (in addition to the individual filtration). Inclusion of additives to limit the degradation of the polymer;

– adherence to the specifications on painting: flame testing, inclusion of additives during extrusion.

11.4. Conclusion

Four years after the launch of the Greenlene project and 18 months after the launch of the 208, it can be considered that this project was, technically, a success.

Many of the technical innovations are classified or in the process of being classified. These innovations could serve for other recycled materials, but some could also be used for virgin materials, the comprehension of the phenomena at play was so far advanced.

Ultimately, though, in spite of the improvements made to its manufacture and transformation, Greenlene remains a material with a more restricted "processing" window than a virgin equivalent.

The upstream process (sorting, concentration, extrusion) still needs to be industrialized in order to truly transform this experiment into a *bona fide* industrial success (though, that said, other Greenlene fenders are now being made). Too many uncertainties about the quality of EOL inputs, the handling of the upstream extrusion processes and reactive extrusion mean that ultimately many batches are rejected before reaching the injection factory to be made into painted parts.

The downstream process (design, assembly, molding and painting) has also been greatly improved upon in the wake of this project. The strong points are the control of the injection process and the understanding of certain paint-related phenomena. Here again, though, points still remain to be clarified and understood in order to confidently envisage the generalization of Greenlene for the widespread manufacture of car fenders (without spending more hours on it than would be spent on a project using virgin material).

At present, it is impossible to predict what Greenlene's future will be. Indeed, significant efforts still need to be invested in obtaining a material that is truly similar to a virgin plastic, from a technical point of view but above all from a financial one. The end customer (i.e. all of us) is not prepared to pay more money for an item (even temporarily) because it is recycled or made of partly recycled material. A great deal of research remains to be done in this area in order to achieve a competitive stance. Otherwise, recycled polymers will remain restricted to "low-grade" applications, the processes in Europe will have difficulty in developing and the resources will continue to be siphoned off to countries where costs are low or sorting is done manually for applications to low-cost manufactured items.

12

Recyclable and Bio-based Materials Open Up New Prospects for Polymers: Scientific and Social Aspects

12.1. Introduction

Polyvinyl chloride (PVC) was invented less than a century ago. It was the first synthetic polymer used for packaging (plastic bottles, punnets) and is still widely used today in construction. Despite consuming a tiny part of petrol and gas (approximately 4% [HOP 09]), polymers have widely contributed to modern conveniences of the 20th and 21st Centuries with regard to packaging, hygiene, medicine, housing, transport, etc. Their low density makes them the components of choice in the automotive and aerospace sectors, providing more fuel to be consumed and new designs. They are used as materials, i.e. they undergo a series of transformations before obtaining the final shape of an object or a part with specific properties required for the thermoplastic polymer (which can deform and be reshaped in heat), or a thermoset polymer (whose chains form a non-fusible three dimensional matrix). Moreover, studies show that some complex high-performing materials can be produced from renewable resources (starch, cellulose, poly(lactic acid) (PLA), etc.) [MUL 13].

Whether polymers are produced from oil or biomass, they possess numerous qualities. The energy required for their production (simultaneous to their polymerization in the case of thermoset polymers) is much lower than that required

---

Chapter written by Valérie MASSARDIER.

for metals, glass and ceramics. In 1998, K. Mulder [MUL 98] already considered that, with regard to their lifecycle, polymers were a good option compared to heavier materials such as glass or steel. In 2001, I.S. Arvanitoyannis suggested that substituting plastic packaging with paper or glass would cause a 100% increase in energy, a 400% increase in the amount of raw materials consumed, a 150% increase in the volume of waste and not to mention a 100% increase in the price of packaging [ARV 01].

Most regular polymers (made of oil) are highly durable and can be recycled, which results in the production of less virgin material and therefore reduces the significant environmental impacts of petrochemicals (depletion of natural resources, transport of oil, pollutions generated by refineries). After use, polymers become increasingly collected, sorted and recycled for ever more ambitious applications. However, do not underestimate the problems associated with the recycling of polymers, responsible for the slow start to this process. At the beginning of the millennium, only 19% of polymers were recycled, which is low compared to the percentage of steel (100%), aluminum (27%) and cardboard recycled (27%) [BOC 06]. In 2011, with 60.9% of post-consumer plastic waste recovered, France was within the European average (59.1%). Both in France and Europe, two-thirds of plastic waste come from the packaging of products with short-lifecycles. The rest comes from building and construction (4.7%), automotive sectors (7%), the electronic and electrical equipment sector (EEE) (4.9%) and finally furniture, household appliances, agriculture, sport and healthcare (20%). In 2011, the overall recycling rate of plastic packaging reached 66.2%, with 23.5% accounting for recycling and 42.7% for the level of energy recovered [ACT 12]. As we will see, research and development studies have opened up new prospects for recycling.

With regard to bio-based polymers, made of biomass (for example by biorefinery processes), they appear to slightly reduce the effects of heavy chemicals. Nevertheless, since they are produced from intensive agriculture, usage of numerous fertilizers and phytosanitary products, their impacts upon the environment are not necessarily much less than their oil-based counterparts [WEI 07]. Their main advantages are much greater when derived from agricultural or forestry waste and appear to reside in their renewability and the fact that they can trap carbon dioxide [ALV 12]. Bio-based does not necessarily mean biodegradable. Moreover, is the biodegradation of fine particles, which can produce toxic metabolites, a good outcome? Without recycling, we would, for example, accelerate deforestation [MUL 13]. By giving materials many lives, even bio-based materials, we can reduce pollution, carbon dioxide emissions [FUJ 12] and the consumption of raw materials.

Originally, research aimed to find new polymers or copolymers as well as to improve their performance and manufacture, many other criteria have since been added (lightening, barrier properties, recyclability, toxicity (emission of volatile organic compounds (VOCs), dissemination of hazardous additives, etc.). One of the current challenges is to prepare for a sharp increase in the production of polymers in a world where oil and fossil fuels are becoming rarer. Thus, J. Ravenstjin predicted that the usage of synthetic polymers would reach 250,000 kt/yr in 2000 and 1,000,000 kt/yr in 2100, which would correspond to a 25% increase in the production of oil. In this whole estimation, polyurethanes should represent 35% of the biopolymer market [RAV 11].

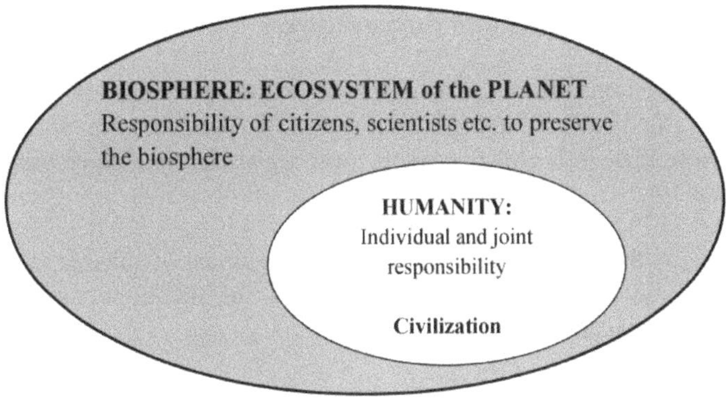

Figure 12.1. *Interactions between human activities and behavior with the biosphere*

In this context, there are two pathways to be explored:

1) after-use recovery of increasing waste flows;

2) usage of renewable resources produced by plants (fibers, starch, etc.) or animals (chitosan extracted from the carapaces of crabs) which can be used as a substitute for oil in practically all applications (Figure 12.1).

Of course, as we will see with recycled and bio-based polyamide blends [LOU 13], it is wise to associate them both. It is also necessary to develop recyclable bio-based materials [MUL 13] and develop the use of natural additives [AMB 11].

It is with sustainable approaches that we have become aware of the strong relationships between humanity and the biosphere, which form part of numerous research and development projects (Figure 12.1). In this context, it is important to consider the overall raw material–process–usage–sustainability sequence, which can lead to compromising situations. With recycling, many criteria essential for its recovery must be satisfied. Firstly, we must start with a relevant eco-design, avoiding combinations of poorly compatible materials during the recovery of waste. Then, as it is often difficult to obtain pure materials, the sorting must provide mixtures that can be widely used, with the help of well-chosen ideas and processes. With regard to these multicriteria and multidisciplinary approaches that lead to complex situations, we must try to find the best compromises, by having properly weighed up the advantages and disadvantages of the proposed scenarios.

12.2. Resources

12.2.1. *Mechanical recycling with energy recovery*

In Europe, end-of-life vehicles (ELVs), waste of electrical and electronic equipment (WEEE) and packaging are subject to legislations trying to improve their recycling. Note that, in a global market, local legislation has repercussions on the whole planet.

The fast growth of recycling helps in the development of suitable industries that provide reclaimed materials that meet ambitious specifications such as those of technical automotive parts [MAS 08, CAS 11].

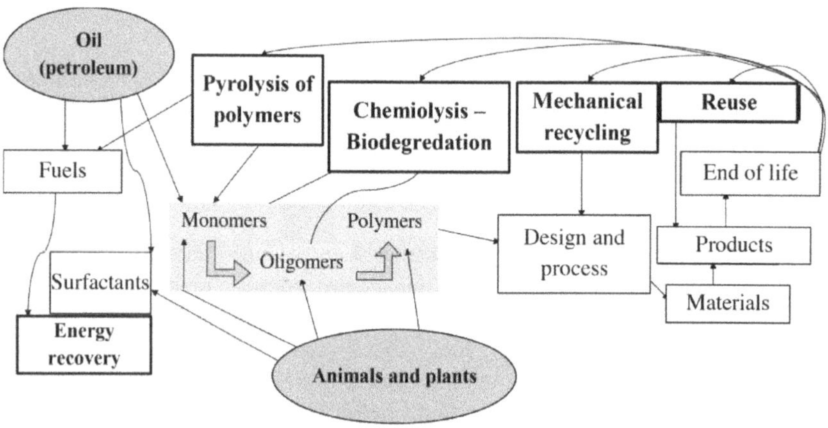

Figure 12.2. *Different stages in the lifecycle of polymers from oil or animal or plant biomass*

Polymers can be recycled using different processes (Figure 12.2), the most economically and environmentally interesting being the reprocessing and remelting of thermoplastics and fine crushing of thermoset plastics (which do not melt and cannot be reused after modest heating) to use them as loads in polymer formulations, for example. In what follows, it is mainly recycled thermoplastics which we will discuss. Whereas it is relatively easy to recycle metals, recycling plastics is much more tricky. In fact, plastics may degrade throughout their lifecycle (sensitivity to

UV, thermomechanic cycles for their reuse). Moreover, they are often used in the form of rather complex formulations to be recovered as materials. Although deconstruction stages, particularly ELV, can increase recycling potential [SCH 13], deconstruction is often considered as poorly profitable. Current sorting technologies separate materials mainly by density (sorting by flotation) and chemical structure (spectroscopic sorting), often without taking into account the nature of formulations (rheological properties (Figure 12.3), nature of additives etc.), which can prove to be restrictive. The purer the material desired, the less is recovered. If we simply sort by flotation, we recover polymers that are more or less mixed. Most polymers are immiscible, which results in poor mechanical properties if the interfaces are not improved using additives, called compatibilizers, which act to uniformly disperse the minority phase in the majority phase (matrix) and improve "dispersed phase–matrix" interfaces to obtain good mechanical properties. We therefore cannot think about mixing them without previously studying the "formulation–process" combination [MAS 02]. We must therefore choose a twin-screw extruder, which will help to disperse the phases by shearing the matter, with an adapted screw profile. A relevantly chosen process and formulation compatibility [MAS 08, CAS 11] will therefore render mechanical properties to be obtained and should be compatible with high value-added applications. The recycling of polymer blends is the next major challenge [HOP 09].

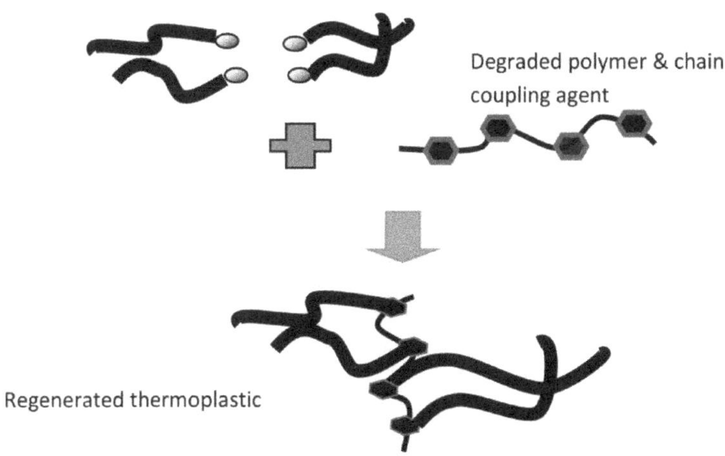

Figure 12.3. *Example of thermoplastic recovery*

For example, because of their close densities, polypropylene and high-density polyethylene can be recovered in the same fraction. However, these polymers often present antagonistic behaviors during the thermomechanic cycles required for their implementation. These cycles lead to breakages in polypropylene chains and reticulating reactions of polyethylene, which can limit their capacity to have several lifecycles. This problem is often complex and solved with relevant "formulation–process" combinations.

In the case of additives such as brominated flame retardants, whose toxicity is proved, spectroscopic sorting allows materials containing these toxic additives to be removed from the recycling process. These will be specifically treated. When the toxicity of chemical products is questioned by the European Union REACH regulation, we find that recycled materials are poorly defined, with additives often appearing on lists of problematic molecules. Furthermore, the chemical structure of organic additives may alter, particularly under the effect of oxidation and UV rays.

Nevertheless, regarding the prospect of a second lifecycle, the properties of recycled polymers, restabilized using additives, are often sufficient for a certain number of applications and recycling seems to be an alternative resulting in fewer negative impacts than the synthesis of virgin resins. Thus, Azapazic *et al.* [AZA 03] show that, with regard to the use of virgin resins, the mechanical recycling of plastic packaging waste (bottles, films) decreases energy consumption, climate warming, acidification and eutrophication of water bodies. With regard to the use of virgin materials, the use of "post-consumer" materials is also advantageous from an economic point of view.

However, with regard to inorganic materials that have the possibility of being recycled an infinite number of times, sorting limitations, ageing, oxidation and the evolution of polymer chains will limit their number of lifecycles. The latter can be increased by diluting the recycled material in the virgin material. We can also recover some polymers such as polyethylene terephthalate (PET) (polymer used to make most drinking bottles, particularly fizzy drinks) which tends to degrade as chains break during recycling operations. We are therefore going to associate chemistry with the process (reactive extrusion) to couple macromolecules with chemical agents that react with both extremities of the chains [LAC 05] (Figure 12.3). Coupling agents that can efficiently regenerate PET by reacting with –OH and –COOH end groups are molecules that include diisocyanates, dianhydrides, bisoxazolines, carbodiimides and phtalimides, for example.

However, when the recycling limits are reached for the material, it is therefore wise to cut chains by pyrolysis or chemiolysis, to recover monomers or oligomers from them. Another rational solution could also be to resort to incineration with energy recovery. So, in the 1990s, there was already much debate within scholarly societies about whether it was necessary or not to develop technologies to recycle polymers; incineration with energy recovery appeared to be a potentially relevant choice, with regard to the extensive use of carbon-based energies.

Moreover, energy recovery is probably the best solution for very contaminated and dispersed plastics (for example, the yoghurt pot which is a multi-layered material, with low mass and whose residual content after consumption can be significant). It is true that the high calorific value of plastics helps regulate the incineration of household waste, improve the quality of effluents (fewer micropollutants and unburned compounds in slag) and conserve oil used in the production of heat and electricity. Most European cities are equipped with urban heating networks, which could become possible due to the incineration of urban waste.

In addition to local communities, cement plants use alternative fuels corresponding to fractions of waste where the organic phase has a high calorific value. Thus, cement manufacturers have agreed to use thermoset composites to hold the raw materials of cement.

For relevance, recycling must become incorporated into lifecycle objects and be anticipated during the design of polymers and objects. Thus, to facilitate recycling, sustainability and recyclability criteria must be considered in preliminary studies. In particular, we try to avoid associating polymers that cannot be recycled together. For example, to make recycling easier, we can try to substitute some materials for others. Therefore, it is interesting to develop nanocomposites or coatings for supple food packaging (required functions: mechanical and barrier properties to gases and aromas) that can replace multilayers that are difficult to recycle. For some applications such as agricultural films for example, biodegradable polymers are interesting, because having already been used as films (replacing mulch), they can also be used as fertilizers. Formulations or associations of very complex materials pose problems for recycling.

With regard to the environmental impacts of energy consumption and emissions of pollutants and carbon dioxide, it is generally accepted that recycling is beneficial whatever the material (glass, steel, aluminum, paper, cardboard). Environmental benefits have also been verified for polymers. In the case of PET, total substitution of virgin PET with recycled PET reduces CO_2 emissions by 27% [HOP 09].

The recycling of plastics must be designed in such a way that it anticipates the end of life and makes progress regarding circular economy objectives. In particular, attempts are being made to recover all by-products and use loop recycling, regarding the scarcity of petroleum products and minimizing environmental impacts, by considering the end of life. Bio-based polymers also offer interesting prospects, with regard to the aims to rationalize resources and limit the impacts of industrial activities.

12.2.2. *Bio-based polymers*

To deal with depleting oil reserves, one possible strategy is to produce bio-based polymers, whose current sustainability is generally lower than their oil-based counterparts, but interestingly come from renewable materials, often from plants [ALV 12]. Polymers made of plants can be used as bags at checkouts, food punnets, surfactants or even as reinforcements (flax, hemp, wood, etc.).

The plant kingdom may have future potential in the fields of chemistry, energy and renewable materials. Non-food crops prevent the annual importation of hundreds of thousands of tons of oil, and, in the field of macromolecules, they can provide natural monomers, surfactants, stabilizers and polymers.

Generally speaking, numerous synthons can be extracted from biomass. One known example is that of ethylene, produced from the dehydration of ethanol which is a very common product of fermentation. Another example is 1,3-propanediol, which is a monomer used as a building block for the production of polymers such as polyesters and polyurethanes. Several industrial processes have studied its production by fermentation with the aim of producing it directly from inexpensive plant raw materials (starch or sucrose). To synthesize polyamides and polyesters, we also aim to produce α,ω-dicarboxylic acids by the biological conversion of esters from vegetable oils.

Biopolymers can be directly obtained [KOB 03] from hemicellulose in corn or starch. Starch, a major polymer from arable crops, is a good candidate for the production of biodegradable packaging or objects. If combined with a compound of wheat straw, its potential applications could increase.

Over the past few years, the production of polymers from renewable resources has rapidly increased at the industrial scale. Cargill-Dow has an industrial plant for the production of biodegradable PLA, which has a high-molar mass [DRU 00]. Major chemical companies and downstream industries are all preparing for these changes [MAS 06].

Non-ionic emulsifying surfactants, such as polyglycerols and polyglycerol esters, are used in formulations in the cosmetic and agri-food industries. These surfactants are produced from glycerol, a co-product of the manufacture process of diester, a biofuel produced from oilseeds for diesel engines. Epoxidized soybean oils are good plasticizers and stabilizers of PVC and may be considered as a substitute for phtalates [BEN 10]. Vegetable antioxidants can also be considered as stabilizers [AMB 11]. Bio-based furanic reactive diluants can be considered as a substitute for styrene [SAD 12].

Materials can also be partly produced from plants such as oil-based plant fiber/polymer composites. The interest in plant fiber resides in their low cost, their low abrasiveness (toward implementation machinery) and particularly their low toxicity compared with synthetic fibers, which are irritants to the respiratory tract [MOH 00]. Hemp, hessian, sisal or flax are largely used in cars in Germany. During 2005, 3.5 kg of natural fibers were used per automobile passenger in Germany, particularly in compression molded composite parts [MON 05].

Generally speaking, increasing numbers of natural polymers and wood-based composites are being marketed, which, in some cases, helps balance economic interest and reduce greenhouse gases [DOR 06].

Although "plant" is now a word associated with "carbon dioxide sinks" and "reduction in the greenhouse effect", developing intensive agriculture (high usage of fertilizers, pesticides, motorized farming machinery, etc.) to provide raw materials to the industry is not a clear alternative, as it is often presented. On the other hand, bio-based polymers must be extracted from plants or animals (chitosan is extracted from the carapaces of crabs), which requires chemical treatment. In most cases, they are also chemically modified to alter their hydrophilic properties. Without chemical treatment, most natural fibers would be too hydrophilic to strengthen polyolefins, for example.

Some studies analyzing the lifecycles of materials have compared oil-based and bio-based polymers. Thus, even though natural polymers are renewable and have an interesting lifecycle with regard to greenhouse gas emissions [DOR 06], they are not necessarily an alternative without disadvantages. By considering four categories of environmental impacts (non-renewable energy consumption, climate warming, eutrophication and acidification of water bodies), M. Weiss *et al.* [WEI 07] showed that packaging made of bio-based polymers was slightly better than with typical oil-based polymers.

The current trend is to recover agricultural and food waste. Thus, certain fibers (palm tree) used in composites [KAD 06] are examples of waste that were not recovered. Biodegradable waste can also be treated and inoculated with selected bacteria to produce solvents (ethanol), monomers (lactic acid, ethylene) which can subsequently polymerize [SAK 04]. Here, we have an example of industrial ecology where the polymer industry recovers plant waste. Hemp is an interesting plant that is fully recovered, grows easily, with very little water, without fertilizers or pesticides. In this case, there is a clear advantage of using natural fibers, especially since the chemicals required for treatment have little environmental impact. Developing the recovery of lignin and cellulose (produced by trees), modified to control hydrophilic properties, may aid the sustainable management of forests and the recovery of small trees following clearing of forests (to select future stumps to be used in construction and furniture). French forests grow fairly naturally, without artificial irrigation systems, and with virtually no use of fertilizers and pesticides. The growth rate is good, but, unfortunately, due to a lack of optimized management, cutting trees back is not carried out enough, which causes trunks to decay (releasing methane). Transforming cellulose and lignin into polymeric materials could bring advantages for surplus biomass that does not require polluting agriculture to grow.

12.3. Social acceptability of recycled and bio-based polymers

According to B. Halvorsen [HAL 12], the most important incentives for recycling household waste are the belief that recycling is good for the environment and that it is a civil right. With respect to these considerations, recycling should therefore ensure that recycled materials do not generate polluting emissions that are harmful to the environment and to health. It was only in 1991 that the United States allowed recycled PET to be used in direct contact with food. Today, advanced decontamination processes help lower the levels of contaminants in recycled PET with regard to their virgin equivalents [WEL 11, WEL 13].

It is also important to know whether poorly defined recycled polymers subjected to temperatures of around 200°C, which does not eliminate all volatile molecules, are particularly dangerous. Whereas toxicology usually studies the toxicity of individual molecules, it is much more complex in reality as there are mixtures of volatile compounds that are inhaled by factory workers, or even by consumers during use. Social acceptability of recycled plastics must therefore rely on scientific studies representative of reality. For this reason, VOCs, emitted during various stages, can be collected and analyzed by chromatography coupled with mass spectrometry for example. The mixtures of VOC collected can therefore be used in toxicological tests [LOU 13] where the actual toxicity of the emitted particles can be assessed. Manufacturing conditions can then be optimized to minimize dangerous emissions. The results of these studies coupled with associating polymer science and toxicology help provide scientific evidence when it comes to discussing the social acceptability of recycled materials.

The use of bio-based raw materials contributes to an increase in food prices, due to competition with industrial applications. Thus, in 2006, a 20% production of biofuels in the United States led to a 53% increase in the price of corn, a 30.5% increase in the price of flour and an 11% increase in the price of soya. This use for non-food purposes can lead to famine in poorer countries [COE 07], and poses ethical problems.

12.4. Formulation examples of blends based on recycled PA and bio-based PA and their toxicological considerations

Two polymers with similar chemical formulae are not necessarily miscible. This is the case for the following:

1) polyethylene and polypropylene;

2) high- and low-density polyethylene;

3) polyamides with different chemical formulae.

Carefully formulated mixtures often provide access to properties that cannot be obtained with a single polymer and there are numerous commercial materials with these associations [UTR 98].

The blends are usually binary and composed of a primary polymer for the most part (representing 70–90% of mass) and a second (20–30% of mass) which constitutes the dispersed phase, often in the form of almost spherical nodules. Generally, we aim to minimize the size of the nodules in the matrix and produce high-quality interfaces between the phases in order to improve the mechanical properties of the blends.

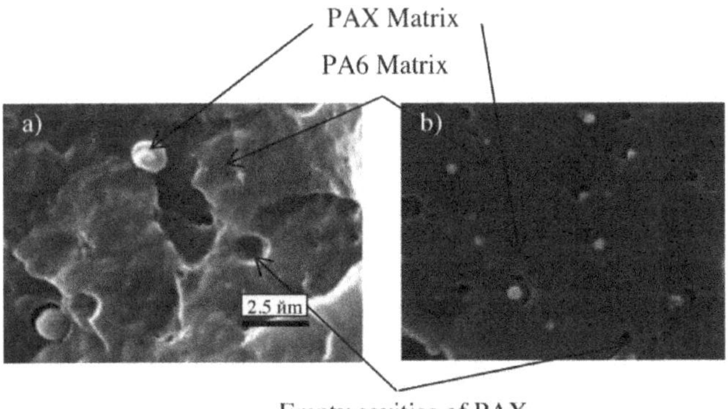

Figure 12.4. *Blend morphologies (75%PA6/25%PAX) non-compatibilized a) and (75%PA6/25%PAX/DGEBA) compatibilized b)*

Bio-based and recycled polymers often have short-lifecycles compared to oil-based virgin resins. We studied bio-based (PAX) and recycled (PA6) polyamide (PA) blends [LOU 13]. Scanning electron microscopy (Figure 12.4) shows that the formulations are composed of 75% PA6 and 25% PAX by mass (denoted PA6/PAX (25/75)) and PAX nodules appear in the PA6 matrix. To refine the morphology and improve PA6/PAX interfaces, we conducted reactive compatibilization to couple the

PA chains with epoxy functional groups. We are interested in three different PA6/PAX blends (25/75), (50/50) and (75/25) to which we added 0.8% mass of bisphenol A diglycidyl ether (BADGE). This significantly enhances the mechanical properties (a 50% increase in elongation at break up to 670%, and in shock resistance up to 15 kJ/m^2, without decreasing these Young's modulus). These good results can be attributed to a decrease in diameters in the dispersed phase as well as an improved interface between PA6 and PAX when BADGE is added.

To ensure that our recycled PA66 was not toxic, we collected VOCs emitted during use, shown in Figure 12.5. The VOC collected were placed in contact with cells of the immune system, under test conditions, and it was found that our VOC did not affect the viability of human macrophages.

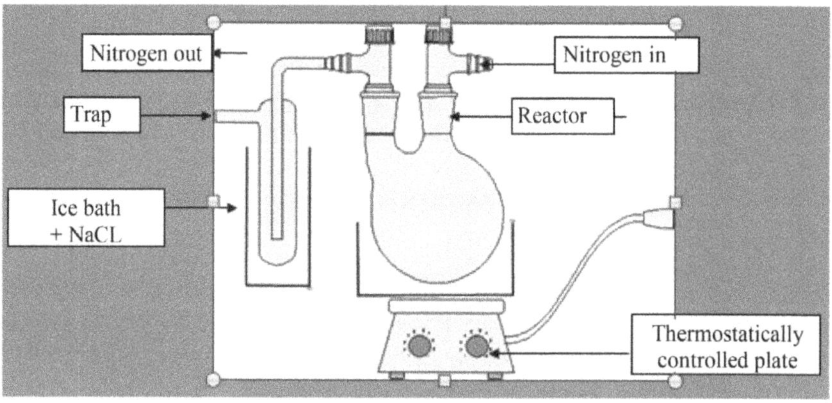

Figure 12.5. *Capture of volatile organic compounds*

12.5. Conclusion

The increasing demand for plastic may not be completely met by the production of virgin resins from fossil resources. The design of polymeric materials, particularly from recycled material or biomass, provides new interesting prospects and helps reduce carbon dioxide emissions in both cases.

We find that an increase in the recovery and recycling of plastic waste helps provide both economic and environmental benefits [HOP 09].

The more recycling is considered (no associations difficult to deal with at end of life), the more it is preferred. While the sorting techniques have become more refined, we may still be faced for a long time to come with the recovery of polymer blends, which is becoming easier each day by combining formulation (compatibilizers which are often copolymers) and the manufacturing process (in particular relevant choice of the extruder profile and screw speed).

Synthons and bio-based polymers can produce polymers whose properties can rival those of oil-based polymers. Unlike their oil-based counterparts, with regard to energy, their production processes are more efficient and theoretically they can be recycled or become compost [ALV 12]. At the same time, ethically and environmentally speaking, they are particularly interesting when made of agricultural waste. The use of agricultural land for purposes other than food is thus prevented, and we do not have to resort to polluting intensive agriculture.

By taking into account environmental and ethical criteria, it helps give plastics a sustainable character. With regard to the objectives set by the circular economy, these may be from agricultural waste (synthons produced by products of fermentation, fibers, etc.), and it is necessary to optimize their end of life (recycling, compost, or even incineration with energy recovery).

12.6. Bibliography

[ACT 12] ACTU ENVIRONNEMENT, http://www.actu-environnement.com/ae/news/valorisation-dechets-plastiques-retard-francais-17127.php4, accessed 23 November 2012.

[ALV 12] ALVAREZ-CHAVEZ C.R., EDWARDS S., MOURE-ERASO R., *et al.*, "Sustainability of bio-based plastics: general comparative analysis and recommendations for improvement", *Journal of Cleaner Production*, vol. 23, pp. 47–56, 2012.

[AMB 11] AMBROGI V., CERRUTI P., CARFAGNA C., *et al.*, "Natural antioxidants for polypropylene stabilization", *Polymer Degradation and Stability*, vol. 96, pp. 2152–2158, 2011.

[ARV 01] ARVANITOYANNIS I.S., BOSNEA L.A., "Recycling of polymeric materials used for food packaging: current status and perspectives", *Food Reviews International*, vol. 17, no. 3, pp. 291–346, 2001.

[AZA 03] AZAPAZIC A., EMSLEY A., HAMERTON I., *Polymers, the Environment and Sustainable Development*, 42, pp. 158–172, John Wiley & Sons, London, 2003.

[BEN 10] BENANIBA M.T., MASSARDIER-NAGEOTTE V., "Evaluation effects of biobased plasticizer on the thermal, mechanical, dynamical mechanical properties, and permanence of plasticized PVC", *Journal of Applied Polymer Science*, vol. 118, no. 6, pp. 3499–3508, 15 December 2010.

[BOC 06] BOCQUET P.Y., *L'usine nouvelle*, N°3032, 2006. http://www.usinenouvelle.com/article/le-dernier-voyagede-la-clio.N54041

[CAS 11] CASSAGNAU P., MANI S., MASSARDIER V., *et al.*, Procédé de fabrication d'une composition à haute fluidité, patent, filed 22/12/11, no. 1 162 328, Plastic Omnium, UCBL, INSA, 2011.

[COE 07] COESNON P., L'Usine Nouvelle, N° 3041, 2007. http://www.usinenouvelle.com/article/l-ethanol-moteur-des-cereales-et-oleagineux.N54913

[DOR 06] DORNBURN V., FAAIJ A., PATEL M., *et al.*, "Economics and GHG emission reduction of a PLA bio-refinery system-combining bottom-up analysis with price elasticity effects", *Resources, Conservation and Recycling*, vol. 46, pp. 377–409, 2006.

[DRU 00] DRUMRIGHT R.E., GRÜBER P.R., HENTON D.E., "Polylactic acid technology", *Advanced Materials*, vol. 12, no. 23, pp. 1841–1846, 2000.

[FUJ 12] FUJII M., FUJITA T., CHEN X., *et al.*, "Smart recycling of organic solid wastes in an environmentally sustainable society", *Resources, Conservation and Recycling*, vol. 63, pp. 1–8, 2012.

[HAL 12] HALVORSEN B., "Effect of norms and policy incentives on household recycling: An international comparison", *Resources, Conservation and Recycling*, vol. 67, pp. 18–26, 2012.

[HOP 09] HOPPEWELL J., DVORAK R., KOSIOR E., "Plastics recycling: challenges and opportunities", *Philosophical Transactions of the Royal Society*, B, vol. 364, pp. 2115–2126, 2009.

[KAD 06] KADDAMI H., DUFRESNE A., KHELIFI B., *et al.*, "Short palm tree fibers: Thermoset matrices composites", *Composites Part A: Applied Science and Manufacturing*, vol. 37, no. 9, pp. 1413–1422, September 2006.

[KOB 03] KOBAYASHI S., UYAMA H., "Biomacromolecules and bio-related macromolecules", *Macromol. Chem. Phys.*, vol. 204, no. 2, pp. 235–256, 2003.

[LAC 05] LACOSTE J.F, BOUNOR-LEGARE V., LLAURO M.F., *et al.*, "Functionalization of PET in the melt state: chemical and rheological aspects", *Journal of Polymer Science: Polymer Chemistry* , vol. 43, pp. 2207–2233, 2005.

[LOU 13] LOUIZI M., MASSARDIER V., TAHA M., *et al.*, "New routes of valorization of recycled and bio-sourced polyamides with a low toxicity process", *Waste and Biomass Valorization*, vol. 4, no. 1, pp. 47–54, 2013.

[MAS 02] MASSARDIER V., "Le recyclage des mélanges de polymères thermoplastiques, Aspects thermodynamiques, chimiques et applications industrielles", *Comptes-rendus de l'Académie des Sciences C.R. Chimie* vol. 5, pp. 1–6, 2002.

[MAS 06] MASSARDIER-NAGEOTTE V., VERNEY V., "Polymères et développement durable, intégration des questions environnementales dans la recherche et l'enseignement", *l'Actualité chimique*, August–September 2006.

[MAS 08] MASSARDIER V., SONNIER R., CASSAGNAU P., *et al.*, patent, filed Procédé de traitement d'un matériau issu de récupération et de broyage, 01/12/08, no. 0 858 167, Plastic Omnium et INSA, patent filed in Europe, extented to USA and India, 2008.

[MOH 00] MOHANTY A.K., MISRA M., HINRICHSEN G., "Biofibres, biodegradable polymers and biocomposites: an overview", *Macromolecular Materials and Engineering*, vol. 24, no. 1, pp. 276–277, 2000.

[MON 05] MONCEL C., *Recyclage Récupération Magazine*, vol. 15, p. 12, 2005.

[MUL 98] MULDER K., "Sustainable production and consumption of plastics", *Technological Forecasting and Social Change*, vol. 58, pp. 105–124, 1998.

[MUL 13] MULHAUPT R., "Green polymer chemistry and bio-based plastics: dreams and reality", *Macromolecular Chemistry and Physics*, vol. 214, pp. 159–174, 2013.

[RAV 11] RAVENSTJIN J., Bio-based polymer: a revolutionary change, http://www.jeccomposites.com/news/composites-news/bio-based-polymers-revolutionary-change, 2011.

[SAD 12] SADLER J.M., NGUYEN A.P., GREER S.M., *et al.*, "Synthesis and characterization of a novel bio-based reactive diluent as a styrene replacement", *Journal of Biobased Materials and Bioenergy*, vol. 6, pp. 86–93, 2012.

[SAK 04] SAKAI K., TANIGUCHI M., MIURA S., *et al.*, "Making plastics from garbage: a novel process for poly-L-lactate production from municipal food waste", *Journal of Industrial Ecology*, vol. 7, nos. 3–4, pp. 63–74, 2004.

[SCH 13] SCHMID A., NAQUIN A., GOURDON R., "Incidence of the level of deconstruction on material reuse, recycling and recovery from end-of life vehicles: an industrial-scale experimental study", *Resources, Conservation and Recycling*, vol. 72, pp. 118–126, 2013.

[UTR 98] UTRACKI L.A., *Commercial Polymer Blends*, Chapman & Hall, London, 1998

[WEI 07] WEISS M., PATEL M., HEILMEIER H., *et al.*, "Applying distance-to-target weighing methodology to evaluate the environmental performance of bio-based energy, fuel and materials", *Resources, Conservation and Recycling*, vol. 50, no. 3, pp. 2017–2028, 2007.

[WEL 11] WELLE F., "Twenty years of PET bottle to bottle recycling: an overview", *Resources, Conservation and Recycling*, vol. 55, pp. 865–875, 2011.

[WEL 13] WELLE F., "Is PET bottle-to-bottle recycling safe? Evaluation of post-consumer recycling processes according to the EFSA guidelines", *Resources, Conservation and Recycling*, vol. 73, pp. 41–45, 2013.

13

Food Packaging: New Directions for the Control of Additive and Residue Migration

13.1. Introduction

Controversy over the actual safety of materials in contact with food is becoming more prevalent in the United States and Europe. This debate has been popularized by magazines and scientific literature with evocative titles – for example: *How Dangerous is Plastic* (*Time Magazine* 12 April, 2010) [WAL 10] – or editorial debates – for example, "Bottled water left in a hot car may cause breast cancer" (*Nature Reviews Endocrinology*, May 2010) [HEA 10]. The mixing of concepts used to protect the consumer (low toxicity, low migration of substances, low consumer exposure, etc.) does not reassure consumers and, for now, fails to prevent health crises: tin coatings [GOU 98], varnish, plasticized plastics [GRO 08], polycarbonate materials in contact with food [NAM 10], printing ink [BEN 08], recycled fiber cardboard [BIE 11], lids of baby food jars [HAM 98], etc. Finally, it tends to lessen the responsibility of the packing supply chain.

Our research helps to demonstrate the risk of food contamination by substances coming from material in direct contact or used in assemblies (e.g. outer packaging) can be reliably predicted by simulation and modeling. A "bottom-up" modeling approach is used (from the molecule to the packaging) together with various techniques including probabilistic approaches and molecular modeling [VIT 05, GIL 09a, GIL 10, DUR 10, VIT 10, FAN 13]. This chapter reviews new directions in the attempt to better control the safety of food packaging and other materials in contact with food. This endeavor forms part of the collaborative research project *SAFEFOODPACK DESIGN* (2011–2013) [INR 11] financed by the National Agency for

Chapter written by Olivier VITRAC and Audrey GOUJON.

Research, with 11 participating partners. This chapter has been written for the satisfactory understanding of both the user of packaging materials as their intent is to pack. To consult a more elaborate version, we can refer to the following chapters in French [VIT 08, VIT 09, VIT 11a] and to the following publications in English [VIT 07a, VIT 07b].

This chapter is split into three independent sections. Section 13.2 presents the ubiquitous or even universal nature of material component migration. It does not give an alarmist point of view, but instead suggests that this risk, just like other food hazards involving chronic chemical contaminants (environmental, components produced by processing of food, natural toxins), requires special attention and careful consideration beyond the regulatory constraints.

Section 13.3 outlines the principles of modeling packaging and food, using two simple but effective examples. They illustrate the concepts of critical steps, or materials or substances critical for food safety via a quantitative failure mode, effects and criticality analysis (FMECA) approach, derived from concepts used by the aviation or electronic industry for critical systems. It is a systematic approach which today facilitates the risk management at the sector scale, of several thousand references or complex assemblies integrating various materials and dozens of substances [NGU 13] via an expert distributable and modifiable system: FMECAengine [VIT 11b].

Section 13.4 is dedicated to the molecular description of transport phenomena in thermoplastic polymer matrices with regard to both diffusion and thermodynamics. Two promising approaches to structure–property associations are introduced, on the one hand, molecular modeling in the framework of Flory–Huggins theory to calculate and predict food-packaging partition coefficients; and on the other hand, free-volume and scaling approaches to estimate activation energies of diffusion in thermoplastics and elastomers as well as the effects of chemical structure on diffusion coefficients. With regard to the latter, suggestions are provided to develop substances of technological interest (e.g. additives, pigments, etc.) with reduced diffusion capacities.

13.2. Migration of packaging components

13.2.1. *Evidence for migration*

Evolving urban life styles (eating out, snacks, microwavable or ready to eat food) inevitably leads to an increase in material surfaces in contact with food, times in contact with materials (Figure 13.1(a)) and therefore exposure to packaging substances from early life stages (e.g. baby food shown in Figure 13.1(b)).

a)

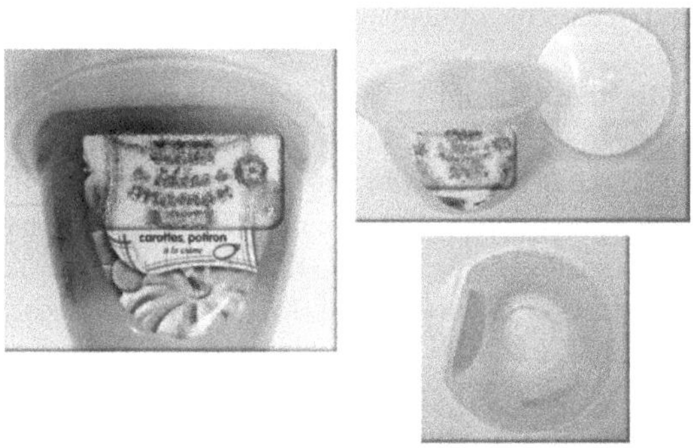

b)

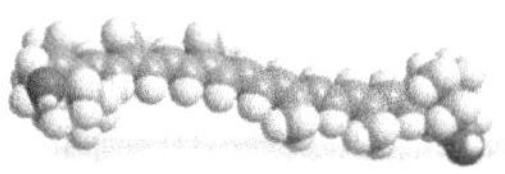

c)

Figure 13.1. *a) Evolving life styles accompanied by the evolution of materials (e.g. composites) and functions (e.g. microwavable); b) coloration of sterile polypropylene (PP) packaging following sterilization at 121°C; and c) example of β-carotene responsible for coloring the PP packaging*

The suspicion surrounding epoxy-phenolic and polyvinyl chloride lacquers of metal packaging, as well as the weight and energetic cost of recycling glass, favors plastic materials. They are used for applications including sterilized food.

Figure 13.1(b) shows the transfer of food matter to the packaging (here, polypropylene (PP)) which involves organic pigments (Figure 13.1(c)) the same size as additives in plastic materials with molar masses between 100 and 800 $g.mol^{-1}$. As described in the final section, this transfer is highly active at storage temperatures, close to the glass transition temperature of PP (–10°C to 0°C) if the product is refrigerated, and sterilization temperatures (120–130°C), close to the melting temperature of PP (180°C). In the domestic environment, a similar migration of pigments, albeit less intense, can be observed when tomato sauce is in contact with popular PP storage containers.

The solubility diffusion of food pigments is the phenomena responsible for this coloration. It is important to highlight that in the opposite direction, from the packaging to food, substances from the material, of approximately the same size of β-carotene or smaller, can also migrate provided that they are not covalently bonded to the polymer matrix. This contamination of food is usually invisible, without being smelled or tasted, but occurs insidiously.

At the end of life, the material may be polluted by environmental substances (e.g. mixing of waste), by secondary use of the material (e.g. a water bottle used to stock a pain solvent or petrol) or even during pick-up (e.g. mixing of shampoo bottles and PET bottles containing fruit juice). These pollutions affect the reuse potential of these materials even after decontamination (solvent washing and vacuum drying of pellets) for subsequent contact with food. Currently, these practices are only authorized for PET [EFS 11, EUR 08].

All materials can potentially, and to varying degrees, contaminate food without being in contact (Figure 13.2(a)) or even after brief contact (Figure 13.2(c)). Figure 13.2(a) shows the no contact transfer of substances from the outer packaging (brown recycled cardboard) into pasta individually packed in cardboard boxes (white and made of virgin fibers) covered in a low-density polyethylene film [BIE 11]. After 65 days in contact, the paraffin concentration (residues of printing ink present in the recycled fibers) in the pasta reaches 6.1 $mg{\cdot}kg^{-1}$. This is extrapolated to 10 $mg{\cdot}kg^{-1}$ after a contact time of two years. The contamination of 33 dried food samples by mineral oils and aromatic polycyclic hydrocarbons packaged in the presence (direct contact or not) of recycled fibers is shown in Figure 13.2(b) [GRO 11]. All food, both liquid and dry food, is potentially affected. Only the partial vapor pressure, chemical affinity and diffusion rates determine the extent of migration. Fresh chopped kiwi in contact with printed serviettes can be significantly contaminated by primary aromatic amines (o-anisidine). In the configuration shown in Figure 13.2(c), the contamination was 5.3 $mg{\cdot}kg^{-1}$ after 2 hours of contact (i.e. migration of 17% for the parts in contact with the kiwi) [HEL 11].

a)

b)

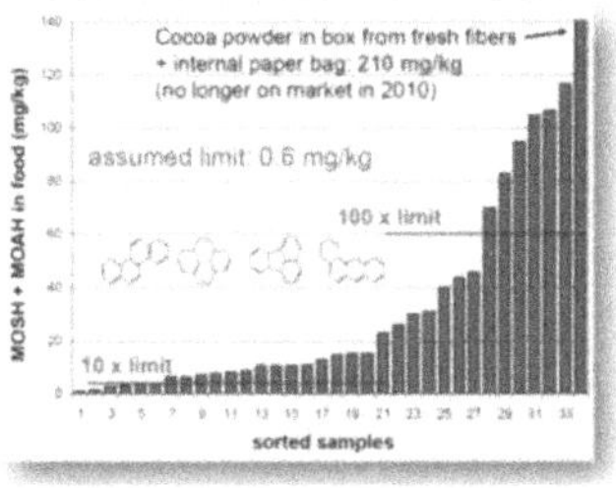

c)

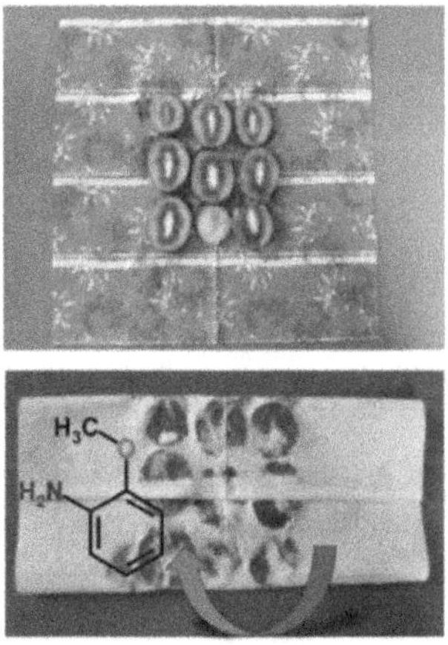

Figure 13.2. *Examples of contamination by materials not meant to be in direct contact with food*

13.2.2. *Food contamination pathways*

The migration of molecules with higher molecular masses (between 200 and 2,000 g·mol^{-1}), i.e. additives, is only possible after prolonged contact (typically several hours to several days at room temperature) between the packaging and food. Different mechanisms are shown in Figure 13.3(a) and categorized in Table 13.1:

food contamination by desorption of packaging constituents (additives, additional technologies), by permeation of gases (volatile contaminants, additives of low-molar masses, ink solvents or polymerization promoters) or by redistribution of contaminants (transfer of printed surfaces onto inner packaging). For substances with sufficient vapor pressure (e.g. mainly molecules with a liquid reference state at contact temperature), contamination can occur with or without contact after passing through a gas space below the food or between two packaging layers. The specific effect of the interactions between the packaging and food is described in Figure 13.3(b). The sorption of food constituents (here, flavorings, but may also be fats) plasticizes the packaging and promotes the release of constituents from the packaging by stimulating polymer relaxation (decrease in glass transition temperature).

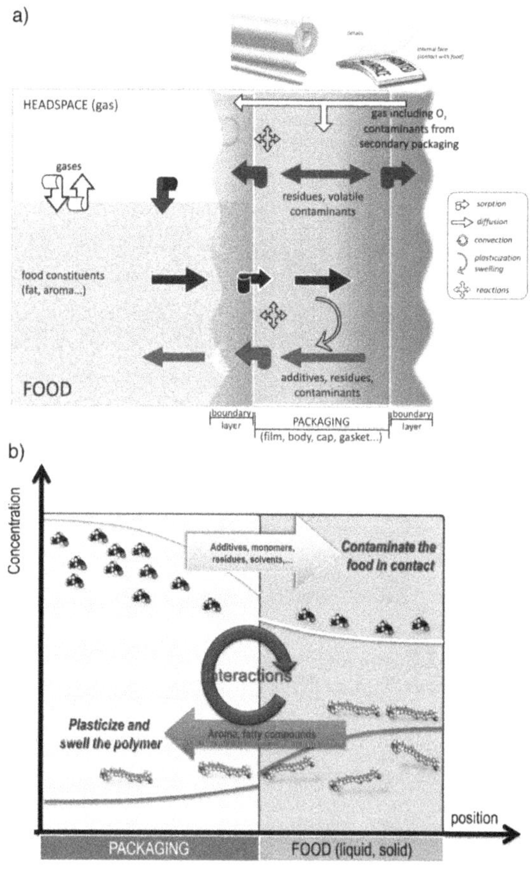

Figure 13.3. *a) Contamination pathways with and without contact; and b) transfers between packaging and food*

Contamination pathway	Description	Examples
In direct contact with food	In direct contact with food during packaging	Monolayer materials
Via the layer in contact with food	The substance passes through one or several materials before coming in contact with food	Multilayer materials, labels, lacquers, motif
Contamination of the layer in contact with food before packaging	Contact between inner and outer sides of packaging (smear, transfer phenomenon or setoff)	Film rolls, packaging or parts of stacked packaging
No contact	Contamination via headspace or permeation of gases	Ink solvents, flavorings, off flavors, contaminants from secondary packaging

Table 13.1. *Main food contamination pathways*

13.2.3. *Aggravating factors of contamination*

Table 13.2 sums up the different technological factors affecting the contamination of food. In the case of migration through contact (Figure 13.3), the level of contamination depends on the contact area, the quality of the contact between the packaging and the food and the contact time. Migration is usually greatest in the case of a liquid food product and is enhanced when the food product spreads out (low-surface tension) onto the polymer material (e.g. liquid fat). In the case where the food product has a high affinity for the packaging, a partial sorption of its constituents (e.g. fats, pigments, flavorings and so on) can be identified. It can modify the physico-chemical properties of the polymer and increase the diffusion and therefore desorption of substances through the packaging. In fact, depending on the quantities absorbed, the polymer may swell and become plasticized, thereby losing its barrier properties.

The kinetic and geometric effects and laws of mass action are described in the second section. The effects of activation by temperature are analyzed in the final section. The effects of polymer plasticization due to sorption of food constituents are shown in Figure 13.3(b).

Evolution of factor	Consequence of contamination	Examples
↗Concentration of contaminant in packaging before being packaged	↗	Production incident
↗Contact time	↗	Use by date, best before date
↗Contact area	↗	Rather wide join Adhesive seal
↗Packing temperature	↗	Hot filling
↗Temperature of conservation	↗	Frozen, refrigerated, room temperature
↗ Thermal stabilization treatment	↗	Pasteurization, sterilization of packaged food
↗Heating up temperature/cooking with packaging	↗	Bain-marie, microwave
Shaping of material	variable	Coextrusion/thermoforming of multilayers, blow molding
↘Size of contaminant	↗	Plasticizer
↗Chemical affinity of contaminant for food product	↗	Fatty food (hydrophobic)
State of polymer	↗↗ above Tg	Hot filling uncontrolled
↗Chemical affinity of food product for polymer	↗	Sorption of food substances into packaging
↗Hydrostatic pressure	↘	Pascalization

Table 13.2. *Main technological factors affecting food contamination*

13.3. Assessing and controlling migration

13.3.1. *European regulation and its limits*

The current regulations of the European Union can be consulted on the European Commission website [EUR 12a]. The regulations of the United States and Japan are independent, but still have the same objectives: to limit the consumer exposure to substances from packaging. In Europe, the inertia principle of packaging is defined in article 3 of the regulation 1935/2004/EC [EUR 04]: "to preclude substances from being transferred to food in quantities large enough to endanger human health or to bring about an unacceptable change in the composition of the food or a deterioration in its organoleptic properties". The main principles and their limitations for a variety of materials are summarized in Figure 13.4. Note that the principles of low migration are only used for plastics [EUR 11a, EUR 11b] (Figure 13.4(a)). For an assembly of materials (a printed multilayer assembly with a cardboard support is shown in Figure 13.4(b)), different principles are applied according to the nature of the components.

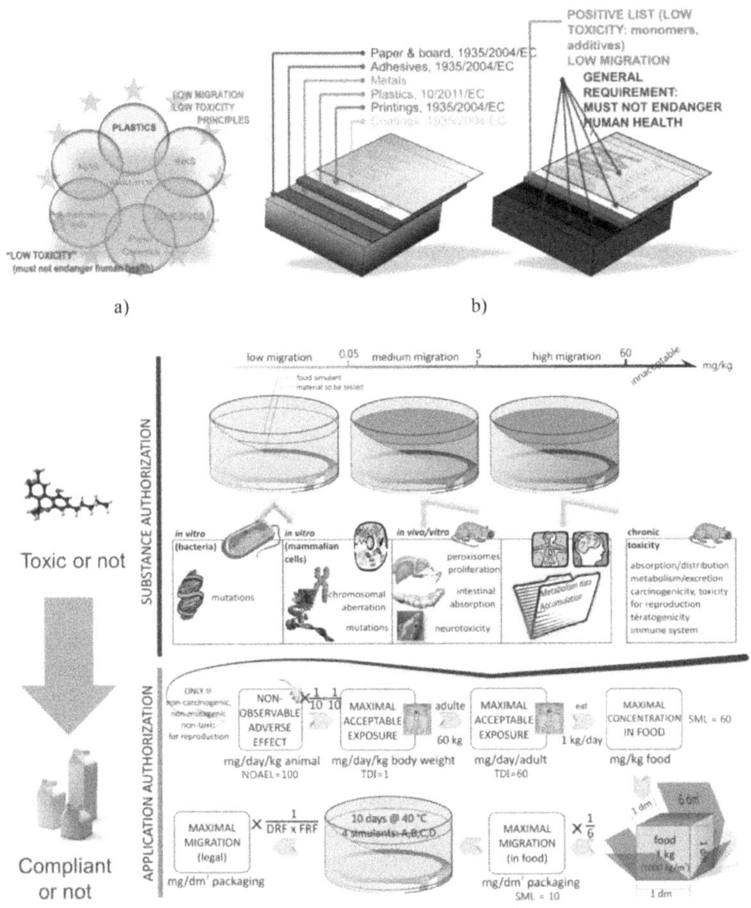

Figure 13.4. *Overview of the principles held for the regulation of materials in contact with food at the European level*

The European framework identifies 17 groups of materials. Only plastics have a positive list of monomers and usable additives for their manufacture and implementation. The principles of substance authorization and the restrictions of use, via specific migration limits (SMLs), are defined in the regulation 10/2011/EC [EUR 11a, EUR 11b] and shown in Figure 13.4(c). The 16 other groups of materials (ink, glue, paper, cardboard, etc.) are not covered by specific measures (no positive list, no overall or SMLs, etc.) and must only be manufactured according to good industrial practices 2023/2006/EC [EUR 06] (Figure 13.4(b)). These good practices must be established for the time being by professional procedures without accurate frameworks. The current principles used in industry are still vague with few restrictions, usually suggesting low toxicity requirements, but do not include the maximum levels of contamination or consumer exposure. In the case of food inks, we highlight the EUPIA effort to develop a guide [EUP 11] and its implementation.

Materials and substances that are not regulated have been particularly at fault in food alerts and product recalls. This is why the inertia principles applied to plastics have been proposed to be generalized to all materials through an FMECA method presented in section 13.3.3. Its main benefit is to integrate within a unique approach assemblies of materials and components, which must comply with different regulations.

The principles of precaution used to authorize a substance and verify the restrictions of use are described in Figure 13.4(c). This process is obligatory for plastic additives and monomers. The authorization of putting a new substance on the market depends on toxicological test results, whose type and significance depend on the level of migration in the food. The process forms part of an independent evaluation the European Food Safety Agency (EFSA). According to the test results, an SML value is proposed, knowing that an overall migration limit of 60 $mg \cdot kg^{-1}$ is required for a group of substances released by plastic. The SML value is based on consumer exposure hypotheses taking into account a theoretical packaging geometry (1 kg of food in 6 dm^2) and a theoretical daily consumption of 1 kg of food for an adult weighing 60 kg. The tolerable daily intake in humans is randomly established to be 1/100th of the non-observed adverse effects level in animals. This type of scenario maximizes the weight-related exposure of adults, but decreases that of children. It is not, however, appropriate for liquids such as water, milk or dairy products.

Since the directive 2002/72/EC (currently repealed and replaced by regulation 10/2011/EC), demonstrating the conformity of plastics can be carried out experimentally, or through calculations such as those in section 13.3.2. Due to the lack of toxicological data on the substance concerned, it is usually considered that a migration less than the theoretical threshold of toxicological concern of 0.15 $mg \cdot kg^{-1}$ does not need special attention.

13.3.2. *Modeling migration*

13.3.2.1. *General principles in the case of monolayer materials*

The contamination of food by substances in packaging involves diffusion into the packaging material, controlled by a diffusion coefficient D (units SI $m^2 \cdot s^{-1}$), and the distribution of substances between the packaging and food, controlled by a partition coefficient $K_{F/P}$, as well as the sampling of substances from the food side as shown in Figure 13.5. The first two properties are explained in detail in sections 13.4.1 and section 13.4.2, respectively.

The equations of transport and mass balance to describe the desorption of substances from a monolayer packaging (Figure 13.5), whose transport properties are uniform and constant over time, are written in one dimension:

$$\frac{\partial C(x,t)}{\partial t} = \frac{\partial}{\partial x}\left(D\frac{C(x,t)}{\partial x} \right)$$

$$j\big|_{x=lp} = -D\frac{\partial C}{\partial x}\big|_{I} = h(C_F\big|_{x=lp} - C_F) = h(K_{F/P}C\big|_{x=lp} - C_F); \frac{\partial C}{\partial x}\big|_{x=0} = 0 \qquad [13.1]$$

$$C_F(t) = C_F(t=0) + \frac{A}{V_F}\int_0^t j\big|_{x=lp} \cdot dt$$

where $C(x,t)$ is the local concentration in the layer in contact with food, C_F is the concentration in the food product or the food simulant, h is the mass transfer coefficient at the interface (units SI $m \cdot s^{-1}$), A is the area of the material in contact with food and V_F is the volume of the food product.

Assuming that the diffusion coefficient is constant and uniform, we can find the analytical solutions to the system of equations [13.1] in [VIT 06] and [GOU 13]. A numerical solver, effective in the presence of one or several layers, is integrated within open-source simulation tool FMECAengine [VIT 11b].

13.3.2.2. *Material balance in the case of multilayer materials*

With regard to multilayer materials (Figure 13.6), the chemical affinities for the contaminants must be taken into account for each layer. A continuous formulation is obtained by replacing the concentration by a continuous potential at the interfaces between each layer: chemical potential or partial pressure. The ratio between the concentration and this potential is determined for each material using the sorption/desorption isotherm of the contaminant in the material under consideration.

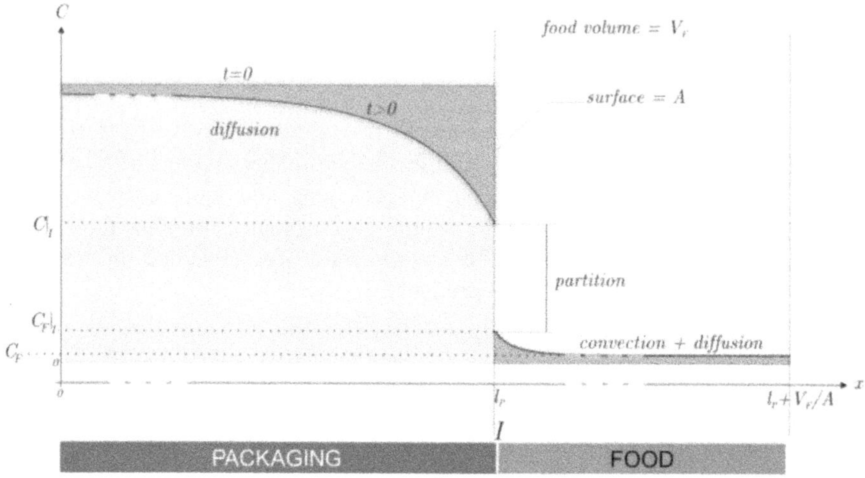

a)

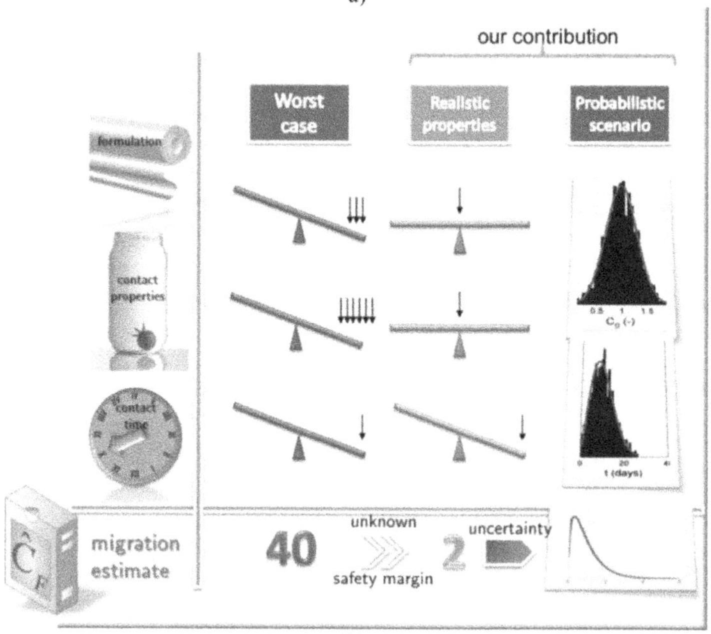

b)

Figure 13.5. *a) One-dimenstional (1D) modeling of transfers between packaging (here monolayer) and food; and b) framework hypotheses used to make decisions based on migration calculations*

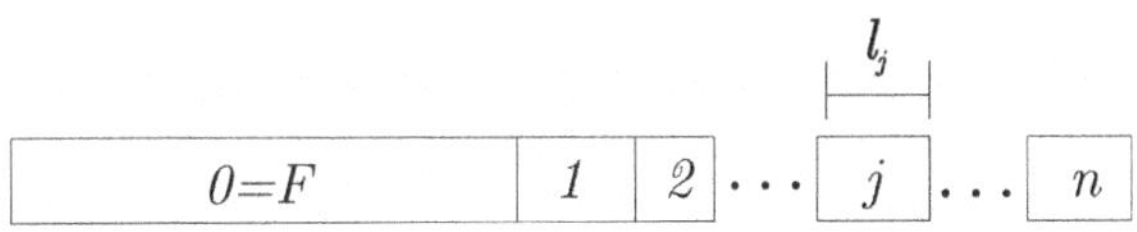

Figure 13.6. *Indexing of n layers of a multilayer material. Layer 1 is the layer in contact with food. The thickness of each layer is denoted l_j. l_0 corresponds to the equivalent size of food and is defined by V_F/A, where V_F and A are food volume and contact area, respectively*

The sorption and desorption equilibrium in each layer is assumed to be reversible and obeys Henry's law. As a result, a vapor pressure $p_j(x)$ of the substance is defined at equilibrium with the quantity locally dispersed in each layer *j* according to:

$$p_j(x) = k_j \cdot C_j(x) = \frac{k_j^H \cdot \rho_j}{M} \cdot C_j(x) \qquad [13.2]$$

where k_j^H and ρ_j are the Henry coefficient of the substance in the layer j and the density of the layer j, respectively. *M* is the molar mass of the substance concerned. Note that k_j^H (with units $J{\cdot}mol^{-1}$) is also the inverse of the solubility of the substance in the layer *j*.

Knowing that no analytical solution exists to this problem in the case of more than two layers or a mixed Robin boundary condition as shown in equation [13.1], we can find a complete numerical implementation in [NGU 13] with a detailed physical formulation in [VIT 11a, VIT 07b]. In order to be concise, only the material balance at equilibrium for an assembly composed of *n* components or layers is presented. This type of formulation coupled with transport equations, initial and boundary conditions, forms *ad hoc* the basis of the FMECA method presented in section 13.3.3.

Assuming that the substance considered is initially present only in the packaging material, the material balance between the food and the packaging is written in the absence of reactions or losses:

$$\sum_{j=1}^{n}\left(\rho_j \cdot C_j\big|_{eq} \cdot l_j\right) + \rho_0 C_0\big|_{eq} \cdot l_0 = \sum_{j=1}^{n}\left(\rho_j \cdot C_j\big|_{t=0} \cdot l_j\right) \qquad [13.3]$$

with:

$$C_j\big|_{eq} = \frac{k_0}{k_j} C_0\big|_{eq}$$

Note that the equilibrium state requires:

$$\left\{ p_j \big|_{eq} \right\}_{j=1,\ldots n} = p_0 \big|_{eq}$$

The concentration at equilibrium in the food is finally given by:

$$C_0\big|_{eq} = \frac{\sum_{j=1}^{n} \left(\frac{\rho_j}{\rho_0} \frac{l_j}{l_0} C_j \big|_{t=0} \right)}{1 + \sum_{j=1}^{n} \left(\frac{k_0}{k_j} \frac{\rho_j}{\rho_0} \frac{l_j}{l_0} \right)} \qquad [13.4]$$

By convention and according to equation [13.4], the choice of $k_0 = 1$ identifies k_j as the partition coefficient between the food and the layer j.

13.3.2.3. *Practical applications of modeling*

Modeling the migration helps demonstrate the conformity of plastic packaging materials with regard to SMLs (see Figure 13.4). This possibility was introduced in 2002 in the directive 2002/72/EC [EUR 02] and is found in the regulation 10/2011/EC [EUR 11a]. However, this is accepted as long as the user can show that the calculated value overestimates the actual value (e.g. Figure 13.7(a)). It has also been shown that, when correctly used, this model helps prevent health crises [GIL 09b] such as that associated with the migration of a photo primer used in printing inks on liquid [BEN 08] or solid food products [CAS 97].

In a more general context, modeling the contamination can be coupled with consumer data and/or household and industrial practices to assess chronic consumer exposure to one or several substances (Figure 13.7(c)) [VIT 11a, VIT 07c].

Typical examples of modeling are shown in Figure 13.7. Figure 13.7(a) compares the experimental kinetics and those predicted by the software SFPP3 [VIT 12] of the migration of an antioxidant (Irganox 1035) from a PP packaging in contact with 95% v/v ethanol. Figure 13.7(b) illustrates the modeling capacity to predict migration of isopropyl thioxanthone [GIL 09b] in milk, which led to a major recall of products in Europe [EUR 12b]. Finally, a calculation of the exposure of French consumers (survey data showing consumption in 5,535 households in 1998) to styrene from yoghurt pots [VIT 07c] is shown in Figure 13.7(c).

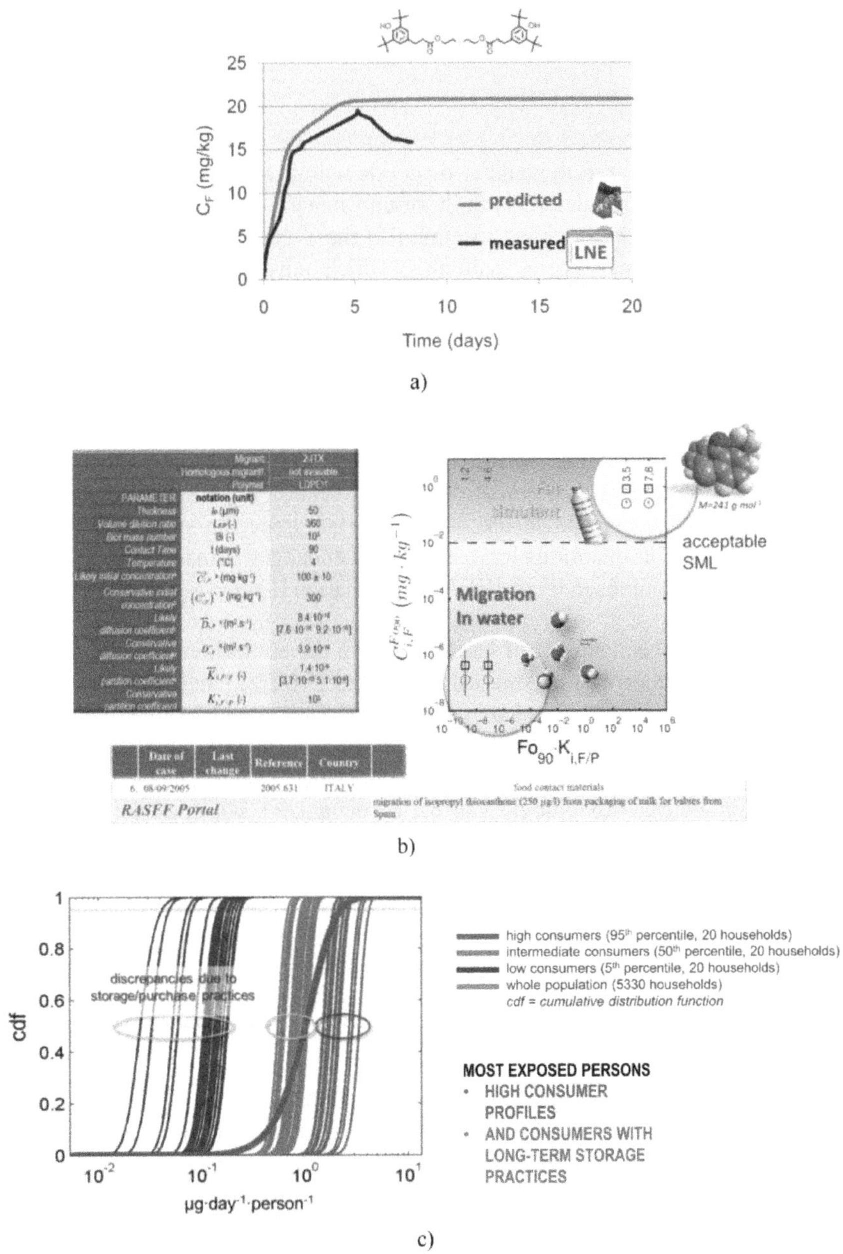

Figure 13.7. *Examples of modeling applications*

13.3.3. *FMECA approaches to design safe packaging*

13.3.3.1. *Harmonized risk assessment*

The European approach to protect consumers was developed along with the desire to promote the free movement of goods within a single market, then in the Union. The resulting regulation is built around materials and their associated dies rather than around a single object to minimize the eventual contamination of food. There are some inconsistencies with this, which must be corrected in practice. Examples include:

– the tolerance to a substance can depend on the material origin (e.g. a UV stabilizer in plastic has restricted use, but this is not the case if the substance is used in ink printed on paper or cardboard);

– the substances used in plastics are subject to a strict assessment of their toxicology and a careful assessment of their associated exposure (Figure 13.4(c)), whereas the substances used in the other materials obey the single law of "[not] transferred to food in quantities large enough to endanger human health" [EUR 04] and the restrictions imposed by the REACH regulation [EUR 07a].

The top-down management of health risks by the authorities was formalized in the 1980s at the request of the American Congress [NRC 83] and was used by many states (Figure 13.8(a)). It aimed to separate risk assessment (subject to scientific expertise) from risk management (a matter of political responsibility). For materials in contact with food, this approach does apply when authorizing or prohibiting a substance for which a toxicological effect is recognized, but does not identify the conditions of use for a substance whose contamination would be unacceptable. In another context, since the 1950s, the American space program has implemented general FMEA standards for the safety management of processes (Figure 13.8(a)). These standards are based on total quality concepts initially proposed by Deming [DEM 66], popularized by the automobile industry. As the agri-food industry was also a supplier of space programs, it had to adopt FMEA methods, in the form of hazard analysis and critical control points (HACCP). These standards are internationally recognized by the joint FAO/WHO Codex Alimentarius Commission [SOM 11] after being recognized as a general standard in the United States [BAU 66, FDA 72] and Europe [EUR 93]. More recently, the United Nations encouraged the combination of horizontal and vertical risk management [FAO 03]: the management of food risks using a the same approach regardless of their nature: physical, chemical, biological or microbiological. These principles form part of the international norm ISO 22000:2005 "Food safety management systems: Requirements for any organization in the food chain" [ISO 05a].

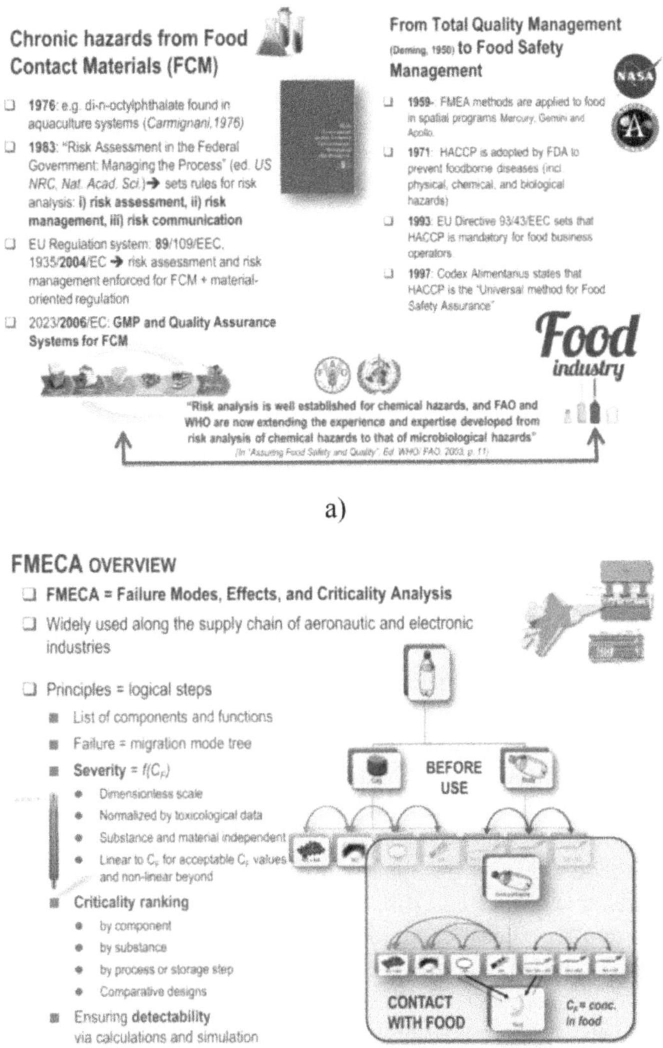

Figure 13.8. *a) Comparison of safety risk assessment and risk management; and b) general principles of FMECA approaches (failure mode, effects and criticality analysis) applied to materials in contact with food*

To meet the requirements of both establishing good industrial practice guidelines for the packaging industry (regulation 2023/2006/EC) [EUR 06] and combining risk management methods for the agri-food industry, an FMECA method was developed [NGU 13] and implemented into the framework of the collaborative research program

SAFEFOODPACK DESIGN [INR 11]. This is summarized in Figure 13.8(b). It is based on the inertia principle of materials based on the regulation 1935/2004/EC [EUR 04]: food contamination must be minimized regardless of the nature of the contaminating materials. However, with an assembly of materials and substances, it involves obtaining dimensionless magnitudes of contamination, called "severity", which take into account the toxicity of each substance. The weighted summation of severity values helps assess the criticality of each design element of a package or substance when it is used in several materials at once.

13.3.3.2. *Evaluating the severity of an industrial stage*

The traditional approaches used so far have not allowed us to assess the consequence of industrial practices upstream of the sector. The proposed approach, which abundantly uses simulation, allows us to rank the risks associated with the practices and conditions of use for these materials: from the industrial storage of these components to the final preparation of food. In comparison, due to cost and efficiency, the current plastic regulation 10/2011/EC [EUR 11a] only assesses the consequences of sterilizing soup in a plastic container in Figure 13.1(b) in the finished product and under accelerated test conditions, and it is unable to reproduce all the critical stages. By defining the contribution of a stage as the variation in contamination when this stage is removed from the technological chart or program, the severity of ant stage can then be tested or compared, whether the food is in contact with the packaging or not.

Figure 13.9 shows the effect of a redistribution of contaminants in ink (here, benzophenone) [EFS 09] during storage of stacked pots before industrial use in the case of pasteurized soup. The concentration profiles and the corresponding concentration kinetics in the food are given in Figure 13.10(b). The Pareto chart of severity for each stage is given in Figure 13.11(a). It clearly shows the impact of storing empty stacked pots (setoff or decal) and long-term storage at room temperature. The calculations were carried out using the free software FMECAengine [VIT 11b] which helps consider sources of doubt or variability during analysis. Figure 13.11(b) shows how the classification of stages evolves when the storage conditions and those for microwave heating are modified.

The common simulation hypotheses are: 1D simulation of transport between components and food; no interaction between substances; no transfer of food constituents into the packaging; no reaction; Henry isotherms; constant diffusion and partition coefficients during the same stage; uniform diffusion coefficients; the simulations may be chained, inherited, branched etc. with arbitrary complexity.

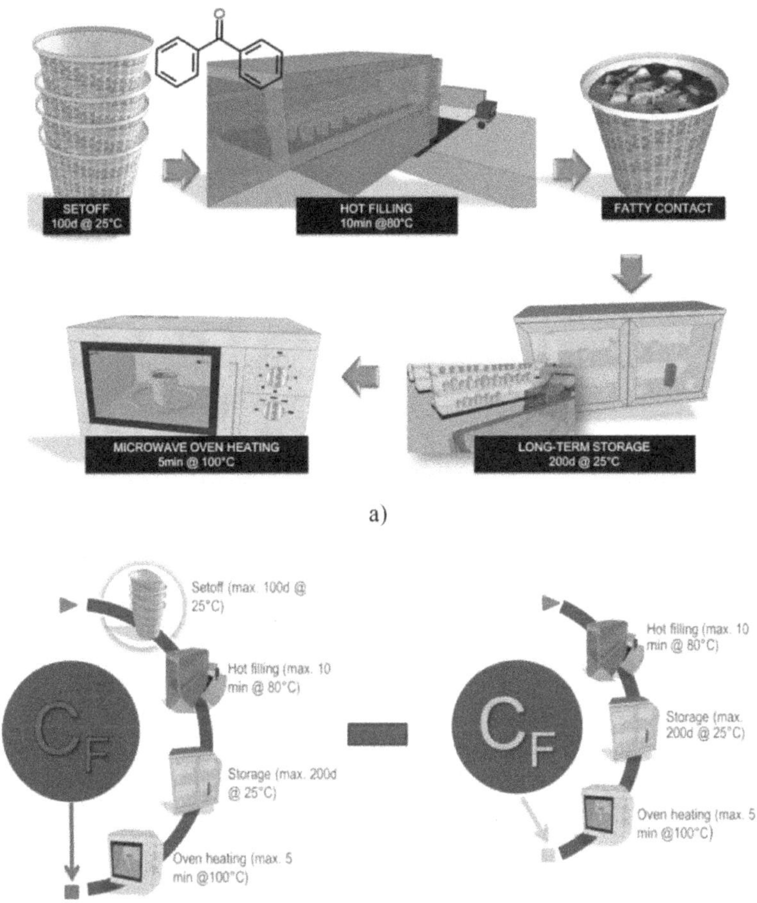

"Setoff" effect assessed for $i = 1$, $M = 4$

$$\text{Severity}\left(\hat{C}_F\left(\text{step}\,i\right)\right) = f\left(\max\left(\underbrace{C_{F_M}\Big|_{1\to 2\to\cdots\to M} - C_{F_M}\Big|_{1\to 2\to\cdots\to M/i}}_{\text{comparison with step } i \text{ removed}},\ \underbrace{C_{F_i}\Big|_i}_{\text{step } i \text{ alone}}\right)\right)$$

b)

Figure 13.9. *a) Diagram showing the production (M = 4 steps) of soup packaged aseptically in a printed polypropylene pot; and b) calculation of the severity of a step (here "Setoff"). The manufacture of the packaging-product couple has four steps: storage of empty packaging in setoff form, hot filling, long storage and microwave heating*

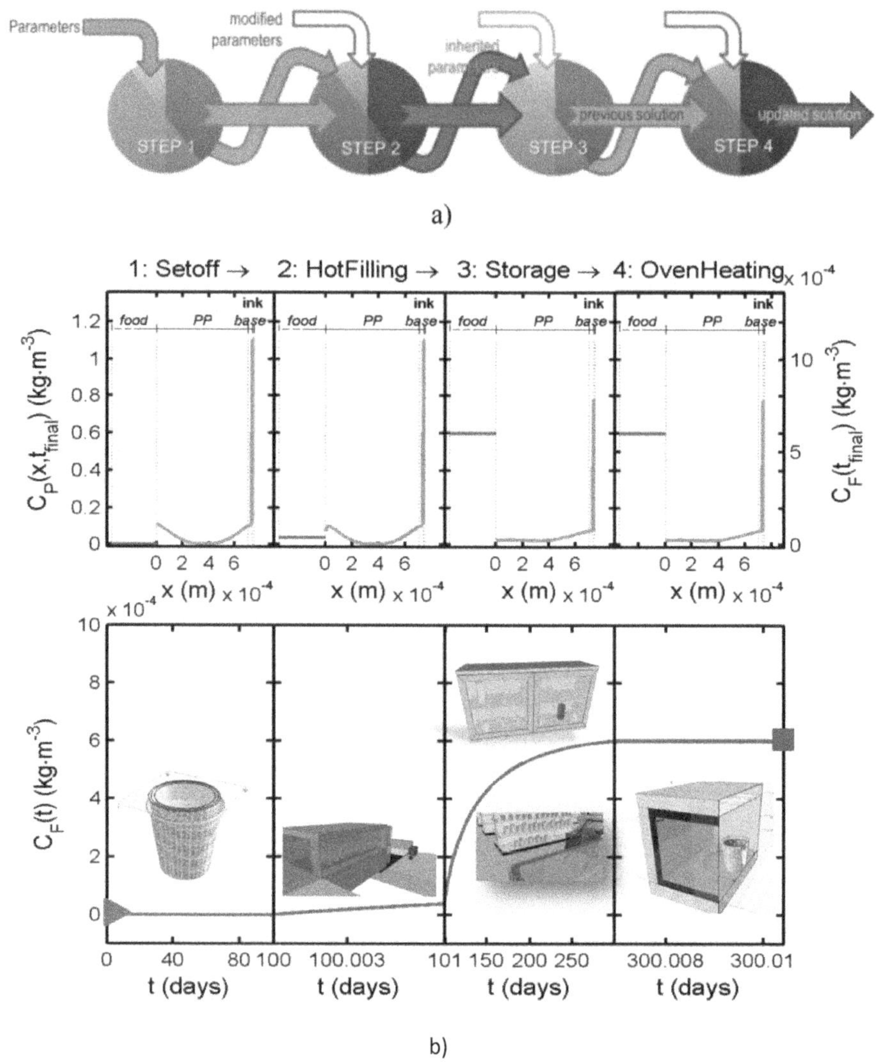

Figure 13.10. *Simulation of the migration of the photo-primer, benzophenone, from printing ink during the four stages described in Figure 13.9; a) diagram produced using free software FMECAengine [NGU 13]; b) profiles of concentration (top) and migration kinetics (bottom). The packaging is composed of two layers: a layer of PP in contact with the food product with a fine layer of ink on top of a base. (For a color version of this figure, see www.iste.co.uk/hamaide/polymers.zip)*

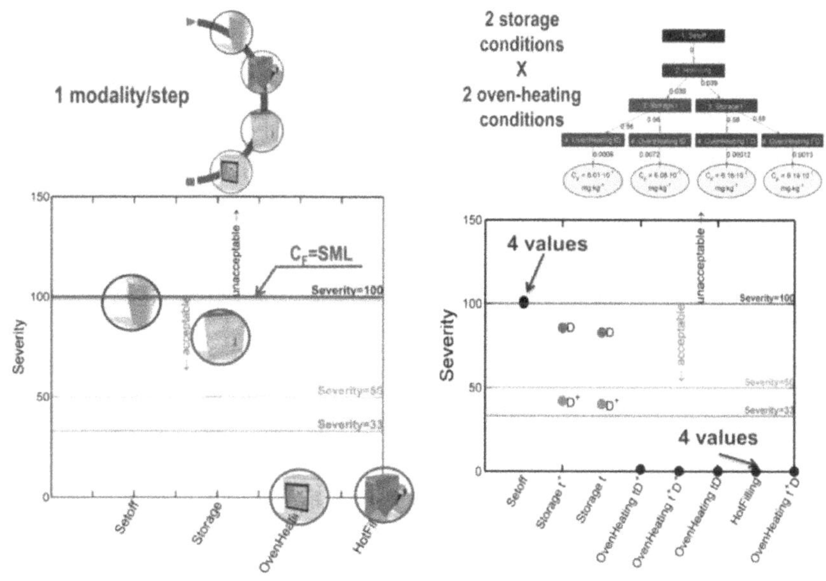

a) b)

Figure 13.11. *Classification of severity of steps described in Figure 13.9: a) for a single modality and b) several modalities (top: combination tree, bottom: corresponding Pareto chart). A severity value of 100 is assigned when the contamination reaches a single regulation (specific migration limit – SML). The combination of 2 × 2 modes consists of short- (t) or long-term storage (t+) and low (D) or high heating (D+). (For a color version of this figure, see www.iste.co.uk/hamaide/polymers.zip)*

13.3.3.3. *Concurrent engineering*

The principles mentioned above can be used to design concurrent packaging. The intent is to add safety considerations as additional engineering constraints to the conventional design approach. The safety of the product is then treated like other mechanical or gas permeability properties and is no longer only tested on the finished product. This approach is shown in Figure 13.12 for the design of an aseptic generic packaging with potential use for fat or dairy products refrigerated or stored at room temperature. Two multilayer oxygen-barrier packaging are compared: bottle or brick (as shown in Figure 13.12(a)). Two contaminants of low-molar mass are studied: a photo-primer in printing ink, benzophenone (migrant X) and a biocide, used to stabilize inks, glues and adhesives, 1,2-benzoisothiazolin-3-one (migrant Y). The "brick" design is subject to set off (since prior to filling, the printed complex was stored in a coiled form) and therefore maximizes the risk of contamination by substances from printing inks. Conversely, the "bottle" design, which comprises two glue layers, maximizes the risk of contamination by substances from glue layers.

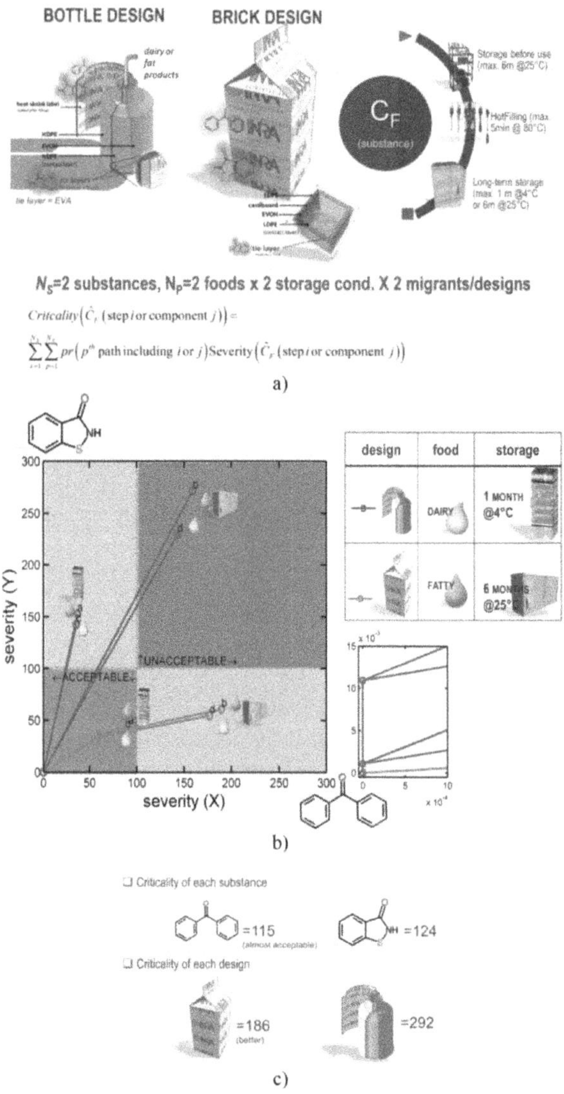

Figure 13.12. *Illustration of the concurrent engineering approach for the design of a packaging adapted to liquid fat or dairy products packaged aseptically and refrigerated or stored at room temperature. (For a color version of this figure, see www.iste.co.uk/hamaide/polymers.zip)*

The severity of substances X and Y during eight technological modality 2 designs × 2 food types × 2 storage conditions is plotted in Figure 13.12(b). The acceptable zone corresponds to a severity lower than 100 (migration lower than the

SML) in the final step: after storage, hot filling and storage. The values greater than 100 provide evidence for one or several design faults. The severities weighted by substance and design, also called criticalities, are shown in Figure 13.12(c). The criticality of each substance is similar, but only the bottle design has a criticality greater than the contamination risk (3 sources of contaminants: 1 ink and 2 glue layers).

Figure 13.12 is reproduced below:

a) designs considered (brick and bottle);

b) comparison of the severity of two ubiquitous substances (migrant X: benzophenone, ubiquitous in printing inks and migrant Y: 1, 2-benzoisothiazolin-3-one, ubiquitous in glues) throughout their technological journey (3 steps: storage of empty packaging, hot filling, storage prior to use) for the eight combinations considered: 2 designs × 2 contacts × 2 storage types;

c) summary of criticalities by substance and design.

It is important to note that the "bottle" design comprises two glue layers and therefore two sources of migrant Y (1, 2-benzoisothiazolin-3-one), whereas the "brick" design contains only one.

13.4. Predicting and controlling migration at the molecular level

This section covers the relations between the chemical structure of the migrants and the main properties controlling the migration of these substances through polymers: partition coefficients and diffusion coefficients. For brevity, only the food-packaging partition coefficients are presented, but the same principles can be applied to polymer 1/polymer 2 partition coefficients. We can refer to data and additional models in [VIT 09].

13.4.1. *Packaging-food partition coefficients*

For a solution *i*, the apparent partition coefficient $K_{i,F/P}^{apparent}$ between a food product (or food simulant), denoted by *F*, and a polymer, denoted by *P*, can be indifferently described by the ratio of mass concentrations at equilibrium:

$$\left\{C_{i,k}^{eq}\right\}_{k=P,F}$$

the activity coefficient ratio:

$$\left\{\gamma_{i,k}^{y}\right\}_{k=P,F}$$

the difference in chemical potential with respect to pure bodies:

$$\left\{\mu_{i,k}^{excess}\right\}_{k=P,F}$$

the difference in isoteric heat of sorption:

$$\left\{\Delta H_{i,k}^{sorption}\right\}_{k=P,F}$$

and sorption entropies:

$$\left\{S_{i,k}^{sorption}\right\}_{k=P,F}$$

For a reversible sorption/desorption isotherm for both media, food and polymer, the thermodynamic equilibrium (equal chemical potentials for both media) imposes:

$$\begin{aligned} K_{i,F/P}^{apparent} &= \frac{C_{i,F}^{eq}}{C_{i,P}^{eq}} = \frac{1}{1-c}\frac{\gamma_{i,P}^{y}}{\gamma_{i,F}^{y}} = \frac{1}{1-c}\exp\left(-\frac{\mu_{i,P}^{excess}-\mu_{i,F}^{excess}}{k_B T}\right) \\ &= \frac{1}{1-c}\exp\left(S_{i,P}^{sorption}-S_{i,F}^{sorption}\right)\exp\left(-\frac{\Delta H_{i,P}^{sorption}-\Delta H_{i,F}^{sorption}}{RT}\right) = \frac{1}{1-c}K_{i,F/P} \end{aligned} \qquad [13.5]$$

where c is the crystallinity of the polymer, k_B is the Boltzmann constant, R the ideal gas constant.

For similar heats of sorption between P and F, the partition coefficients are poorly activated by temperature. Also note that equation [13.5] takes into account the crystallinity of the polymer, with the apparent affinity for food increasing with crystallinity. The expression therefore assumes that the crystalline zones cannot be accessed by diffusing additives. If we disregard the quantity of additives potentially trapped in the crystalline phases, the previous assumption is due to the diffusion mechanisms of large solutions in the polymer matrices (see section 13.4.2): the relaxation of polymer chains is necessary for the movement of additives and it is negligible in crystalline regions. A second partition coefficient, denoted by $K_{i,F/P}$ between the food and the amorphous parts of polymer P, appears in equation [13.5].

This coefficient can be predicted by molecular modeling using an off-lattice Flory–Huggins simulation approach [GIL 09a, GIL 10, VIT 10].

The Flory–Huggins theory generalizes the theory of regular solutions by taking into account the entropic effects associated with the dissimilarity of sizes in the mixing of molecules. The excess chemical potential with regard to the pure bodies is written:

$$\frac{\left\{\mu_{i,k}^{excess}\right\}_{k=P,F}}{k_B T} = Ln\gamma_{i,k}^{v} = \left(1-\frac{1}{r_{i,k}}\right)\phi_k + \chi_{i,k}\phi_k^2 \approx \left(1-\frac{1}{r_{i,k}}\right) + \chi_{i,k} \qquad [13.6]$$

where ϕ_k is the mass fraction of phase $k = P,\ F$ ($\rightarrow 1$ for the infinitely diluted solution i), r_k is the size ration between the reference molecule and the solution, and $\chi_{i,k}$ is the Flory–Huggins interaction coefficient (enthalpic contribution, without units). The term $(1\text{-}1/r_{i,k})$ represents the entropic contribution. In practice, $r_{i,P}\rightarrow 0$ and $r_{i,F}$ takes a non-zero value either approximated using the ratio of molar volumes or is accurately determined from radical distributions [GIL 10, VIT 10]. The coefficient $\chi_{i,k}$ is determined from pair-to-pair interactions between molecules at the atomic level and number of neighboring molecules [GIL 09a].

Some examples of comparisons between experimental and calculated values of partition coefficients are shown in Figures 13.13 and 13.14 for numerous molecules and several polar solvents: methanol, ethanol and water–ethanol mixtures. The arrows pointing downward in Figure 13.13 represent the corrections when taking into account the radial distribution of the solution in F in the assessment of $r_{i,F}$. The solutions (denoted i in equations 13.5 and 13.6) are as follows: 1: decane; 2: dodecane; 3: tetradecane; 4: hexadecane; 5: octadecane; 6: eicosane; 7: docosane; 8: tetracosane; 9: octacosane; α: dodecanol; β: tetradecanol; γ: hexadecanol; δ: octadecanol; a: camphor; b: diphenyl oxide; c: diphenylmethane; d: D-limonene; e: DL-menthol; f: eugenol; g: isoamyl acetate; h: linalyl acetate; j: phenylethyl alcohol; A: BHT; B: Chimassorb 81; C: erucamide; D: Irgafos 168; E: Irganox 1035; F: Irganox 1076; G: Irganox 245; H: Irganox 3114; J: Irganox PS802; K : stearic acid; L: Tinuvin 326.

Figure 13.14 illustrates, more specifically:

a) the effect of the mass fraction in ethanol on the partition coefficients;

b) the robustness of the choice of water–ethanol mixture as a liquid simulant of milk.

The modification to the European regulation [EUR 07b] corresponds to the replacement of distilled water by 50% ethanol.

Calculating partition coefficients using molecular modeling reduces the cost of studying the relevance of certain solvents simulating complex components. Thus, we have shown that choosing water as the initial simulant of a product such as milk, containing 3% fat, was not appropriate [GIL 10]. The choice of this component in the past has been responsible for the accidental contamination of baby milk by the photo-primer isopropyl thioxanthone. Using a 50% ethanol–water mixture was proposed in the directive 2007/19/EC [EUR 07b] as an alternative. The importance of this choice is described in Figure 13.14(a) and is compared with predicted results. Molecular modeling [GIL 10] clearly suggests that only 95% ethanol or higher allows partitions similar to those of a dairy product to be obtained. This choice is confirmed by the simulation in Figure 13.7(b) which obtains contamination values close to those measured by the authorities, responsible for the product recall [GIL 09b].

a)

b)

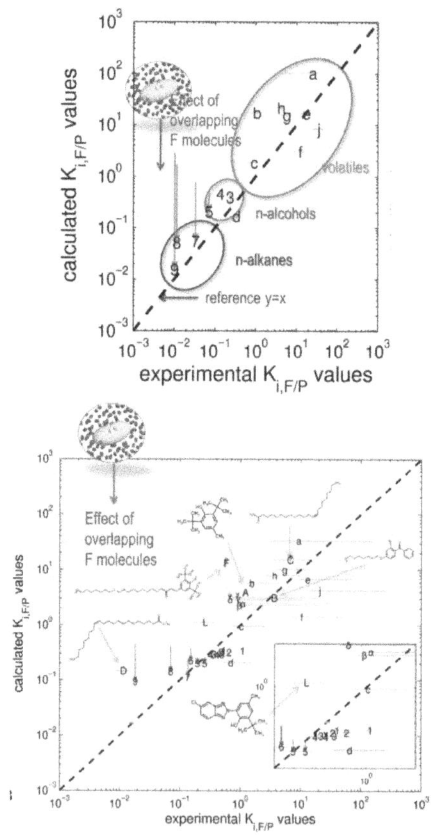

Figure 13.13. *Comparison between partition coefficients values, $K_{i,F/P}$, calculated at 313 K according to equations [13.5] and [13.6] [GIL 10] and the values measured between P = polyethylene (amorphous) and a) F = methanol; b) F = ethanol*

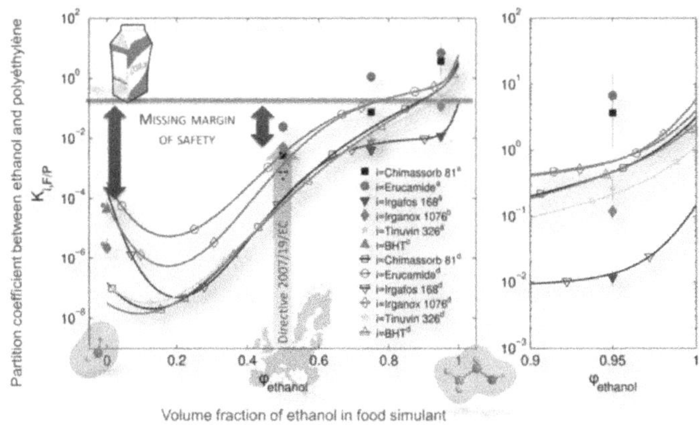

a)

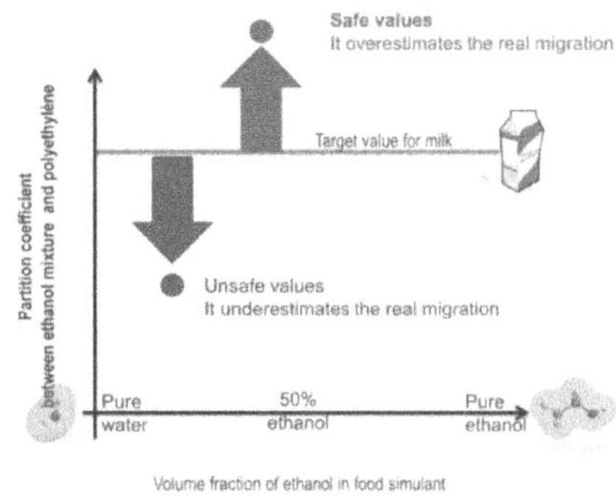

b)

Figure 13.14. *Partition coefficients, $K_{i,F/P}$, between a water–ethanol binary mixture and polyethylene (amorphous) at 313 K [GIL 10]. The symbols represent experimental data and the outlined shapes represent calculated values. (For a color version of this figure, see www.iste.co.uk/hamaide/polymers.zip)*

13.4.2. *Diffusion scaling laws*

Understanding the dynamics of migration phenomena requires the establishment of scaling laws and diffusion mechanisms according to the size (Figure 13.15(a)) and shape of substances (Figure 13.15(b)). There are two typical situations: substances resembling macromolecules (e.g. oligomers and plasticizers) and substances that do not resemble macromolecules (e.g. hindered phenolic antioxidants). Since substances from materials are larger than the pocket of free volume between the macromolecules, their movement requires the collective movement of several segments of macromolecules, and it is therefore highly affected by the state of the amorphous phase and the temperature. Figure 13.15(b) illustrates the volumic dependence of Van der Waals for approximately ellipsoid molecules. This dependence is even greater when the amorphous phase is vitreous.

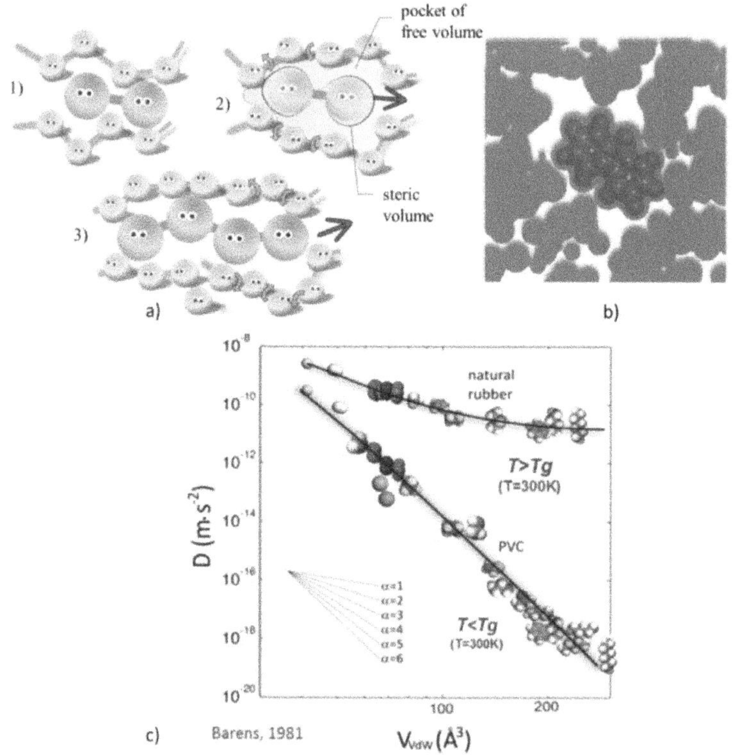

Figure 13.15. *Diffusion mechanisms and scaling*

Figure 13.15 illustrates:

a) temporary encapsulation of a linear molecule between the macromolecular segments (adapted from [MAU 81]);

b) a map of free volumes around a UV stabilizer (Chimassorb 90) in a polyethylene matrix at 298 K. Only the longitudinal section according to the main axes of the dissolvent is presented. The effect of the polymer matrix (natural rubber and polyvinyl chloride) on the diffusion coefficients of small molecules (<200 g·mol^{-1}) at 300 K is analyzed on a log–log scale according to the volume of Van der Waals (V_{VdW}) in the diffusant (Figure 13.15(c)). The results are adapted from [BER 81].

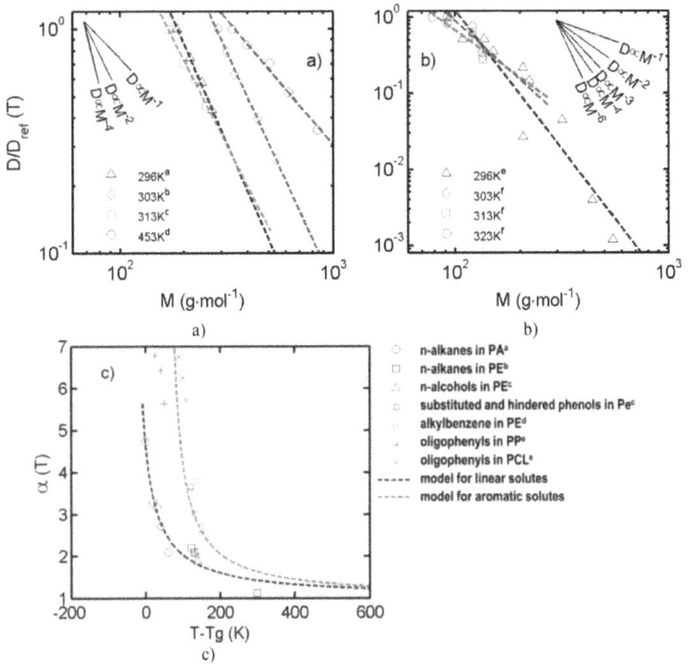

Figure 13.16. *Scaling law associated with the diffusion of homologous solutions [FAN 13]. (For a color version of this figure, see www.iste.co.uk/hamaide/polymers.zip)*

Generally speaking, diffusion coefficients D can be expressed according to a power law of the molar mass [DUR 10, VIT 09, VIT 06, FAN 13]:

$$D \propto M^{-\alpha(T-Tg)} = M^{-1} M^{-\Delta\alpha(T-Tg)} \qquad [13.7]$$

where α is a temperature-dependent exponent, more specifically dependent on the difference in glass transition temperature $T–Tg$ in the rubberized state. $\Delta\alpha$ is the positive deviation of exponent 1 predicted by the Rouse model [ROU 53, DUR 10]. The theoretical values of α can only be obtained for a real homologous series based on the repetition of the same pattern. The values of α for linear and aromatic molecules in aliphatic polymers are shown in Figure 13.16(a) and (b), respectively. Note that the experimental values of α are expected for molten polymers and elastomers (e.g. when T≫Tg) [FAN 13]. The scaling exponent α decays with $T–Tg$ as in:

$$\alpha(T-T_g) = 1 + \Delta\alpha(T-T_g) = 1 + \frac{K_\alpha}{T-T_g+K_\beta} \qquad [13.8]$$

where K_α and K_β are constants. K_α is approximately constant for aliphatic polymers and only K_β depends on the sequence of molecules considered. Figure 13.16 compiles numerous results from literature obtained under controlled conditions and for homologous solutions. The details are given below:

A) flexible solutions;

B) semi-rigid solutions in an aliphatic polymer (polyethylene). The diffusion coefficients (D) are normalized with regard to the value D_{ref} of the first molecule in the sequence and are represented on a log–log scale according to molecular mass, M, for:

a) *n*-alcohols (C7–C18) in LDPE [KOS 86];

b) alkyl-phenols (C6–C18) in LDPE [MOL 94];

c) *n*-alkanes (C12–C18) in LDPE [VIT 07d];

d) *n*-alkanes (C24–C60) in HDPE [MEE 07];

e) substituted and hindered phenols in LDPE;

f) alkyl-benzene in LDPE [DOO 92].

C) dependence of coefficient α on the T–Tg difference for linear diffusants (flexible) and aromatic diffusants in aliphatic polymers. The symbols represent experimental determinations from:

a) [KWA 03];

b) [MOL 94, VIT 07d, MEE 07];

c) [KOS 86];

d) [DOO 92];

e) [FAN 13].

Continuous models are obtained by adjusting equation [13.8] with K_α= 144 K, K_β= 40 K (linear solutions) and K_α= 156 K, K_β= –51 K (aromatic solutions). The experimental data + and × are external validation data not included in the adjustment.

The good correspondences between the experimental data and equation [13.8] is represented in Figure 13.16(c). The figure shows a transfer of 91 K between flexible and rigid aromatic molecules. In other words, when the temperature decreases toward the glass transition temperature of the polymer, the rigid molecules "sense" the transfer restrictions well before the flexible molecules. Similarly, heating a polymer beyond the glass transition temperature is responsible for activating the diffusion of substances of greater masses. This effect caused by the high thermodilatation of thermoplastic polymers is particularly responsible for the significant migration of food pigments under sterilization conditions, as shown in Figure 13.1(b).

Note that equation [13.8] imposes both a deviation in Arrhenius's law (for a large temperature range) and a dependence on the activation energy, *Ea*, on the logarithm of molecular mass (i.e. a variation by a factor of 2–3 is expected in the range 100–1 000 g·mol^{-1}):

$$Ea(T,M) = R\frac{\partial D(T,M)}{\partial 1/T} = K_\alpha \frac{RT^2}{(T-T_g+K_\beta)^2} Ln\left(\frac{M}{M_0}\right) \qquad [13.9]$$

where M_0 is a reference molar mass used to determine α.

Figure 13.17 shows that incorporating a flexible segment close to the center of mass of a rigid molecule allows us to obtain a much larger decrease in diffusion coefficients (*D*) than the effects of size described in Figure 13.16. In the case of diphenyl alkanes, *D* decreases by 10 when 1–3 carbons are incorporated between two aromatic rings. This is associated with an anchor effect of aromatic rings, which are temporarily "trapped" by the absence of free volumes facilitating the reorientation or tilting of the aromatic ring [FAN 13].

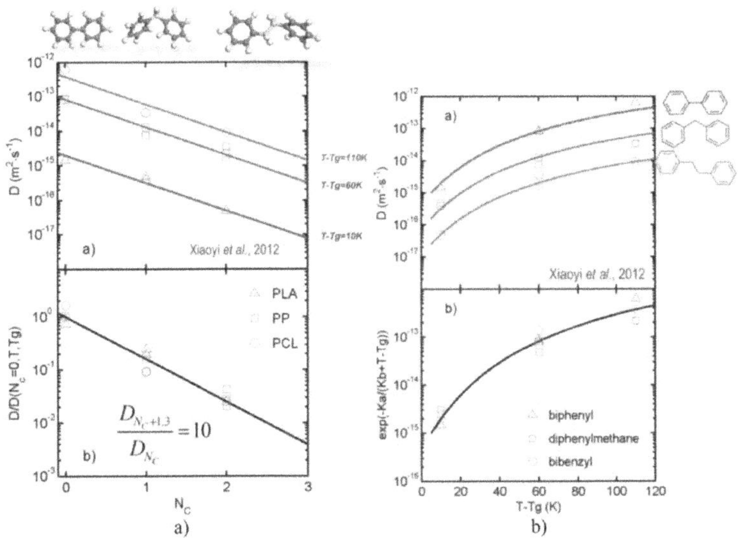

Figure 13.17. *Scaling law associated with the diffusion of diphenyl alkanes. Effects of a) length of the flexible segment (N_C = 0, 1, 2) and b) the T–Tg difference [FAN 13]. (For a color version of this figure, see www.iste.co.uk/hamaide/polymers.zip)*

13.5. Conclusion

This chapter shows that the risk of contamination to food by materials in contact can be assessed from:

1) a description of the transfer phenomena between different packaging and food elements;

2) material composition data (occurrence of substances, ranges of concentrations).

This observation allows us to consider an integrated risk management, pooled across the sector (Figure 13.18) which integrates all the packaging components, the industrial practices and the technological journey of the food package ensemble. The approach proposes, via a quantitative FMECA approach, responses, scenarios and consequences of material transfers, predictable or not, which could become major health crises. On an industrial scale, it is particularly applicable in the regulation 2023/2006/EC [EUR 06] and the norms ISO 22000-200 [ISO 05a, ISO 05b]. Depending on the sector participants, this chapter proposes three levels of corrective actions to minimize the risk of contamination:

1) to modify the formulation of materials (minimize the quantities or choose substances with a weaker migratory power);

2) to develop polymer additives or alternate rigid and flexible segments;

3) to modify the elements of packaging design or use.

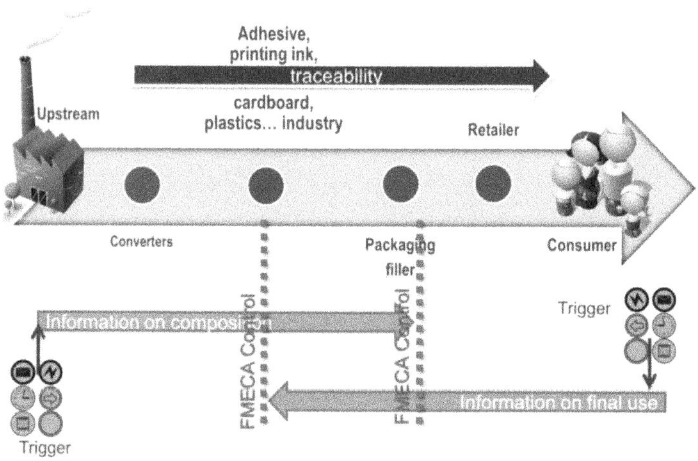

Figure 13.18. *Proposal of mutual management of safety of packaging and materials in contact in contact with food*

The food packaging sector is sketched in Figure 13.18. It includes the chemical industry, material formulators and transformers, agri-food industry, the distributors as well as the control and health monitoring authorities. In the preventative approach to manage the contamination of food by materials, the flow of material from upstream to downstream must be accompanied by a flow of two-dimensional information in line with industrial practices. From upstream to downstream, it is essentially composition data (categories and identities of substances, maximum quantities used). From downstream to upstream, the data must include a range of information about the final use, assemblies, storage conditions and transformations undergone by the packaged food.

13.6. Bibliography

[BAU 66] BAUMAN H.E., *Food Technology*, vol. 28, p. 30, 1974.

[BEN 08] BENETTI C., ANGELETTI R., BINATO G., *et al.*, *Analytica Chimica Acta*, vol. 617, p. 132, 2008.

[BER 81] BERENS A.R., *Pure Appl. Chem*, vol. 53, p. 365, 1981.

[BIE 11] BIEDERMANN M., INGENHOFF J.E., BARBANERA M., *et al.*, *Packaging Technology and Science*, vol. 24, p. 61, 2011.

[CAS 97] CASTLE L., DAMANT A.P., HONEYBONE C.A., *et al.*, *Food Additives and Contaminants*, vol. 14, p. 45, 1997.

[DEM 66] DEMING W.E., *Some Theory of Sampling*, Courier Dover, New York, 1966.

[DOO 92] DOONG S.J., HO W.S.W., *Industrial and Engineering Chemistry Research*, vol. 31, p. 1050, 1992.

[DUR 10] DURAND M., MEYER H., BENZERARA O., *et al.*, *Journal of Chemical Physics*, vol. 132, p. 194902, 2010. Available at http://scitation.aip.org/content/aip/journal/jcp/132/19/10.1063/1.3420646

[EFS 09] EFSA, *EFSA Journal*, vol. 7, no. 1, p. 104, 2009.

[EFS 11] EFSA, *EFSA Journal*, vol. 9, no. 2, p.184, 2011.

[EUP 11] EUPIA, EuPIA Guideline on printing inks applied to the non-food contact surface of food packaging materials and articles, 2011.

[EUR 93] EUROPEAN COMMISSION, Council Directive 93/43/EEC of 14 June 1993 on the hygiene of foodstuffs, 1993.

[EUR 02] EUROPEAN COMMISSION, Official Journal of the European Union, L220, 18, 2002.

[EUR 04] EUROPEAN COMMISSION, Official Journal of the European Union, L338, 4, 2004.

[EUR 06] EUROPEAN COMMISSION, Official Journal of the European Union, L384, 75, 2006.

[EUR 07a] EUROPEAN COMMISSION, Official Journal of the European Union, L136, 50, 2007.

[EUR 07b] EUROPEAN COMMISSION, Official Journal of the European Union, L91, 17, 2007.

[EUR 08] EUROPEAN COMMISSION, Official Journal of the European Union, L86, 9, 2008.

[EUR 11a] EUROPEAN COMMISSION, Official Journal of the European Union, L12, 1, 2011.

[EUR 11b] EUROPEAN COMMISSION, Official Journal of the European Union, L328, 8, 2011.

[EUR 12a] EUROPEAN COMMISSION, Health and Consumers: Food Contact Materials. Legislative lists, http://ec.europa.eu/food/food/chemicalsafety/foodcontact/legisl_list_en.htm, accessed on 19 July 2012.

[EUR 12b] EUROPEAN COMMISSION, Rapid Alert System for Food and Feed-RASFF (Search criteria : "THIOXANTHONE" or "PRINT" for the period of 01/01/1980-31/12/2012), https://webgate.ec.europa.eu/rasff-window/portal/?event=SearchForm&cleanSearch=1, accessed on 25 March 2012.

[FAN 13] FANG X., DOMENEK S., DUCRUET V., *et al.*, *Macromolecules*, vol. 46, 874, 2013.

[FAO 03] FAO, WHO, Assuring food safety and quality: guidelines for strengthening national food control systems, FAO Food and Nutrition Paper 76, Rome, 2003.

[FDA 72] FDA, *Proceedings of the 1971 National Conference on Food Protection*, (eds. U.S. Dept. of Health, Education, and Welfare, Food and Drug Administration, Public Health Service), HHS publication 72-2015, U.S. Government Print Office, Washington DC, 1972.

[GIL 09a] GILLET G., VITRAC O., DESOBRY S., *Industrial and Engineering Chemistry Research*, vol. 48, p. 5285, 2009.

[GIL 09b] GILLET G., VITRAC O., TISSIER D., *et al.*, *Food Additives and Contaminants Part a-Chemistry Analysis Control Exposure and Risk Assessment*, vol. 26, 1556, 2009.

[GIL 10] GILLET G., VITRAC O., DESOBRY S., *Industrial and Engineering Chemistry Research*, vol. 49, p. 7263, 2010.

[GOU 98] GOUBET I., LE QUERE J.L., VOILLEY A.J., *Journal of Agricultural and Food Chemistry*, vol. 46, p. 1981, 1998.

[GOU 13] GOUJOT D., VITRAC O., *Journal of Chemical Engineering Science*, vol. 99, p. 2, 2013.

[GRO 08] GROB K., *New Food*, vol. 1, p. 48, 2008.

[GRO 11] GROB K., "Residues of food contact materials in food", *International Fresenius Conference*, Cologne, Germany, 17–18 October 2011.

[HAM 98] HAMMARLING L., GUSTAVSSON H., SVENSSON K., *et al.*, *Food Additives and Contaminants*, vol. 15, p. 203, 1998.

[HEA 10] HEATH V., *Nature Reviews Endocrinology*, vol. 6, p. 237, 2010.

[HEL 11] HELLING R., "Residues of food contact materials in food", *International Fresenius Conference*, Cologne, Germany, 17–18 October 2011.

[INR 11] INRA, SAFEFOODPACK DESIGN, 2011, http://modmol.agroparistech.fr/SFPD/, accessed on 2 March 2012.

[ISO 05a] ISO 22000, International Standard. ISO 22000, Food safety management systems: Requirements for any organization in the food chain, 1st ed., September 2005.

[ISO 05b] ISO 22000, Technical specification ISO/TS22004, Food safety management systems: Guidance on the application of ISO 22000:2005, 1st ed., November 2005.

[KOS 86] KOSZINOWSKI J., *Journal of Applied Polymer Science*, vol. 31, p. 2711, 1986.

[KWA 03] KWAN K.S., SUBRAMANIAM C.N.P., WARD T.C., *Polymer*, vol. 44, p. 3061, 2003.

[MAU 81] MAURITZ K.A., STOREY R.F., GEORGE S.E., *Macromolecules*, vol. 23, p. 441, 1990.

[MEE 07] VON MEERWALL E.D., LIN H., MATTICE W.L., *Macromolecules*, vol. 40, p. 2002, 2007.

[MOL 94] MÖLLER K., GEVERT T., *Journal of Applied Polymer Science*, vol. 51, p. 895, 1994.

[NAM 10] NAM S.H., SEO Y.M., KIM M.G., *Chemosphere*, vol. 79, p. 949, 2010.

[NGU 13] NGUYEN P.M., GOUJON A., SAUVEGRAIN P., *et al.*, *AIChE Journal*, vol. 59, p. 1183, 2013.

[NRC 83] NRC, Risk assessment in the Federal Government: Managing the process, (eds. Committee on the Institutional Means for Assessment of Risks to Public Health, Commission on Life Sciences, National Research Concil), National Academy Press, Washington, 1983.

[ROU 53] ROUSE J.P.E., *J. Chem. Phys*, vol. 21, p. 1272, 1953.

[SOM 11] SOMOGYI A., HATHCOCK J., BIESALSKI H.K., *et al.*, *Regulatory Toxicology and Pharmacology*, vol. 60, p. 161, 2011.

[VIT 05] VITRAC O., HAYERT M., *Aiche Journal*, vol. 51, p. 1080, 2005.

[VIT 06] VITRAC O., HAYERT M., *Industrial and Engineering Chemistry Research*, vol. 45, p. 7941, 2006.

[VIT 07a] VITRAC O., CHALLE B., LEBLANC J.C., *et al.*, *Food Additives and Contaminants Part a-Chemistry Analysis Control Exposure and Risk Assessment*, vol. 24, p. 75, 2007.

[VIT 07b] VITRAC O., HAYERT M., "Design of safe food packaging under uncertainty", in BERTON P. (ed.), *Chemical Engineering Research Trends*, Nova Publishing, New York, p. 251, 2007.

[VIT 07c] VITRAC O., LEBLANC J.C., *Food Additives and Contaminants: Part A*, vol. 24, p. 194, 2007.

[VIT 07d] VITRAC O., MOUGHARBEL A., FEIGENBAUM A., *Journal of Food Engineering*, vol. 79, p. 1048, 2007.

[VIT 08] VITRAC O., JOLY C., "Physique chimie TI053, chimie des milieux complexes 42529210, Article AF6930", *Techniques de l'Ingénieur*, 22, 2008.

[VIT 09] VITRAC O., JOLY C., "Physique chimie TI053, chimie des milieux complexes 42529210, Article AF6931", *Techniques de l'Ingénieur*, 32, 2009.

[VIT 10] VITRAC O., GILLET G., "An Off-Lattice Flory-Huggins Approach of the Partitioning of Bulky Solutes between Polymers and Interacting Liquids", *International Journal of Chemical Reactor Engineering*, vol. 8, no. 1, June 2010. Available at: http://www.degruyter.com/view/j/ijcre.2010.8.1/ijcre.2010.8.1.2094/ijcre.2010.8.1.2094.xml

[VIT 11a] VITRAC O., JOLY C., "Mathématiques pour l'ingénieur TI052, Applications des Mathématiques 42102210, Article AF1446", *Techniques de l'Ingénieur*, 40, 2011.

[VIT 11b] VITRAC O., FMECA engine open source project, https://github.com/ovitrac/FMECAengine, accessed on 2 March 2012, 2011.

[VIT 12] VITRAC O., SFPP3, http://modmol.agroparistech.fr/SFPP3/SFPP3_migresives/, accessed on 19 July 2012.

[WAL 10] WALSH B., "The Perils of Plastic", *Time Magazine*, 7, 2010.

14

Biodegradability and/or Compostability?

14.1. Biodegradation

14.1.1. *Introduction*

All organic materials, based on the chemistry of organic carbon, can be gradually degraded by diverse pathways (physical, chemical and biological) until they are reduced to molecules as simple as water, carbon dioxide, methane and various inorganic compounds (nitrogen, phosphorous, sulfur, etc.), initial constituents of organic matter. In this respect, we can say that all organic materials are biodegradable, provided that they have sufficient time to degrade: from several days to several weeks for starch, from several weeks to several months for PBAT or PHA, and several hundred years for PET or PE.

From an environmental point of view, to prevent accumulations and dispersions of particles, it is attractive to use polymers whose end of life is characterized by a rapid biodegradation.

The "accurate" measurement of the speed and degree to which a material biodegrades[1] is still very controversial. The only way to remain impartial is to refer to standards determined from laboratory research.

Chapter written by Guy CÉSAR.

1 Very dependent on external factors such as the type of incubation medium, temperature, oxidation, humidity, types of microbial populations and others.

14.1.2. *The definitions of biodegradation*

There are many definitions of biodegradation, all of which include the same main principles.

According to the French norm NFU 52001 *Biodegradable materials in agriculture and horticulture – mulching films – AFNOR 2005*, the definition of biodegradation is as follows:

> The biodegradation of a material occurs via the succession of concomitant physical, chemical and biological phenomena, in all cases leading to the reorganization of biomass and the release of CO_2 (and/or CH_4), water, energy (in the form of heat), eventually producing new organic molecules and possible inorganic residues.

The "*guide to vocabulary in the domain of degradable and biodegradable polymers and plastics. CEN/TR 15351:2006*" proposes the following definitions:

– biodegradation: the degradation of a polymer system due to phenomena resulting from the action of cells;

– aerobic biodegradation: biodegradation under aerobic conditions;

– anaerobic biodegradation: biodegradation under anaerobic conditions;

– biodegradability: ability to undergo biodegradation;

– biodegradable: a polymer system that can be biodegraded;

– degree of biodegradation: fraction of an original polymer system that is biodegraded, measured via phenomena or specific techniques sensitive to the formation of inorganic compounds and biomass;

– maximum degree of biodegradation: maximum degree to which biodegradation that can be reached under selected experimental conditions;

– theoretical degree of biodegradation: theoretical value of the degree of biodegradation corresponding to the total conversion of the organic material initially present in the polymer system into inorganic matter and biomass. In fact, this is also called bioassimilation.

14.1.3. *Stages of biodegradation*

In what follows, to better understand the phenomena, we will study the successive stages, even though they are generally concomitant.

14.1.3.1. *First stage: fragmentation of the material*

According to CEN/TR 15351:2006, fragmentation is the dissociation of a polymer into small particles via any mechanism. Under the action of rather severe erosive anthropogenic and/or natural phenomena, materials in the process of biodegradation are torn (broken) into shreds (pieces) of several decimeters to several millimeters, without a significant impact upon their physico-chemical properties (Figure 14.1).

Figure 14.1. *Fragmentation of covering plastic films: a) PBAT-starch and mulch, b) PE-starch used in agriculture*

14.1.3.2. *Second stage: degradation of the material*

According to *CEN/TR 15351:2006*, degradation is the undesirable modification of the original properties, due to the chemical cleavage of macromolecules forming a polymer system, whatever the mechanism that cleaves the chain.

This stage involves the modification of the material physico-chemical properties following the chemical cleavage of their macromolecules. Six fundamental factors are involved in this degradation:

– humidity and pH of the degradation medium;

– light, and more specifically, UV radiation;

– temperature, which speeds up degradation as it increases;

– oxygen, vital for oxidative degradation[2];

– the presence of exogenous enzymes.

Degradation may be abiotic, if it is purely physico-chemical and/or biotic if at any stage living organisms intervene such as bacteria, fungi and/or algae.

2 The combination of several factors defines degradation as thermo-oxidative, photo-oxidative or even thermo-photo-oxidative.

Generally speaking, under natural conditions, two types of degradation go hand in hand.

Whether biotic or abiotic, degradation is essentially hydrolytic or *(photo/thermo)-* oxidative (*other modes exist such as acid attacks by Thiobacillus).* For example, for the hydrolysis of poly(lactic acid) (PLA), the reaction is very slow at low temperatures (20–25°C), but very fast at temperatures around 60°C (Figure 14.2). We can also mention the abiotic degradation of poly(ethylene) by photo- and thermo-oxidation (Figure 14.3) which results in a drastic decrease in molar masses, a loss of mechanical properties of the material, an increase in the carbonyl index up to 1713 cm^{-1} and the progressive appearance of degradation molecules, such as ketones, alcohols, acids and so on.

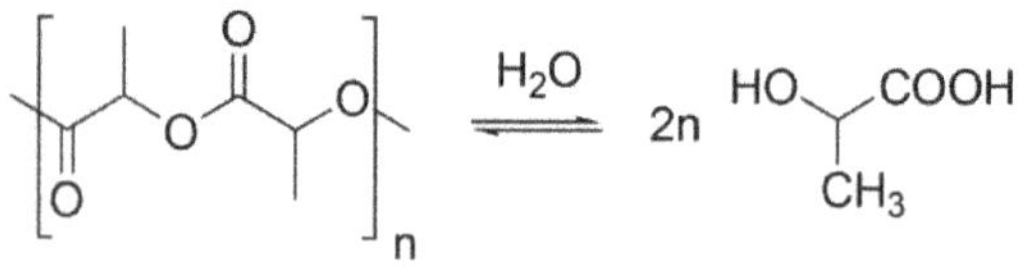

Figure 14.2. *Hydrolysis of poly(lactic acid)*

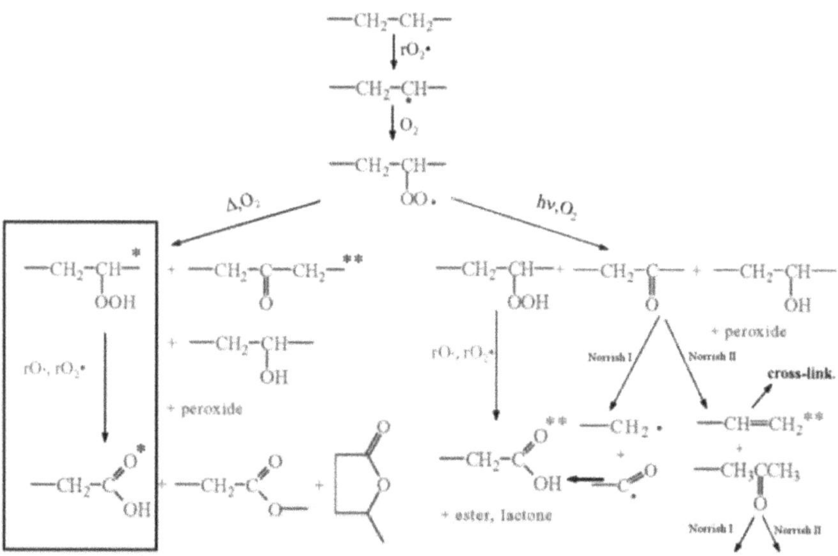

Figure 14.3. *Example of the abiotic degradation of poly(ethylene) by photo- and thermo-oxidation (according to Jacques Lemaire CNEP)*

14.1.3.3. *Third stage: bioassimilation of material*

According to *CEN/TR 15351:2006*, bioassimilation is the conversion of a polymer system into biomass. It is the initial stage of the degradation process. In the final stage, biodegradation and bioassimilation are often synonymous and used interchangeably, wrongly in our opinion. The macromolecules are reduced to sufficiently small molecules that can penetrate living cells and become integrated into anabolic (construction, maintenance and reorganization of biomass) and catabolic (source of energy) physiological cycles.

The simplified equations of catabolic bioassimilation can be written as follows:

a) *in the presence of oxygen (under aerobic conditions):*

$$C_6H_{12}O_6 + 6\ O_2 \rightarrow 6\ CO_2 + 6\ H_2O + \text{energy} \quad [14.1]$$

(production of carbon dioxide)

b) *in the absence of oxygen (under anaerobic conditions):*

$$C_6H_{12}O_6 + 3\ H_2O \rightarrow 3\ CH_4 + 3\ H_2O + 3\ CO_2 + \text{energy} \quad [14.2]$$

(production of methane and CO_2)

$$3\ H_2CO_3 + 12\ H_2 \rightarrow 3\ CH_4 + 9\ H_2O + \text{energy} \quad [14.3]$$

(production of methane)[3]

c) *with the balanced equation:*

$$C_6H_{12}O_6 + 3\ H_2O + 12\ H_2 \rightarrow 6\ CH_4 + 9\ H_2O + \text{energy} \quad [14.4]$$

In practice, the reaction is almost never completed and bioassimilation under anaerobic conditions is almost always accompanied by a production of CH_4 and CO_2, with the amount of CH_4 being produced varying according to the circumstances.

There are three key characteristics of the final stage of biodegradation:

1) the production of CO_2 (under aerobic conditions);

2) the consumption of O_2 (under aerobic conditions);

3) the production of CH_4 and CO_2 (under anaerobic conditions).

It is important to note that the characteristics of the final stage are directly proportional to the quantity of organic material bioassimilated.

3 In strict anaerobiosis, methanogenic bacteria can reduce CO_2 (or HCO_3^-) to methane.

14.1.4. *How to measure biodegradation*

14.1.4.1. *Estimation a priori (non-normative methodology)*

By knowing the exact nature of the material concerned, it is possible to know *a priori* if it will easily biodegrade or not.

For example, groups of NO_2, halogens, SO_3H_2, CF_3 and quaternary carbons will slow the rate of biodegradation, whereas groups of COOH and OH increase the biodegradability regardless of the molecular structure of the compound. Ester and amide groups favor biodegradation, as they are easily hydrolyzed in the environment to form carboxylic acids and amines [SCO 82, ALE 94].

Starch, cellulose and sugars quickly biodegrade (several days or weeks). (Co)polyesters (PBAT, PBAT-starch, PHA, PBS, PBSA) biodegrade less easily (several months). However, some materials almost do not biodegrade; this is the case for PET, polyolefins (PE, PP) and PVC which take several hundred years to biodegrade. PLA degrades very quickly in compost (several weeks) but a lot longer at room temperature (several years).

It is useful to know these elements prior to more specific tests. In particular, this allows us to question the validity of a biodegradation test before concluding when the response observed is different to that obtained *a priori*.

14.1.4.2. *Estimation by measuring weight loss*

The principle is simple. The samples tested, previously weighed in their initial state, are exposed to previously established biodegradation conditions (landfill, compost or in natural water bodies, in the open air, etc.).

The norms NF EN ISO 846 of August 1997 and ISO 15985 of December 2004 (Appendix B) propose methods to calculate the loss of mass for sampled incubated under aerobic and anaerobic laboratory conditions.

As stated in ISO 846, these methods "are not intended to determine the biodegradability of plastics". They do, however, form an initial approach as they do not require complex facilities or high costs. The errors made during the assessments are often due to overestimations caused by the difficulty to distinguish between degradation and biodegradation and/or difficulty correctly cleaning the samples before final weighing and/or the heterogeneity of incubation soils. In addition, to assess the dynamics of (bio)degradation, it is necessary to multiply the repetitions to obtain intermediate points between the initial phase and the final phase of the test[4].

---

4 One material with a minimum of 3 repetitions and 10 intermediate points = 30 tests!

14.1.4.3. *Estimation by image analysis (non-normative methodology)*

This method, valid only for films (preferably colored), allows us to estimate (bio)degradation by image analysis (Figure 14.4).

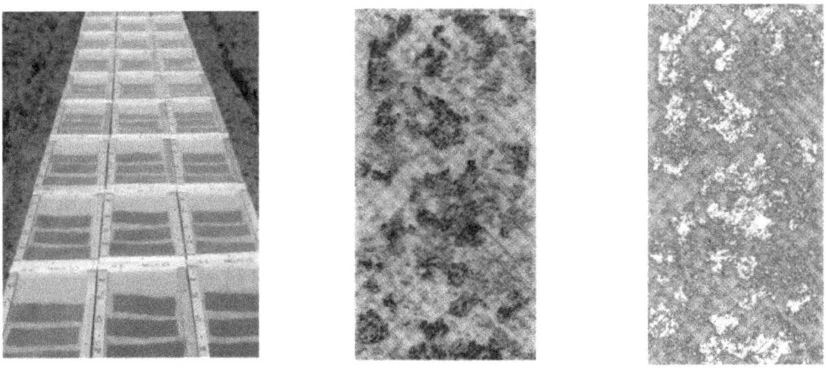

a) b) c)

Figure 14.4. *a) Different types of films (pure PBAT, PBAT-Starch, PE with additives, pure PE) are buried in the ground to carry out weight loss analyses and image analyses after biodegradation; b) after biodegradation the samples are cleaned and scanned; and c) the scanned images are analyzed. The black parts corresponding to any remaining plastic become white and the pixels are counted. The ratio of "white pixels/total pixels × 100" gives the rate of (bio?)degradation*

The samples are exposed to biodegradation conditions according to the same modalities as those defined for estimations by measuring weight loss.

The samples are scanned during sampling, and the images are then processed using image analysis software which separates the zones still with plastic from the rest.

This method is quick, but a certain number of difficulties are encountered:

– the quality of the image analysis obtained will largely depend on the resolution of the scanner;

– the difficulty in properly cleaning the films from the incubation soil leads to overestimations in the rate of (bio)degradation;

– intermediate points can only be obtained by multiplying the tests;

– the difficulty to distinguish between degradation and biodegradation, especially if degradation into very fine particles occurs, with these particles being lost during cleaning or scanning.

14.1.4.4. *Estimating the rate of (bio)degradation of mulching films (protocol by simulation in humid chamber) (non-normative methodology)*

This technique is recommended by the test stations SERAIL (Brindas) and CTIFL (Balandran). It is restricted to films used in agriculture. The aim is to simulate in a humid chamber the degradation of films partially covered by soil. The controls are strictly visual, on top of the soil, below the soil and at the soil–air interface (Figure 14.5).

Figure 14.5. *Each strip of film is half buried in damp soil and half remains on the surface. The following are examined over time: the samples buried in soil, the samples on the surface and the soil–air interfaces (testing the degradation of mulching films was developed by SERAIL, September 2002)*

The state of the sample was assessed every 10 days according to a rating scale 0–5, from not degraded to completely degraded.

Although the method has the advantage of being "visual", it does have some disadvantages:

– the evaluation is qualitative and not quantitative;

– there may be confusion between "degradation" and "biodegradation".

14.1.4.5. *Estimation by observing the development of fungal micro-organisms and or/bacteria on the polymer surface*

Sterilized parts of the polymer (or even calibrated specimens) are placed on a sterile agar medium inoculated with known strains of bacteria and/or fungi. It is also possible to simply put the plastic polymers in contact with agricultural soil or compost. The development of bacteria and/or fungi is then observed as well as the state of the polymer (visual assessment, calculation of the loss of mass, tensile testing, etc.) compared with the samples on the non-inoculated and the sterile media (Figure 14.6). This strictly empirical method was not intended to show the biodegradability of materials in contact[5]. The norm NF EN ISO 846 describes this methodology, but these kinds of tests are insufficient to show the biodegradability of the tested material, as they are too prone to subjective interpretation. However, they can be useful to show and reinforce other types of tests (particularly respirometric).

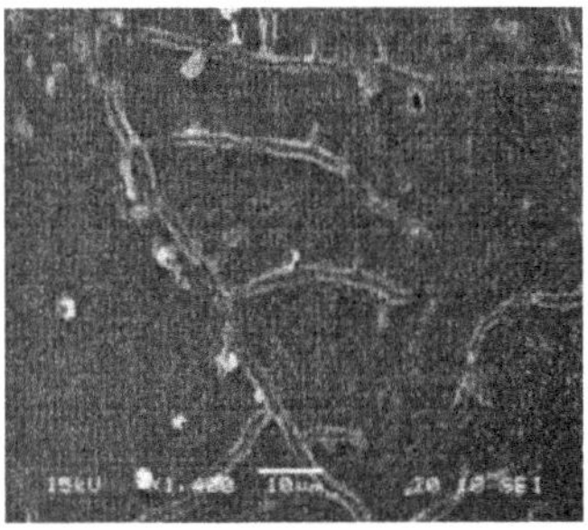

Figure 14.6. *Microphotograph of a scanning electron microscope SEM (1,400 ×) of mycelial filaments developing on the surface of an additive LDPE film after 42 days buried in soil. Can we therefore conclude that there is biodegradation of the material? (Photo Professor Emo Chielini at University of Pisa)*

14.1.4.6. *Estimation by ATP assay (non-normative methodology)*

The principle of this method, developed by the firm BIOTHEMA and its ATP "Biomass Kit HS", involves evaluating against a control the populations of active bacterial cells and/or fungi from perfectly identified strains.

In the presence of luciferase, ATP is determined by luminometric assay according to equation [14.5]:

$$\text{ATP} + \text{D-luciferin} + O_2 \rightarrow \text{AMP} + \text{PPi} + CO_2 + \text{light} \qquad [14.5]$$

where ATP: adenosine triphosphate; AMP: adenosine monophosphate; PPi: inorganic pyrophosphate.

The active cell count by ATP assay is directly proportional to the amount of nutritional substrate consumed, a polymer in our case.

5 NF EN ISO 846 itself specifies that this norm is empirical and does not intend to determine the biodegradability of plastics.

The CNEP [LEM 12] recommended this method to demonstrate the biodegradability of pre-treated additive polyolefins. Biodegradability is said to have occurred if after 180 days, the ATP assay gives a value of at least 100 times greater than the content of ATP determined during the initial incubation with the selected micro-organisms: Rhodococcus, Nocardia, Aspergillus, Mortierella and Cladosporium.

The method is simple to apply, but it requires that the tests are repeated as many times as samples carried out by the ATP assays. It is not possible to establish the dynamics or the final rate of biodegradation with this method. However, interesting results are often obtained that appear to show that pre-treated PE with additives promotes bacterial activity and thus a faster biodegradation of PBAT-starch or PLA [LEM 12]. The method uses selected micro-organisms according to unspecified criteria. There is therefore a bias, as the whole microbial spectrum is not covered, and it is not representative of natural phenomena, which could explain these results that are often difficult to explain.

14.1.4.7. *Estimation by enzymatic assay (non-normative methodology)*

This method involves placing a polymer sample of an initial known weight in the presence of one or several previously chosen enzymes, and comparing it to a control. After a certain amount of time, the sample is removed, cleaned, dried and weighed. The loss of weight allows us to estimate the degree of "attack" on the sample by the enzymatic solution. Many authors, concluding that there is degradation of the tested polymers, have used this technique (Table 14.1).

Authors	Type of polymer studied	Enzyme used
Alejandra Rodriguez-Contreras *et al.* in 2012 [ALE 12]	PHA	Lipase AK from pseudomonas fluorescens and Lipase Lipopan 50 BG produced by Thermomices Lanuginosus
Zhang *et al.* [JIA 11]	PHB	PHB depolymerase produced by Ralstonia pickettii
Tanaka *et al.* [TAN 07]	PHB	PHB depolymerase produced by Ralstonia pickettii
Yem *et al.* [YEM 05]	PLA	α-amylase *which is a digestive enzyme in saliva and pancreatic juices, produced industrially from a culture of Bacillus subtilis*
Chang Lang *et al.* [CHA 05]	Viscose (Rayon and Lyocell)	Cellulase produced by Trichoderma viride

Table 14.1. *Examples of enzyme tests found in literature*

Simple to carry out, this technique is however dependent on the enzyme chosen and does not conclude whether there is definite biodegradation of the sample tested when placed under "natural" conditions. Most authors remain cautious and speak only of degradation.

By choosing the "best enzyme(s)", it is always possible to "show" (bio)degradability of any organic compound. It is therefore necessary, in our opinion, to remain wary with regard to the conclusions deduced from such tests.

The previous sections (14.1.4.1–14.1.4.7) review a variety of tests (non-exhaustive) likely to demonstrate the possible biodegradation of polymers. None of these tests prove that there is biodegradability of the material concerned. The following sections describe methods present convincing arguments, if not definite evidence, to conclude biodegradability.

14.1.4.8. *Estimation by respirometric analyses*

This involves methods whose aims are to determine the rate and speed of biodegradation, or more specifically the bioassimilation of polymers.

The final stage of biodegradation ultimately occurs under aerobic conditions with the consumption of O_2 and release of CO_2 and under anaerobic conditions with the release of CH_4 (methane).

These uptakes and releases are directly proportional to the quantity of organic nutrients containing carbon made available for micro-organisms.

If the initial quantities of organic nutrients available for micro-organisms are known, the quantities of CO_2 (or CH_4) emitted or O_2 absorbed must be accurately measured to deduce the rates of biodegradation.

The principle is as follows:

– the tests are performed in triplicate and include one control test containing only the incubation medium;

– one cellulose test composed of incubation medium + micronized cellulose which help verify the validity of the current test;

– one test of the material composed of incubation medium + the material.

It is considered that the quantities of gas emitted (or absorbed) by the incubation medium and the tested product are additives[6].

Therefore, the actual quantity of gas emitted (or absorbed) by a single tested material will be equal to the difference between that observed of the test minus that of the control.

Finally, the net quantity of gas will be prepared to the theoretical quantity of gas that can be emitted (or consumed) and the rate of biodegradation deduced.

The measurement is regularly repeated until no gas is emitted (or absorbed)[7], a phase usually referred to as "plateau".

There are many norms that define the conditions of measuring bioassimilation (Table 14.2). We will only mention a few principles. In practice, only the aerobic tests measuring the CO_2 released are practical, because in most cases the manufacturers of biodegradable plastics must be compatible with the norms EN 13432, EN 14995 and NF U 52001, which essentially refer to the aerobic measurements of released CO_2.

These methods simulate, with good estimations, the phenomena that take place in a natural medium, liquid or solid (stagnant water or currents, agricultural or forest soils, or a purification or composting facility). These tests therefore have a universal nature that makes them indispensable.

---

6 This is not, however, always true, as synergic or ecotoxic effects may be observed.

7 For example, if the control emits 150 mg of CO_2, the test which emits 225 mg of CO_2, whose maximum theoretical amount of CO_2 that can be emitted is 1,250 mg of CO_2, then the percentage of biodegradation is equal to (225 mg – 150 mg)/1,250 mg × 100 = 6%. The resulting curves are generally sigmoid (except for some exceptions) which can be smoothed using the Hill equation with three constants $[y = ax^b/(c^b + x^b)]$, whose asymptote is determined by constant a, the half-time of biodegradation by coefficient c and the radios of the sigmoid curve by coefficient b. The derivative function of this curve $[y = abc^b x^{(b-1)}/(c^b + x^b)^2]$ is an asymmetric bell-shaped curve that intersects zero and gives biodegradation in units of time, analogous to the degree of instantaneous microbial activity of the incubation medium. By cancelling out the second derivative of the Hill equation $[x = (((c^b(b-1))/(b+1))^{(1/b)})]$, we indicate the point where the rate of biodegradation (thus bacterial activity) is at its maximum.

Anaerobic tests		Aerobic tests		
Incubation on liquid medium	Incubation on solid medium	Incubation on liquid medium	Incubation on solid medium	
Water + digested sludge from anaerobic digestion plants *ISO/DIS 14853*	Digested sludge from anaerobic digestion plants *ISO/DIS 15985* *ASTM D 5511-94*	Water + reactive bacteria from purification plants *NF U 52001* *NF EN ISO 14851* *NF EN ISO 14852*	Actual soil *NF U 52001* *ASTM D5988-96*	Compost *NF U 52001* *NF EN 13432* *NF EN 14995* *NF EN ISO 14855* *NF EN 14046*
Measurement of biogases released (CH_4 + CO_2)	Measurement of biogases released (CH_4 + CO_2)	Measurement of CO_2 released or O_2 consumed	Measurement of CO_2 released or O_2 consumed	Measurement of CO_2 released
35°C ± 2°C	52°C ± 2°C	38 °C ± 1°C	28°C ± 2°C	58°C ± 2°C
Max 90 days	Max 15 days extra time possible	Max 183 days	Max 365 days	Max 183 days
		Validation		
Cellulose ≥70% when stationary phase is reached	Cellulose ≥70% after 15 days	Cellulose ≥70% after 45 days	Cellulose ≥70% after 183 days	Cellulose ≥70% after 45 days
		Interpretation		
Material ≥90% of the value of cellulose at the end of the test	Material ≥90% of the value of cellulose at the end of the test	Material ≥90% of the value of cellulose at the end of the test	Material≥60% of the value of cellulose at the end of the test	Material ≥90% of the value of cellulose at the end of the test
		Areas of application		
Solid or water soluble polymers likely to enter anaerobic digestion plants or landfill site	Solid polymers likely to enter anaerobic digestion plants or landfill site	Solid or water soluble polymers likely to enter purification plants (PVA, for example)	Polymers destined for landfill (agricultural films, for example)	Polymers whose end of life is in an industrial composting facility (packaging, for example)
Tests hardly used		Tests most frequently used (except O_2 consumed which is hardly used)		

Table 14.2. *Different types of tests of biodegradability*

14.1.4.8.1. Respirometric test on a solid anaerobic medium with the measurement of emitted biogases

A diagram of this test, drawn according to the norm ISO/DIS 15985, is represented in Figure 14.7.

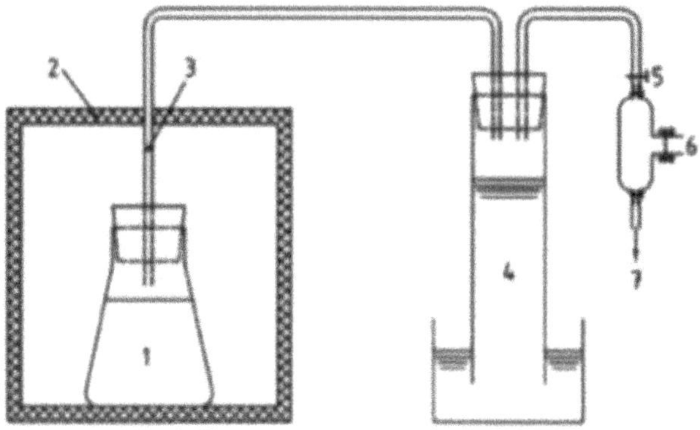

1. Digester
2. Incubator
3. Gas exit
4. Gas collector
5. Valve
6. Gas sample
7. Outflow

Figure 14.7. *Diagram of an anaerobic digestion apparatus of high solid content according to the norm ISO/DIS 15985 (the gas collector must contain water of pH* < 2 *to prevent the emitted* CO_2 *from dissolving, the volume of* $(CH_4 + CO_2)$ *is measured in (4), converted into normal conditions by* $PV/T = P_1V_1/T_1$*, then expressed in moles using the relation* $n = pV/RT$ *before being converted into mg of C by being multiplied by 12,011).*

14.1.4.8.2. Respirometric test on a liquid anaerobic medium with the measurement of emitted biogases

The diagram of this test, drawn according to the norm ISO/DIS 14853, is represented in Figure 14.8. By applying n = pV/RT, we determine n (in moles of the $CH_4 + CO_2$ mixture), we obtain the mass of C by multiplying by 12,011 and by analyzing the liquid phase, we determine the dissolved organic C which is added at the end of the process to the biodegradation curve. At the end of the test, the figures are similar, the intermediate and final rates of biodegradation are determined. The graphical representation of this is given in Figure 14.9.

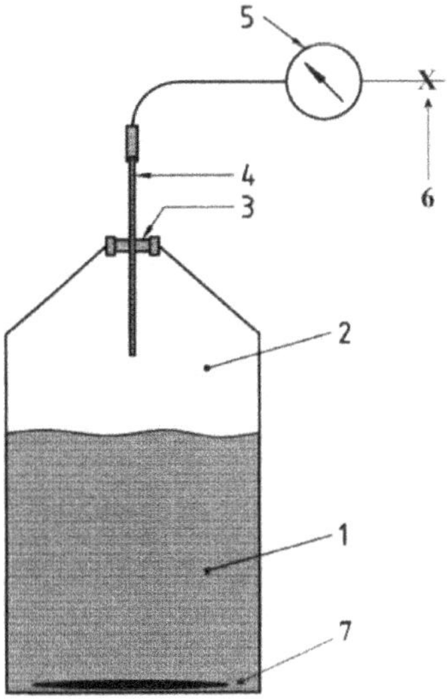

Figure 14.8. *Diagram of a biodegradation apparatus according to the norm ISO/DIS 14853), 1: inoculum; 2: headspace (10% of the total volume when regular gas discharges are programmed, but it may be 30% is a single measurement is carried out at the end of the test); 3: airtight seal; 4: syringe needle attached to the manometer allowing pressures of between 0 and 2,000 hPa to be measured; 5: manometer; 6: valve allowing the outlet of gases formed and the return to nominal pressure (atmospheric); 7: magnetic bar (optional)*

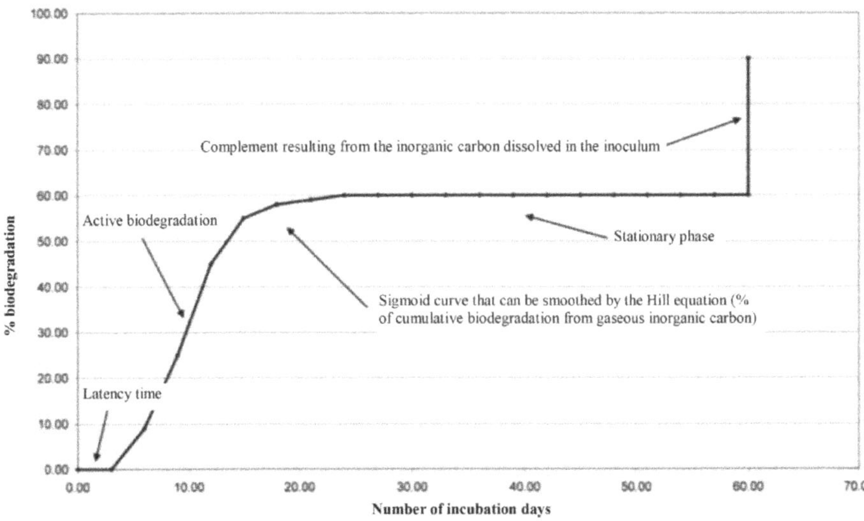

Figure 14.9. *Example of typical biodegradation curve (the first part of the curve resembles the Hill sigmoid)*

14.1.4.8.3. Respirometric test on a solid aerobic medium with the measurement of oxygen consumed

A diagram of this test, drawn according to the norm ISO/DIS 17556.2, is represented in Figure 14.10. The CO_2 trap is composed of KOH pellets. The production of oxygen is triggered electrolytically. The manometer measures the vacuum caused by the respiration of micro-organisms and triggers the electrolysis of water). The production of oxygen required to compensate for the vacuum produced is immediately translated into $mg.dm^{-3}$ of oxygen injected into the respirometer and is recorded by the computer. It is directly proportional to the amount of electricity used, but is specific to each respirator and is provided by the manufacturer, for an enthalpy of dissociation of water of 285 $kJ.mol^{-1}$, an electrolyzer yield of 75–95%, using between 8 and 12 $mg.dm^{-3}$ of oxygen produced at 0 °C and 1 bar. This operation can also be completely manual (although tedious) with less sophisticated respirometers. The system described may be replaced by vacuum measurement of biological oxygen demand (BOD).

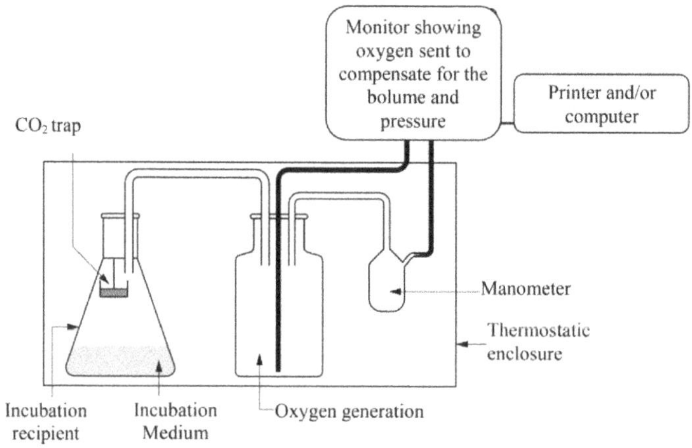

Figure 14.10. *Diagram of a manometric respiration test according to the norm ISO/DIS 17556.2*

14.1.4.8.4. Respirometric test in a liquid aerobic medium with the measurement of CO_2 released[8]

The CO_2 released is captured in a base of a known concentration. The excess base is back titrated by a strong acid in the presence of a universal indicator. The chemical equations used are as follows:

8 Also called a "Stürm test", after the name of the author who described the method for the first time in 1979, applying it to petroleum derivatives.

$$2\ NaOH + CO_2 \rightarrow Na_2CO_3 + H_2O \qquad [14.6]$$

$$Na_2CO_3 + BaCl_2 \rightarrow 2\ NaCl + BaCO_3 \text{ (milky white precipitate)} \qquad [14.7]$$

$$NaOH + HCl + \text{thymolphtaleine} \rightarrow NaCl + H_2O \qquad [14.8]$$

(titration of excess NaOH. Change from blue color to milky white).

Or even:

$$Ba\ (OH)_2 + CO_2 \rightarrow BaCO_3 + H_2O \qquad [14.9]$$

$$Ba(OH)_2 + HCl + \text{thymolphtaleine} \rightarrow BaCl_2 + 2H_2O \qquad [14.10]$$

(titration of excess $Ba(OH)_2$. Change from blue color to milky white).

Incubation usually takes place at 38°C ± 1°C, despite the norm recommending temperatures of between 20 and 25°C ± 1°C. The reference material (positive control) will be micronized cellulose used in chromatography. Figure 14.11 shows an open-circuit "Stürm" set-up. The system can be entirely automated, but with a higher overall cost of apparatus.

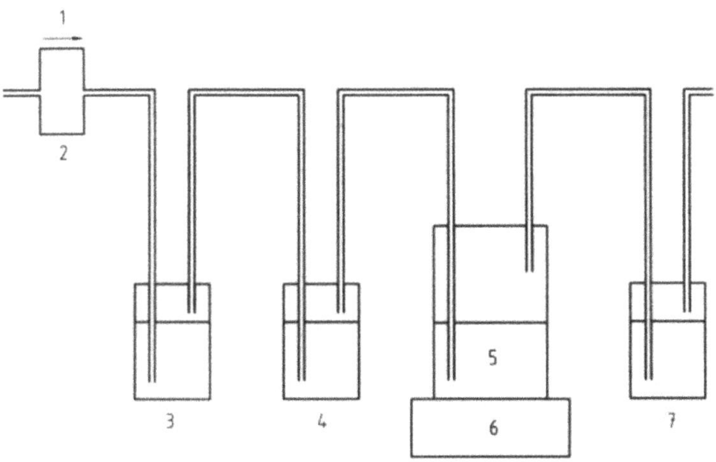

Legend :
1 Compressed air
2 Control of flow
3 Carbon dioxide trap (for example by a gas washing bottle containing alkali)
4 Indicator of carbon dioxide [$Ba(OH)_2$]
5 Test recipient
6 Mixer
7 Carbon dioxide trap [$Ba(OH)_2$ or NaOH]

Figure 14.11. *Diagram showing open-circuit "Stürm" set-up (according to the norm ISO 14852)*

14.1.4.8.5. Respirometric test on an aerobic solid medium by measuring CO_2 released

The incubation medium is real soil at a temperature of 28°C ± 1°C[9].

These tests can be applied to all organic matter destined for landfill, for which one must assess the degree and speed of biodegradation. For example, biodegradable polymers or those considered as such, including paper, PLA, PBAT, PBS, PBSE, PHA, etc. or various combinations of these polymers.

The principle is based on the titration of CO_2 released, resulting from the respiration of micro-organisms using the carbon compound as a nutrient.

The CO_2 is previously recovered from a known solution of NaOH to form Na_2CO_3 according to equation [14.6]; it is then precipitated using a solution of excess $BaCl_2$ as described in equation [14.7]. The remaining unused NaOH is then titrated by HCl in the presence of thymolphtaleine used as an indicator. After each titration, a fresh solution of NaOH is used. The time interval between each titration will depend on the biodegradation speed of the tested polymer.

Clean conventional 1 liter jars were used. Each jar contained 1 vial containing the sample to be tested firmly placed in soil, 1 humidifying vial containing distilled water and 1 vial containing the exact quantity of the known titrated solution of NaOH. The tests are carried out in triplicate against the zero control and the micronized cellulose (Figure 14.12).

The incubation medium is compost at a temperature of 58 °C ± 2°C

While the incubation medium is no longer soil, but compost, the technique is similar to that which we have just described, with some differences (Figure 14.13). The method recommended in the norm is in fact only an adaptation of the open-circuit Stürm test. Just like the Stürm test, this system can be entirely automated (Figure 14.14). In this case, the costs are much greater. (It costs at least €60,000 for a battery with 12 units allowing the testing of 2 samples at the same time = 3 controls + 3 celluloses + 3 samples 1 + 3 samples 2).

9 Adapted from the norm ASTM D5988-03 and appendix F of the norm NF U52-001 which help limit the consumption of reagents as well as the area of lab benches used and the costs by allowing multiple tests to be carried out at the same time.

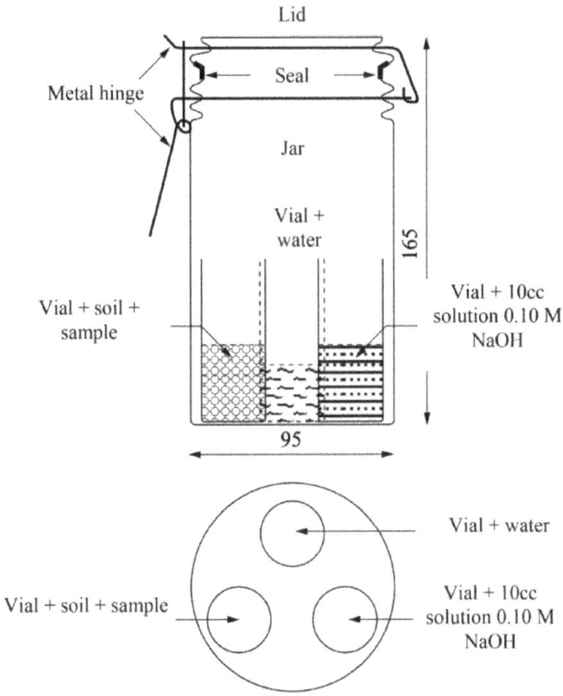

Figure 14.12. *Diagram of biodegradation according to Master II course in the ecodesign of polymers and composites at the University of South Brittany (UBS), Lorient, France*

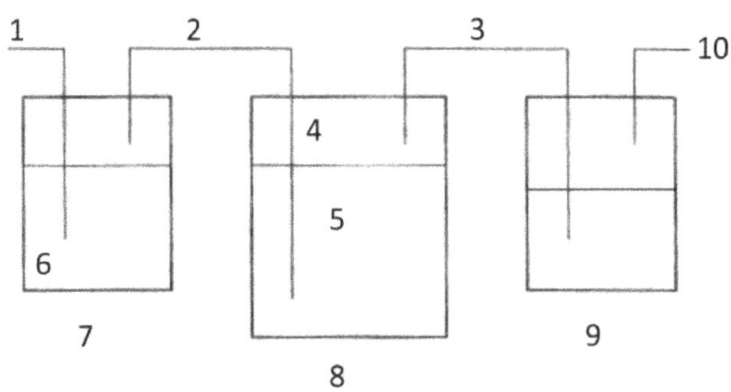

Figure 14.13. *Diagram according to ISO 14855; 1: air entry; 2: passage of air free from CO_2; 3: passage of air loaded with CO_2 through the gas washing bottle containing a basic solution (NaOH or $Ba(OH)_2$); 4: air loaded with CO_2 in the headspace; 5: incubation medium + test material (approximately 600 g, or ±1 liter); 6: solution to trap CO_2 (NaOH 10 moles/liter); 7: vial for capturing CO_2; 8: incubation vial (volume = 3 liters); 9: vial for capturing emitted CO_2; 10: exit of air free of CO_2*

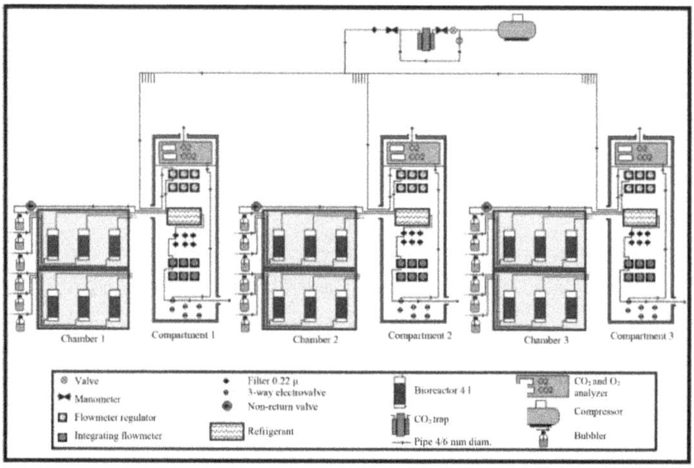

Figure 14.14. *Diagram of an entirely automated set-up of the respirometric test on compost (Siemens)*

Another system for measuring the biodegradability by incubation in compost, based on successive weighing, was proposed by a Japanese company. The technique used is an adaptation of the norm ISO 14855-2. Microbial oxidative degradation analyser (MODA) was designed, tested and manufactured by a consortium composed of the Department of Pharmaceutical Sciences at the University of Shizuoka, the Fuji Industrial Research Institute, the administrative center of Shizuoka, Saida Ironworks Ltd., the University of Michigan state and the Department of Chemical Engineering of Michigan at Lansing (Figure 14.15).

The incubation was carried out on compost agitated once a week, and aerated (air free from CO_2 with a flow rate of 2.40 $dm^{-3}.h^{-1}$) at a defined temperature (58°C ± 2°C). Water and ammonia released are captured by a solution of H_2SO_4. The chambers of SiO_2 and $CaCl_2$ act as additional desiccants, eliminating all traces of water, from either water in the circulating air, or from the chemical reaction between NaOH and CO_2. CO_2 is captured by a chamber full of NaOH pellets whose weight gradually increases in accordance with the biodegradation of the material. Intermediate weight measurements of the chamber helps establish how the dynamics of biodegradation change over time. The system, in triplicate, is completed by a positive control containing micronized cellulose and a control containing only the incubation medium. This system is costly[10] and consumed high quantities of reagents, but is simple to carry out.

10 It costs 50,000 euros for 6 chambers = 2 zero controls, 2 cellulose controls, 2 tests = 1 test in 2 repetitions.

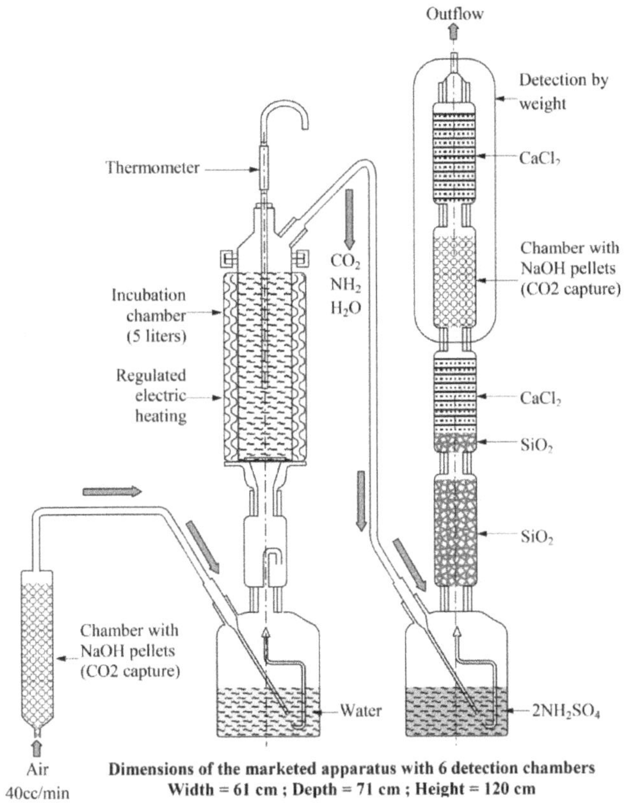

Figure 14.15. *Diagram showing an analyzer of microbial oxidative degradation*

14.1.5. *Specific study of polyolefins added to pro-oxidants*

The oxidation of a polyolefin is a normal phenomenon, but extremely slow.

These reactions can be accelerated by adding pro-oxidant additives to polymers, usually transition metals, such as carbamic compounds of iron or nickel, or stearic compounds such as stearates of cobalt or manganese, etc.

In practice, these phenomena cause a loss in the mechanical properties of the polymer which becomes brittle. From an analytical point of view, a decrease in molar masses is observed as well as an increase in IR absorbance up to 1,713 cm^{-1} characteristic of the presence of carbonyl groups (Figure 14.16). How biodegradation occurs depends on many factors such as type of material, its concentration in the incubation medium, type of incubation medium, temperature. If the responses are often similar to Hill curves, synergic interactions may be observed (which is often the case for PHA) as well as ecotoxic interactions (observed with PVOH and additive PE) between the test material and the incubation medium. In all cases, extrapolations using the equations obtained must be completely avoided, as it is very difficult to predict the evolution of biodegradation from several points obtained in a short period of time.

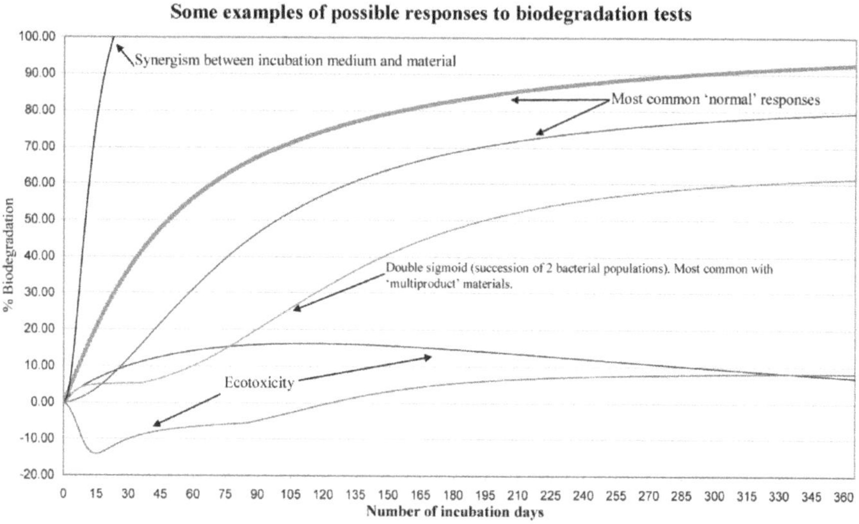

Figure 14.16. *Examples of possible biodegradation test analyses carried out by SERPBIO*

Many organic molecules are released into the immediate environment of the polymer during degradation (carboxylic acids, ketones, ketoacids, alcohols, etc.)[11]. It is well-known and accepted that biodegradation of such materials can only occur when degradation is well under way (sufficiently small molar masses ≈ 500 g.mol^{-1} or carbonyl indices[12] at 1,713 cm^{-1} greater than 0.02). In fact, under "natural" conditions, degradation is always too low for biodegradation to occur a tat sufficient rate. These biodegradable materials are not "in accordance with the current norms" or a "good example of environmental management". To "improve" the results of biodegradation under laboratory conditions, the materials are often subject to, for

11 Sigbritt Karlson, 1996.

12 The carbonyl index is determined for plastic sheets. It is equal to [(absorbance measured t_n – initial absorbance at t_O)/thickness of film in microns]. Note that at index 0.01, the material has lost all its mechanical properties.

several weeks, oxidative pretreatments under well-ventilated conditions, at high temperatures and intense UV light. In this case, it becomes difficult, if not impossible, to establish weak correlations between laboratory tests and tests in the natural environment.

14.1.6. *Examples*

Table 14.3 provides several examples of biodegradation observed under laboratory conditions. Attention must be drawn to the large differences observed under the chosen biodegradation conditions. The standard deviations of the values obtained must also be taken into account as they can be large and vary from one test to another and/or from one laboratory to another. It is not uncommon to observe standard deviations of approximately 10–15%.

Material	Incubation medium	Incubation temperature (°C)	t1/2 (days) (a)	Biodegradation in 183 days (b)	Biodegradation in 365 days(c)
Cellulose	Compost	58	10	96	–
Cellulose	Water with added bacteria from purification stations	38	16	96	–
Cellulose	Seawater	25	139	–	51
Cellulose	Real soil	28	50	–	93
PBAT	Real soil	28	56	–	64
PBAT+starch	Real soil	28	80		80
PBAT+PLA	Water with added bacteria from purification stations	38	14	100	–
PBAT+PLA	Real soil	28	40	–	62
PCL	Real soil	28	153	–	67
PCL + 10% nanomaterials	Real soil	28	153	–	79
PE	Real soil	28	No biodegradation	–	0
PE with added pro-oxidants with preoxidation (d)	Real soil	28	20	–	17

PE with added pro-oxidants with preoxidation (e)	Water with added bacteria from purification stations	37	80	30	–
PE with added pro-oxidants without preoxidation	Real soil	28	No biodegradation	–	0
PHBV	Seawater	25	116	–	97
PHBV	Beach sand	25	102	–	100
PLA	Compost	58	37	100	---
PLA	Seawater	25	167	–	24
PPC	Compost	58	13	33	–
PU (biodegradable)	Compost	58	11	100	–
Viscose Lyocell (*values extrapolated from* Chang Lang Yoon) [CHA 05]	Water with added bacteria from purification stations	30	16	66	–
Viscose Rayon (*values extrapolated from* Chang Lang Yoon) [CHA 05]	Water with added bacteria from purification stations	30	13	88	–

(a) Time needed for 1/2 biodegradation in days
(b) Biodegradation after 183 days as percentage (maximum duration of test)
(c) Biodegradation after 365 days as percentage (maximum of test)
(d) Product exposed to bad weather for several weeks
(e) Product left for several weeks in free air and at room temperature (F. Silvestre, Agrice 2006)

Table 14.3. *Some values, mostly recorded from our experiments*

14.2. Composting

14.2.1. *Compostability*

Compostability involves the fermentation of agricultural and/or urban residues mixed or not with topsoil. The compostability is the ability of a material to be compostable.

The terms "compostable" and "compostability" are not defined in many dictionaries, but can be found in technical norms, particularly in Europe (NF EN 13432). This world recognized European standard defines the conditions that must be satisfied for a packaging (according to the European directive 94/62/CE) to be declared compostable under industrial conditions.

14.2.2. *Norm NF EN 13432*

The norm NF EN 13432 is composed of four main parts.

14.2.2.1. *Characterization of the material*

– Identification of components (each organic component, if its proportion exceeds 1% mass it must comply with NF EN 13432 and the sum of all organic constituents with proportions less or equal to 1% mass cannot exceed 5% mass).

– Determination or harmful substances (e.g. analysis of trace elements, which must comply with acceptability levels).

– Determination of organic carbon.

– Determination of dry solids.

– Determination of volatile compounds (*max. = 50% mass*).

14.2.2.2. *Biodegradability*

14.2.2.2.1. Aerobic biodegradability

This test is always required. It is carried out under laboratory conditions by incubating the test material against cellulose, in compost at 58 °C for a maximum of six months. The validity of the test is assured if cellulose degrades by at least 70% in 45 days. The material is considered to be biodegradable if it reaches 90% of the value for cellulose when the level of cellulose plateaus or after a maximum of six months.

14.2.2.2.2. Anaerobic biodegradability

This test is optional, and almost never requested in practice. The material is considered to be biodegradable if it reaches 50% of the value of cellulose when the level of cellulose plateaus or after two months at 35°C. Two examples are shown in Figures 14.17 and 14.18.

14.2.2.3. *Disintegration*

Aerobic decay is always required. It occurs under actual industrial composting conditions or in a pilot plant reproducing industrial composting conditions as accurately as possible (1% mass or volume of compost). The material is exposed for a maximum of 12 weeks under industrial composting conditions. After 12 weeks, we verify that at least 90% of the products of disintegration pass through a sieve with 2 mm mesh. This test can easily separate materials composed of well-known compounds by their biodegradation qualities (e.g. PLA of 50 µm in thickness will disintegrate easily, but 1 cm thick PLA will not disintegrate properly).

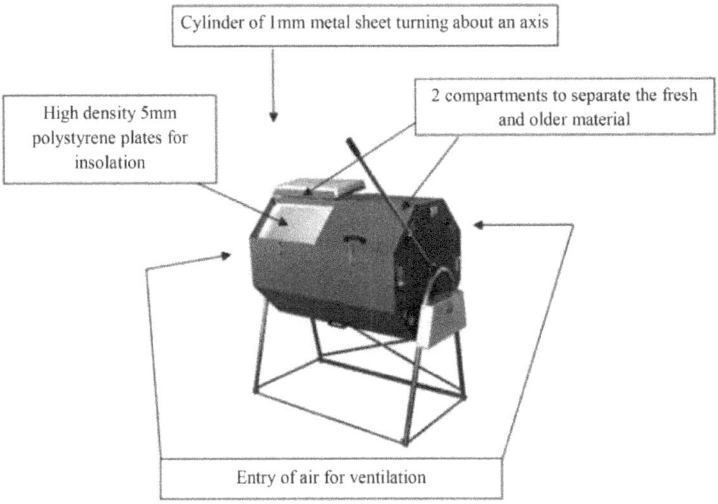

Figure 14.17. *Example of the manual composting pilot plant JK 400 (€10,000/unit) as proposed by Sytevom in Franche-Comté and installed in Faverney college. http://www.sytevom.org*

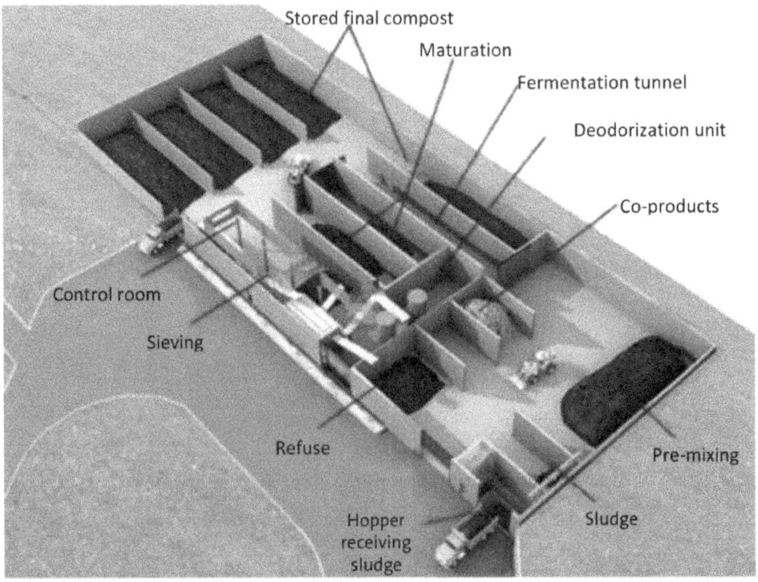

Figure 14.18. *Example of an industrial composing plant (€4,000,000). According to the Environmental Technology center of Pays de Châteaubriant. http://www.pays-chateaubriant.fr/*

Anaerobic disintegration is an optional test, which is almost never requested. It is carried out under actual anaerobic digestion conditions or in a pilot plant that reproduces these same conditions. The material is exposed for a maximum of five weeks to a combination of anaerobic digestion and aerobic stabilization. After five weeks, we verify that at least 90% of the products of disintegration pass through a sieve with 2 mm mesh.

14.2.2.4. *Ecotoxicity*

The tests are performed using a 10% fermentate mixture of weight of the material concerned in the compost for 12 weeks. The fermentate obtained is then mixed by 25 and 50% volume or weight in a culture substrate. One monocot (barley) and 1 dicot (lettuce) are sown on these preparations, and then the germination rates and biomass are measured. There is conformity when at least 90% of the values obtained using the control (pure compost) is reached.

14.3. Ten questions about biodegradability and compostability

14.3.1. *Can certain materials (polyolefins with additives) be sometimes called "compostable"? Is this true?*

No! They may eventually pass several tests required by NF EN 13 432, but in no case the biodegradability test.

14.3.2. *Is it acceptable to say that "my product is biodegradable"?*

No. From an environmental point of view, this kind of statement has no meaning. It is always necessary to refer to the conditions determining this "biodegradability".

Incubation medium? Temperature? What level of biodegradation was reached and during how much time? Is the biodegradation absolute or relative to cellulose?

14.3.3. *Is it acceptable to say that "my product is compostable"?*

No, we must always specify the compostability conditions, i.e. by saying "my product is compostable according to norm NF EN 13432" or even "my product complies with norm NF EN 13432 and can be declared biodegradable by composting under industrial conditions" or even by consulting a known mark of conformity that refers to the same norms *ad hoc*.

14.3.4. *Is there a norm for compostability under "domestic" conditions?*

No. The allocated marks of conformity are only based on "private" specifications, which are non-consensual despite being based on the standards for compostability under industrial conditions.

14.3.5. *Are the materials said to be oxobiodegradable actually biodegradable?*

No. They do not comply with any current biodegradability norm. However, after their oxidative phase (*which can only occur when there is sufficient oxygen, heat and light*), they may show some biodegradability, slow, although faster than those with added polyolefins.

14.3.6. *My material is not packaging, can it still conform to NF EN 13432?*

No. The norm NF EN 13432 is strictly restricted to packaging as defined by the European directive 94/62/CE. However, compliance can be obtained with NF EN 14995, a certified copy of NF EN 13432, but it is restricted to all materials except packaging. Failure to observe this regulation could lead to prosecution for false advertising.

14.3.7. *Can I say that my material conforms to NF EN 13432 and/or to NFU 52001 without consulting a certification body?*

Yes. Applying these norms is not compulsory. However, you will encounter fraud control and mistrusting clients if your material is not consistent with your declarations.

14.3.8. *Can I "officially" declare that my material is biodegradable based on one or several (non-) respirometric and (non-) standardized tests of biodegradation?*

No. All of these types of tests are of interest when providing evidence for hypotheses of biodegradation and there is often no minimum threshold to meet. According to common sense, it is easy to take the results of these tests as proof. An "official" declaration of biodegradation can only be proved if it results from a set of standardized tests, defining the thresholds to be reached, especially respirometric, chemical and ecotoxicological tests, etc. For example, it will then be possible to declare that "my product is biodegradable in landfill sites, as it conforms to the norm NF U 52001" or even "my product is biodegradable via industrial composting techniques, as it complies with the norm NF EN 13432 or NF EN 14995" or even by consulting a well-known mark of conformity that refers to the same norms *ad hoc*.

14.3.9. *Is a biodegradable material necessarily compostable?*

No, since compostability particularly depends on the thickness and/or shape of the material considered. For example, PLA can easily biodegrade in the laboratory following incubation in compost at 58°C, but is completely incapable of passing the disintegration test if its thickness is too large.

14.3.10. *Is a material, compostable under industrial conditions, biodegradable?*

Yes, on the sole condition that you specify that the biodegradability test was carried out on compost at 58°C.

14.4. Conclusion

Only biodegradation tests, preferably normative, can conclude that a given material is biodegradable, but they cannot conclude, in any case, its eco-compatibility. In an attempt to approach this, other tests must complement those that determine the biodegradability of materials, such as chemical analyses and ecotoxicity studies. This is the focus of other norms based on those devoted to studying biodegradability and are supplemented by new tests. Thus, we answer the question posed in the article title; in fact, biodegradability tests are ultimately only part of certification standards, whether this may be NF EN 13432, or NF U 52001 or all other equivalent norms.

There are numerous biodegradability tests, which are often long, costly, and may vary from one laboratory to another, and often with significant variations within the same laboratory. Declaring that a material is biodegradable without specifying exactly under which conditions the tests were performed, does not make sense. Therefore, saying that PE or PP is biodegradable without specific details is correct, but this does not hold any environmental validity.

Many studies are currently underway to limit the duration and variability of tests and ensure more generalizable results, for example, for cultures on inert media with standardized microbial cocktails, but still without concrete results.

In composting, 12-week tests proposed by the norms do not satisfy many composting industrialists working with methods that can obtain compost after only four weeks. Current studies are also trying to limit the duration of the tests.

Ecotoxicity tests required by the norms are criticized, because the scope of the tests is not extensive enough.

Despite their limitations, the standardized requirement currently in force can be considered as a good compromise, as they prevent prohibitive costs associated with significant experimental constraints, as well as limiting the environmental constraints, when managing the end of life of materials classed as biodegradable or compostable.

14.5. Bibliography

[AGR 06] AGRICE P., Etude des mécanismes de dégradation, d'écotoxicité et d'accumulation des biopolymères utilisables en agriculture. Recherche et évaluation des conséquences de la présence éventuelle de micro-fragments issus de cette dégradation dans les sols et les plantes, n° 0201021, Rapport Tache N° 2, Françoise Silvestre, 23 March 2006.

[ALE 12] ALEJANDRA R.-C., MARGARITA C.-M., MARÍA M.-C., "Enzymatic degradation of poly(3-hydroxybutyrate-co-4- hydroxybutyrate) by commercial lipases", *Polymer Degradation and Stability*, 11 January 2012.

[ALE 94] ALEXANDER M., *Biodegradation and Bioremadiation*, Academic Press, 1994 and 1999.

[CES 08] CÉSAR G., Cours de biodégradation de Master II, écoconception des polymères et composites de l'Université de Bretagne Sud, Lorient, 2008-2013.

[CHA 05] CHANG LANG Y., CHUNG HEE P., YUN-KYUNG K., *et al.*, "Biodegradability of viscose Rayon and Lyocell fibers", *Journal of the Korean Society of Clothing and Textiles*, vol. 29, nos. 3–4, pp. 470–477, 2005.

[FON 13] FONTANELLA S., BONHOMME S., BRUSSON J.M., *et al.*, "Comparison of biodegradability of various polypropulene films containing pro-oxidant additives based on Mn, Mn/Fr or Co", *Polymer Degradation and Stability*, vol. 98, pp. 875–884, 2013.

[KAR 97] KARLSSSON S., "Dicarboxylic acids and ketoacids formed in degradable polyethylenes by zip depolymerisation through a cyclic transition state", *Macromolecules*, vol. 30, pp. 7721–7728, 1997.

[LEM 12] LEMAIRE J., DELORT A.M., FROMAGEOT D., *et al.*, "Hydrobiodégradables et oxobiodégradables", *Cahiers techniques. Plastiques et caoutchoucs*, 898, November 2012.

[ROD 12] RODRIGUEZ-CONTRERAS A., CALAFELL-MONFORT M., MARQUÉS-CALVO M.S., "Enzymatic degradation of poly(3-hydroxybutyrate-co-4-hydroxybuturate) by commercial lipases", *Polymer Degradation and Stability*, vol. 97, no. 4, pp. 597–604, 2012.

[SCO 82] SCOW K.M., "Rate of biodegradation", LYMAN W.J., REEHL W.F., and ROSENBLATT D.H. (ed.), *Handbook of Chemical Property Estimation Methods*, McGraw Hill Book Co., New York, 1984.

[TAN 07] TANAKA T., YABE T., TERMACHI S., *et al.*, "Mechanical properties and enzymatic degradation of poly[(R)-3-hydroxybutyrate] fibers stretched after isothermal crystallization near Tg", *Polymer Degradation and Stability*, vol. 92, no. 6, 1016–1024, 2007.

[YEM 05] YEM G.Y., MOHD YUSOF A.M., MOHD ISHAK Z.A., *et al.*, "Water absorption and enzymatic degradation of poly(lactic acid)/rice starch composites", *Polymer Degradation and Stability*, vol. 90, no. 3, pp. 488–500, 2005.

[ZHA 11] ZHANG J., KASUYA K., HIKIMA T., *et al.*, "Mechanical properties, structure analysis and enzymatic degradation of uniaxially cold-drawn films of poly[(R)-3-hydroxybutyrate-co-4-hydroxybutyrate]", *Polymer Degradation and Stability*, vol. 96, no. 12, pp. 2130–2138, 2011.

14.5.1. *Reference material*

ASTM D 5209-92, Standard test method for determining the aerobic biodegradation of plastic materials in presence of municipal sewage sludge.

ASTM D 5511-94, Standard test method for determining anaerobic biodegradation of plastic material under high solid anaerobic digestion conditions.

CEN/TC 223, Amendements du sol et supports de culture.

ISO 10634:1995, Qualité de l'eau. Lignes directrices pour la préparation et le traitement des composés organiques peu solubles dans l'eau en vue de l'évaluation de leur biodégradabilité en milieu aqueux.

ISO 11074-1:1996, Qualité du sol. Vocabulaire. Partie 1 : termes et définitions relatifs à la protection et la pollution du sol.

ISO 11734:1995, Qualité de l'eau. Evaluation de la biodégradabilité anaérobie ultime des composés organiques dans les boues de digesteurs. Méthode par mesurage de la production de biogaz.

ISO 14855-2, Détermination de la biodégradabilité aérobique ultime des matériaux plastiques dans des conditions contrôlées de compostage. Méthode par analyse du dioxyde de carbone libéré. Partie 2 : mesurage gravimétrique du dioxyde de carbone libéré lors d'un essai de laboratoire.

ISO 16929, Plastiques. Détermination du degré de désintégration des matériaux plastiques dans des conditions de compostage définies lors d'un essai à échelle pilote. Norme internationale, AFNOR November 2002.

ISO 846:1997, Plastiques. Evaluation de l'action des micro-organismes.

ISO/DIS 14853:1999, Plastique. Evaluation de la biodégradabilité anaérobie ultime en milieu aqueux. Méthode par détermination de la production de biogaz.

ISO/DIS 15985:1999, Plastiques. Evaluation de la biodégradabilité anaérobie ultime et de la désintégration dans des conditions de digestion anaérobie à teneur élevée en solides. Méthode par analyse du biogaz libéré.

ISO/DIS 17556.2, Plastics. Determination of the ultimate aerobic biodegradability in soil by measuring the oxygen demand in a respirometer or the amount of carbon dioxide evolved.

ISO/TR 15462:1997, Qualité de l'eau. Sélection des essais de biodégradabilité.

JOURNAL OFFICIEL DE L'UNION EUROPÉENNE, OJL, 219, 7.8.98, p. 39, décision de la Commission du 7 avril 1998 établissant les critères écologiques pour l'attribution de l'écolabel communautaire aux amendements du sol.

NF EN13432, Exigences relatives aux emballages valorisables par compostage et biodégradation. Programme d'essai et critères d'évaluation de l'acceptation finale des emballages. Norme française et européenne homologuée. Cette norme comporte les techniques respirométriques à mettre en œuvre, soit en explication directe soit en renvoi à d'autres normes. AFNOR November 2000.

NF U52-001, Matériaux biodégradables pour l'agriculture et l'horticulture. Produits de paillage. Exigences et méthode d'essai. Cette norme comporte les techniques respirométriques à mettre en œuvre, soit en explication directe soit en renvoi à d'autres normes.

OECD, Guidelines for Testing of Chemicals 208 : Terrestrial Plants, Growth Test ; Organisation for Economy Co-operation and Development.

ÔNorm S 2201, Compostable biogenic waste. Quality requirements.

prEN 12578, Amendements du sol et supports de culture. Spécifications. Liste de produits.

prEN 12880, Caractérisation des boues. Détermination des résidus secs et contenu aqueux.

Wl 261 074, Evaluation de la désintégration des matériaux d'emballage lors d'essais à usage pratique dans des conditions de compostage définies.

Wl 261 085, Evaluation de la biodégradabilité aérobie ultime et de la désintégration des matériaux d'emballage dans des conditions contrôlées de compostage. Méthode d'analyse du dioxyde de carbone libéré.

15

The Regulation by Law of Nanosciences and Nanotechnologies

15.1. Introduction

Nanoscience[1] and nanotechnologies[2] are recognized by analysts as possible sources of a fourth industrial revolution[3]. By 2015, nanotechnologies could have an annual global income of approximately 1,000 billion euros[4] and help create, directly and indirectly, approximately 10 million jobs from now until 2014[5]. Nanotechnology applications for the health[6], energy[7] and environmental[8] sectors

Chapter written by Eric JUET.

1 Nanoscience refers to "the study, the discovery and understanding of matter at the nanometric scale (2.1) where properties and phenomena associated with the size and structure, unlike those associated with molecules individual atoms or base materials, can present themselves". ISO/TS 80004-1:2010, Title: Nanotechnologies. Vocabulary. Part 1: Core terms.

2 Nanotechnologies are "the application of scientific knowledge to manipulate and control matter at the nanometric scale in order to make use of the properties and phenomena associated with the size and structure, unlike those associated with molecules individual atoms or base materials". ISO/TS 80004-1:2010, Title: Nanotechnologies. Vocabulary. Part 1: Core terms.

3 Since 2004, they have been recognized by the European Commission as a key element in promoting its economic competitiveness strategy for Europe. See statement published by the Commission, *Vers une stratégie européenne en faveur des nanotechnologies*, COM(2004) 338 Final, 12.05.2004, p. 3.

4 *Etude prospective sur les nanomatériaux*, D. Carlac'h and Y. Hemery, a study carried out by "development and advice" on behalf of MINEFI/DIGITIP/SIMAP, 2004, p. 17.

5 *The Nanotech Report 2004*, Lux Research, 2004.

6 Creation of new tools for prevention (e.g. nanopatch vaccines), diagnosis (use of nanoparticles as fluorescent markers) and treatment (nano-structured and vectorized capsules).

7 A significant improvement is expected in the yield of photovoltaic tiles or even fuel cells. Eventually, using a bottom-up method, objects could be manufactured without waste.

8 Treating air and water pollution using nanoporous filters and improving catalysis processes exiting chimney pipes, etc.

should respond (or at least partially) to several challenged faced by modern societies: strains on our healthcare systems, the depletion of natural resources and the deterioration in the quality of our environment. However, health, environmental and ethical concerns, associated with nanomaterials, have gradually arisen and been supported by various scientific studies and expert reports[9]. The highly significant potential impact (positive and negative) of nanosciences and nanotechnologies on the functioning of society justifies intervention of the law[10]. This intervention is designed to ensure the responsible development of emerging technologies such as NBIC[11]. The concept of responsible development mirrors that of regulation which, in a legal context, is understood as the public or private action to ensure equilibrium of the system through the use of legal instruments (i.e. with legal effects) whether legally binding or not[12]. In what follows, our developments will focus on regulatory actions to help control risks posed by nanotechnologies[13]. We notice that, despite many challenges, particularly scientific and technical, encountered by public and private stakeholders (see section 15.2), the regulation of nanosciences and nanotechnologies has become more extensive over the past few years (see section 15.3).

15.2. Obstacles in the legal regulation of nanomaterials

The legal regulation of nanosciences and nanotechnologies has been hindered at the same time by scientific and technical challenges as well as factors associated with legal complexity.

9 For more specific details about the risks, we can refer to the expert report by the French agency for Food, Environmental and Occupational Health Safety (ANSES): *les nanomatériaux: effets sur la santé de l'homme et sur l'environnement*, July 2006 and no. 96 of the National Consultative Ethics Committee for health and life sciences: *Questions éthiques posées par les nanosciences, les nanotechnologies et la santé*, February 2007.

10 Note that the law can be shown concisely such as a system of rules of behavior, sanctioned by the state, forming a social regulation.

11 NBIC for nanotechnologies (N), biotechnologies (B), computer science (I) and cognitive sciences (C).

12 Note that the equilibrium concept (protecting man and the environment/promoting economic competitiveness) is a subjective concept, but with objective limits. In fact, at the European level, particular emphasis on the general objective of not making it impossible to achieve other general objectives set by the Treaty and in particular "promoting a higher level of competitiveness". In fact, according to CJUE, "institutions must deal with different objectives [...], not focusing too much on a single objectives to the point that it becomes impossible to meet the other objectives", CJCE, 06.12.1984, SA Biovilac NV against the European Economic Community, Aff. 59/83, Rec. p. 4057, issue no. 16.

13 We do not cover the regulatory actions orientated toward promoting nanotechnologies and in particular the activities involved in the construction of a center of excellence or even the actions favoring industrial innovation.

15.2.1. *Scientific and technical challenges*

The legal regulation of nanosciences and nanotechnologies was initially hindered by the existence of problems with metrology and the partial inadaptability of current risk assessment methods described for chemical substances (to which nanomaterials can be equated). In metrology, shortcomings were identified in the detection, quantification and characterization of nanomaterials. To address these shortcomings research has been carried out (which will be discussed later) that help make adapted measuring equipment available, as well as reference materials (essential for calibrating measuring instruments)[14]. As well as metrological challenges, there is also difficulty in evaluating the health and environmental risks associated with nanomaterials. Current risk assessment methods of chemical substances, which are applicable to nanomaterials due to the consensus at the European level that nanomaterials are similar to chemical substances, are assessed by specialists not used to dealing with nanomaterials[15]. In order to be able to appropriately assess specific risks posed by nanomaterials, additional information about the properties of substances, as well as that specifically required by regulation may be necessary. In addition, guidelines for tests adopted for chemical substances (information on the properties of a substance can be obtained by applying standardized testing methods) may also need to be modified (OECD guidelines for the testing of chemicals, ISO technical documentation, etc.[16]).

In an attempt to overcome the scientific and technical challenges mentioned above, research activities with international cooperation were undertaken. Research on the safety of nanomaterials began in Europe with the implementation of FP1, which required 2.5 million euros for the field of impacts of nanomaterials on health

---

14 The first certified reference nanomaterial is a silica nanoparticle (ERM-FD100) presented in February 2011 by IRMM (*Institute for Reference Materials and Measurements*). A second reference material became available in February 2012, made if silica particles in an aqueous matrix (ERM-FD304). Note that the interpretation of data by measuring equipment currently has difficulties with regard to "the need to account for 'background noise', the specificity of responses obtained depending on the type of nanoparticle produced, their agglomeration kinetics, etc.". CEA Press record: *Nanosafe: recherches dans les domaines de la production et de l'utilisation sécurisées des nanomatériaux*, p. 9, 2008.

15 In particular, these are conclusions of the Scientific Committee on Emerging and Newly Identified Health Risks (CSRSEN or SCENIHR) presented in a 2006 report. See *modified opinion (after public consultation) on the appropriateness of existing methodologies to assess the potential risks associated with engineered and adventitious products of nanotechnologies*, SCENIHR, 2006.

16 The majority of test methods described in the regulation 440/2008/EC of the Commission of 30th May 2008 establish test methods that conform with the regulation (EC) no. 1907/2006 of the European Parliament and Council with regard to the registration, assessment and authorization of chemical substances, as well as the restrictions applied to these substances (REACH), OJEU no. L 142/2008, pp. 1–739, are based on OECD guidelines.

and the environment[17]. Fundraising has increased in FP6 with 25 million euros allocated to projects on safety issues raised by nanomaterials[18]. With regard to the current implementation of FP7, 50 million euros have already been allocated by the EU to research aiming to expand the knowledge base of the risks associated with nanomaterials[19]. These European funds, as modest as they may seem (with regard to the 3.5 billion euros allocated to the research priority "Nanosciences, nanotechnologies, materials and new production technologies") have a powerful leverage effect on national spending and private investments in research into the safety of nanomaterials. In France, research into the risks posed by nanotechnologies truly began following the launch of the *National programme in nanosciences and nanotechnologies* (PNANO) in 2005, controlled by the National Agency for Research (ANR)[20]. This program, reintroduced under different names[21], has included from the outset cross-cutting thematic component *Instrumentation and metrology for nanotechnologies*. As of 2007, it was improved with a section on *Societal impact and regulation* (ethical, social and health aspects, risks of nanotechnologies on health and the environment). National and European research into the safety of nanomaterials fall within the limits of international collaboration with the aim of pooling research efforts into the safety of nanomaterials, to guarantee the comparability of scientific data and standardize regulatory methods to control risks posed by nanotechnologies. Two international organizations, ISO and the OECD, have many technical standards (or legal-technical in the case if OECD) that focus on the safety aspects of nanomaterials. Within the OECD, issues regarding nanomaterials have been dealt since 2006 by the Working Party on Manufactured Nanomaterials (WPMN) which is associated with the committee of chemical products. Research has also developed since 2007 within the working party of nanotechnology, which is associated with the Committee for Scientific and Technological Policy[22]. OECD standardization to control the risks posed by nanotechnologies is essentially carried out by WPMN.

---

17 330,000 euros of which is allocated to the project Nanosafe (*risk assessment in production and use of nanoparticles with development of preventive measures and practices codes*) during 2003–2004, 1 million euros for Nano-Pathology during 2002–2005 and 1.1 million euros for Nanoderm during 2003–2006.

18 Nanosafe2 (SAFE *production and use of NANO materials*) had 7 million euros during 2005–2009. The program aimed to develop "tools and methods for detection and chemical and reaction characterization to promote better risk assessment and management for the secure industrial production of nanomaterials". http://www.ineris.fr/centredoc/resume-vsfr-nanosafe2.pdf.

19 The research project NanEx, which aims to develop scenarios of exposure to manufactured nanomaterials was, for example, awarded 1 million euros during 2009–2010.

20 National research agency.

21 Its implementation is supported by the national network of nanoscience and nanotechnology (R3N). In 2009, the program PNANO became *P3N: nanosciences, nanotechnologies and nanosystems program*, then in 2010 *P2N: nanotechnologies and nanosystems program.*

22 Work by GTN is presented as general reports covering only emerging political questions in science, technology and innovation associated with the responsible development of nanotechnologies.

The WPMN aims to "promote" the international cooperation regarding the health and environmental safety of nanomaterials to help increase the rigorous assessment of the safety of nanomaterials"[23]. Its research is organized into eight projects[24], one of which is dedicated to developing safety tests on representative samples of nanomaterials. In the case of this project, 13 nanomaterials have been selected based on their presence in trade, their tonnage and the availability of standardized samples[25]. The OECD publishes various documents such as guidelines (*Guidance manual for the testing of manufactured nanomaterials*), solutions (*resolution on implementing strategic approach to international chemical management*) or even expert reports (*Preliminary Analysis of Exposure Measurement and Exposure Mitigation in Occupational Setting: Manufactured Nanomaterials*). The WPMN carries out similar activities to that of the technical committee *nanotechnology* (TC 221), created in 2005 by ISO. TC 229 addresses four categories or norms developed by joint working groups:

– terminology and nomenclature;

– surveying and characterization;

– health, safety and environment;

– specifications for identified nanomaterials.

Two technical groups, which have since been established, are working on:

– consumers and social dimensions of nanotechnology;

– as well as nanotechnologies and sustainable development[26].

23 OECD project on the safety of manufactured nanomaterials, Environment Department, OECD, 2008. The document is available at the following address: www.oecd.org/dataoecd/32/3/41615970.ppt.

24 Projects concerning the construction of a database of research on the safety of nanomaterials (Project 1), identification of research strategies for nanomaterials (Project 2), development of safety tests on a representative nanomaterial sample (Project 3), publication of redeveloped guidelines for safety testing of chemicals (Project 4), cooperation of voluntary plans of action and regulatory programs (Project 5), cooperation of risk assessments (Project 6), assessment and validation of alternative methods in nanotoxicology (Project 7) and, finally, supply of guidance documents on measuring and limiting exposure (Project 8).

25 The manufacture of the first reference nanomaterial by IRMM allowed the creation of *Repository of representative nanomaterials* by IHCP (*Institute for Health and Consumer Protection*). The creation of *Repository of representative nanomaterials*, by the European Union, responds to the demand for scientists involved in the development of risk assessment methods adapted to the specificity of nanomaterials to make samples of standardized nanomaterials available. In October 2011, 23 nanomaterials featured in the *Repository*.

26 Note that on the same date, at the European level, the technical committee CEN TC 352 *nanotechnologies* was created.

Normative documents adopted by the ISO include guides, specifications, technical reports, technical specifications and/or even international norms[27]. Despite a rather broad field of intervention, in practice, ISO appears to focus on terminology and metrology (characterization and sampling techniques). This forms the basis of a convergent approach to nanomaterial testing (in which OECD proves to be particularly active)[28].

Legal issues together with scientific and technical complications restrict legal intervention.

15.2.2. *Legal issues*

The absence of a common language has for a long time hindered the development a legal framework for the risks posed by nanotechnologies. Until the Commission adopted its recommendation of October 2011 regarding the *definition of nanomaterials*[29], there was no model definition of the term *nanomaterial*. Many definitions, formed by expert organizations, then began to contend[30]. What is more, these definitions, which often differ significantly, were not suitable for use for

---

27 For example ISO/TR 13121:2011. Nanotechnologies. Risk assessment associated with nanomaterials or even ISO/TS 11308:2011. Nanotechnologies. Characterization of single-walled carbon nanotubes by thermogravimetric analysis.

28 The collaboration of work by ISO and OECD (and other organizations such as UNEP) is assured by different mechanisms including cross involvement (ISO, UNEP or FAO, benefits from an observer status within NMWG – the European Union benefits from a particular member status within NMWG-OECD and IHCP have the status "liaison organization" within TC 229). The work program TC 229 also indicates an "obligation to account for current activities carried out by other relevant international organizations and governing bodies such as OECD". Finally, note the existence of an inter-organization program for the sound management of chemicals (IOMC) to facilitate communication between seven international organizations (including UNEP, FAO and OECD) about chemical substances (including in the form of nanomaterials).

29 Recommendation 2011/696/EU of the Commission of 18 October 2011 regarding the definition of nanomaterials, OJEU no. L 275/2011, pp. 38–40.

30 For example, BSI in its 2005 document *PAS 71:2005. Vocabulary. Nanoparticules* defines a nanomaterial as "a material with one or several external dimensions, or an internal structure interne, at the nanometric scale, and which exhibits new properties that differ from those of the same material without nanoscale features. Note: the new characteristics may include improved resistance, chemical reactivity or conductivity" (translation by author). SCENIHR in a 2007 opinion on *The scientific aspects of the existing and proposed definitions relating to products of nanosciences and nanotechnologies* specifies that a nanomaterial was "any kind of material that is composed of distinct functional parts, many of which with one or several dimensions of approximately 100 nm or less" (translation by author).

regulatory purposes[31]. The efforts to form a common language within a regulatory context materialized in the 2011 Commission recommendation which defined in point 2 a nanomaterial as: "a natural material, accidentally formed or manufactured that contains free particles, in the form of an aggregate or agglomerate, with at least 50% of particles, in the number size distribution, having one or several external dimensions between 1 and 100 nm"[32]. The only reference to the size criterion, which is in line with other international approaches[33], is justified by the Commission in an October memo in the following words: "size is the only universally applicable criterion, clear and measurable that can be used to identify materials"[34]. This definition is meant to be used by EU member states, agencies and economic operators "when they adopt and implement legislative and political actions and programs concerning nanotechnology products"[35]. What is more, the Commission requires that it "acts as a reference to determine whether a material should be considered as a 'nanomaterial' in legislation and policies of the European Union"[36].

Another legal issue is associated with the application of the precaution principle. This principle has been nationally defined, since 1995, using the Barnier law which indicates that "the absence of certainty, based on current scientific and technical knowledge, must not delay the adoption of effective and proportionate measures aiming to prevent a risk of serious and irreversible damage to the environment at an economically acceptable cost"[37]. In Europe, the CJEU uses a more flexible definition specifying that "when there is uncertainty regarding the existence or risk to public

31 The conditions for use for regulatory purposes are as follows: a definition based on current scientific knowledge (to facilitate its appropriation by those involved including the scientific community), a specific and clear definition (to allow manufacturers to correctly portray their legal obligations), an understandable definition (to allow the population to understand the information) and, finally, a broad definition allowing the protection of both human health and the environment (one of the objectives fixed by regulation authorities).

32 Point 2 of the recommendation 2011/696/EU, previously cites.

33 For example, ISO characterizes nanomaterials in its document ISO/TS 80004-1 of 2010 Nanotechnologies. Vocabulary. Part 1: Defines "core" as "a material with an external dimension at the nanometric scale or with an internal structure or surface structure at the nanometric scale". The nanometric scale was defined in the same document as "the length scale range from approximately 1 nm to 100 nm".

34 MEMO/11/704, *Questions and answers on the Commission Recommendation on the definition of nanomaterial*, 2011.

35 Point 1 of the recommendation 2011/696/EU, previously cited.

36 Considering point 4 of the recommendation 2011/696/EU, previously cited.

37 Article L110-1 environment code. The precaution principle is explained in the requirements in article 5 of the environment charter which requires that "when damage occurs, although uncertain in the state of scientific knowledge, which may seriously and irreversibly affect the environment, public authorities shall ensure, by applying the precautionary principle, and in their areas of responsibility, the implementation of risk assessment procedures the adoption of provisional and proportionate measures to prevent the occurrence of damage".

health, institutions may take protection measures without having to wait for these serious risks to fully arise"[38]. The precaution principle responds to the question *when* is it that the public authorities *must* (or in Europe *are able to*) adopt measures to protect human health and the environment? This requires, as a constant jurisprudence, a minimum scientific justification stating that there is a serious risk to human health or the environment. This minimum scientific justification appears to have already been provided, in the field of nanotechnologies, at the beginning of the 2000s with publication of several alarming toxicological and ecotoxicological reports on nanomaterials[39]. The precaution principle, justified by the scientific context, however cannot lead to the implementation of arbitrary measures by the public authorities[40]. The attention captured by the proportionality principle during the implementation of the precaution principle helps prevent this threat to measures that are not associated with the risk. The proportionality principle imposes the following on public authorities: "to compare the measures that can arrive at the desired result (protection of human health and the environment) and to retain the measure presenting fewer disadvantages (affecting the freedom of trade)"[41]. The definition protective measures in nanoscience and nanotechnology is a legally difficult exercise complicated by the need to account for the level of protection desired and the essential foundation in a certain economic rationality. Support given to health, security and environment (HSE), ethical, legal and social aspects (ELSA) researches and the referral to experts (in particular, to be clear about the actual risks associated with nanomaterials) are the initial manifestations of the precautionary principle practiced by the public authorities in nanoscience and nanotechnology.

These actions are the foundations for legal regulation measures which were implemented following observations in different expert reports of certain weaknesses in existing regulations to ensure adequate control of the risks posed by nanotechnologies[42].

---

38 CJCE, 12.07.1996, Commission *v.* United Kingdom, Aff. C-180/96, Rec. p. I-3 903, point 99.

39 Many of these scientific studies are cited in the 2006 AFSSET report *nanomateriaux: effets sur la santé de l'homme et sur l'environnement*, Saisine no. 2005/010, July 2006.

40 Even if the jurisprudence acknowledges a discretionary power to public authorities for the environment and public health (accentuated by the context of the scientific uncertainty).

41 *Droit de l'environnement et protection de la santé*, under the direction of P. Billet, M. Durousseau, G. Martin and I. Trinquelle, p. 246, L'Harmattan, 2009.

42 At the European level, SCENIHR, acting on behalf of the Commission, have produced, since 2005, several reports emphasizing the shortcomings of the regulation on controlling chemical substances for the proper control of the risk posed by nanotechnologies.

15.3. The legal regulation of nanomaterials

The legal regulation of nanomaterials (its risk management component) is characterized by a flexibility and progressivity manifested by both voluntary measures, technical adaptations of legislations and non-technical modifications of sector regulations to control the risks posed by nanotechnology[43].

15.3.1. *Legal regulation initiated by voluntary measures*

The time difference between carrying out HSE–ELSA research and the production of HSE–ELSA results and then the exploitation of these results has increased the development of voluntary measures designed to instill confidence in nanotechnologies and associated products. In 2007, the BAUA-VCI (association of German chemical industries) took the initiative to produce a guide of good practices for the production and use of nanomaterials in the workplace[44]. A year later, this guide, which intended to better protect workers exposed nanomaterials, inspired the AFSSET to define recommendations of good practices and use of nanomaterials[45]. These recommendations, described in the 2008 report dedicated to safety at work[46], equate nanomaterials to hazardous chemicals. For these materials, AFSSET (in compliance with the provisions of the Labor Code applied to hazardous chemicals) recommends applying STOP principles[47], (Substitution measures, Technical measures, Organizational measures, personal Protective measures) establishing the hazard communication of risks posed by *nano-objects*, assessing the environment of premises (to assess the residual exposure by inhalation of operators to

43 The decision to make a standardized adaptation, and not develop a specific legislation, is associated with the observation made by the Commission in its communication of 2008 on "the regulatory aspects of nanomaterials" that globally "the current legislation broadly covers the risks associated with nanomaterials and that these ricks can be managed using the existing legislative framework. However, changes may have to be made according to new available information". Communication from the Commission to the Council, at the European Parliament and the European Economic and Social Committee in *Aspects réglementaires des nanomatériaux*, COM(2008) 366 final, 17.06.2008, p. 4.

44 *Guidance for Handling and Use of Nanomaterials at the Workplace*, BAUA-VCI, 2007. The guide is available at the following address: http://www.baua.de/en/Topics-from-A-to-Z/Hazardous-Substances/Nanotechnology/pdf/guidance.pdf;jsessionid=BD5714CB9F77F4BCFFC EB0F95E1BC89E.1_cid253?__blob=publicationFile&v=2.

45 AFSSET was also motivated by a BSI (British Standardisation Institute) guide on good practices, published in 2007. See PD 6699-1:2007, *Nanotechnologies-Part 1: Good practice guide for specifying manufactured nanomaterials*, BSI, 2007.

46 AFSSET report on nanomaterials: safety at work, Saisine no. 2005/010, July 2008.

47 *Substitution* (by a less dangerous chemical substance), *Technologies* (work processes reducing the risk by using isolated, filtration, ventilation, systems etc.), *Organization* (reducing exposed workers, specific signaling, reserved storage zones, etc.) *Individual protection* (adapted work clothing, respiratory masks, etc.).

nanomaterials) or even reinforced medical check-ups[48]. In order to reassure the citizen-consumer about "nano" products, some companies have also marketed certification systems for the risk management and monitoring of nanotechnologies. For example, CENARIOS® certification, resulting from the collaboration of a Swiss and a German company, has helped ensure that EHS risks associated with nanotechnological products and processes are assessed based on the current knowledge[49]. Other voluntary schemes, such as the *Voluntary Reporting Scheme* launched by DEFRA (*Department for Environment, food and Rural Affairs*) in the United Kingdom in September 2006, involved a more specific commitment to better understand the extent to which the population was exposed to nanomaterials[50]. In Europe, the Commission also used voluntary measures by choosing to publish in 2008 a "code of conduct for responsible research in nanoscience and nanotechnology"[51]. This code constructed guidelines encouraging an open and responsible approach to research in nanosciences and nanotechnology within the EU. It promotes durability, precaution, inclusion, excellence, innovation and responsibility among member states, employers and funding bodies of research[52].

However, these voluntary measures have been deemed insufficient by the authorities to ensure adequate protection of humans and the environment. Since 2008, they have been accompanied by technical adaptations of REACH regulations, which were rapidly adopted by non-legislative procedures.

---

48 These risk prevention measures are described in a document published in 2012 by INRS in particular advising about signs, air filtration mechanisms, collection mechanisms, storage facilities, cleaning and transport processes, processing of waste etc. See *Nanomatériaux: prévention des risques dans les laboratoires*, INRS, 2012. The document is available at the following address http://www.inrs.fr/accueil/produits/mediatheque/doc/publications.html?refINRS=ED%206115.

49 For more details, can refer to the presentation leaflet at the following address: http://www.innovationsgesellschaft.ch/images/publikationen/Factsheet_CENARIOS_english_arial2.pdf.

50 The low response level obtained in the English experiment has orientated the French authorities toward implementing the obligatory declaration of substances in the nanoparticle form.

51 Recommendation of the Commission on the code of conduct for responsible research in nanoscience and nanotechnology, C(2008) 424 final, 07.02.2008.

52 A controversial aspect of the code concerns responsibility in the statement of principles for research. The members of the European project Nano Code insist that the European code of conduct is more specific in its formulation. They state that scientists are only responsible for good scientific practice, and not what will be done with their work by other people in the future. This proposition is communicated in Nano Code *Master Plan: Issues and Options on the Path Forward with Europe a Commission Code of Conduct on Responsible N&N Research*, Nano Code, November 2011. See link http://www.innovationsgesellschaft.ch/media/archive2/20111116_News_NanoCode_MasterPlan.pdf.

15.3.2. *Legal regulation prolonged by technical modifications to the REACH legislation*

Since 2006, SCENIHR has indicated (in its expert reports) some shortcomings in the REACH regulation[53] regarding its application to chemical substances in the form of nanomaterials. In response, in 2008, the European authorities technically modified the REACH legislation. Implementing the regulation 987/2008/EC of 8 October 2008 and modifying appendixes IV and V of REACH[54] is the first step of the technical adaptation process carried out by the Commission. Carbon and graphite are removed from appendix IV of REACH on the grounds that "the corresponding Einecs and/or CAS numbers are used to identify forms carbon or graphite at the nanometric scale, which do not meet the inclusion criteria in this appendix"[55]. This was organized by the European executive after non-conclusive discussions about this the issue within CASG Nano (*REACH Competent Authorities subgroup on nanomaterials*)[56]. Along with the changes to the technical appendixes of REACH, work has been done to modify the technical guides of ECHA (*European Chemicals Agency*). These guides help implement REACH by describing good practices of how

53 Regulation 1907/2006/EC of the European Parliament and Council of 18 December 2006 concerning the registration, assessment and authorization of chemicals, as well as the restrictions applied to these substances (REACH), OJEU no. L 396/2006, pp. 1–123.

54 Regulation 987/2008/CE of the Commission of 8 October 2008 modifying appendices IV and V of regulation (EC) no. 1907/2006 of the European Parliament and Council concerning the registration, assessment and authorization of chemicals, as well as the restrictions applied to these substances (REACH), OJEC no. L 268/2008, pp. 14–19.

55 Considering three of the regulation 987/2008/EC, previously cited.

56 CASG Nano is one expert sub-group created in March 2008 of the current CARACAL (*Competent Authorities for REACH and CLP*). CARACAL is a "group of experts who advise the European Commission and the ECHA on issues associated with REACH and CLP". It was founded as the *European Commission Working Group on the Practical Preparations for REACH* in May 2004. In September 2007, it was renamed *REACH Competent Authorities (REACH CA)* and, in March 2009, *Competent Authorities for REACH and CLP (CARACAL)*. CASG Nano aims to provide information about how to manage nanomaterials to conform to REACH and CLP. An early document described the state of discussions within the CASG Nano in such a way that REACH was applied to nanomaterials and published in 2008 by the Commission (see *Follow-up to the 6th Meeting of the REACH Competent Authorities for the implementation of Regulation (EC) 1907/2006 (REACH)*, European Commission, Doc. CA/59/2008 rev. 1, 2008). A second document made available to the public in 2009, released indications on the classification, labeling and packaging of nanomaterials to conform to REACH and CLP (*Classification, labelling and packaging of nanomaterials*, European Commission, Doc. CA/90/2009 rev. 2, 2009). Additional publications provided by CASG Nano are expected on the registration of nanomaterials or even on the communication throughout the supply chain.

to comply with REACH-CPL[57]. Since 2009, CASG Nano has been associated with IHCP (*Institute for Health and Consumer Protection*) to produce RIP-oN (REACH *Implementation Projects on Nanomaterials*) which are preparatory for the development of these technical guides on the particularities of nanomaterials. The first RIP-oN *REACH Implementation Project – Substance Identification of Nanomaterials (RIP-oN 1)*, published in March 2011, has not managed to reach a consensus for the issue with the identification of nanomaterials. Many manufacturers consider that size is not a criterion for substance identification[58]. This stance allows under REACH the current definition of nanomaterials as "a particular form of a raw parent substance". This practice unfortunately does not seem to (with the current available information) help differentiate the quality of risk assessment with the self-registration of nanomaterials[59]. Two other RIP-oN titled *Specific Advice on Fulfilling Information Requirements for Nanomaterial sunder REACH (RIP-oN 2)* and *Specific Advice on Exposure Assessment and Hazard/Risk Characterisation for Nanomaterial sunder REACH (RIP-oN 3)*, published in June 2011 give a consensus on the necessary changes to the risk assessment methodologies applied to chemicals in the form of nanomaterials and to the information required by REACH about the intrinsic properties of nanomaterials. These two RIP-oN should soon help ECHA update these technical guides on "the required information and the assessment of chemical safety". Discussions are under way to further amend the REACH regulation to include the particularities of nanomaterials (Nano Reach). They focus on the

57 CPL refers to Regulation 1272/2008/EC of the European Parliament and Council of 16 December 2008 regarding the classification, labeling and packaging of substances and blends, modifying and repealing the directives 67/548/EEC and 1999/45/EC and modifying the regulation (EC) no. 1907/2006, OJEU no. L 353/2008, pp. 1–1355.

58 The *EU Guidance to the Commission Regulation (EC) No 450/2009 of 29 May 2009 on active and intelligent materials and articles intended to come into contact with food*, European Commission, November 2011 requires that the list must include "identity: nom, synonym, CAS number, CE number, size of particles, composition, other specifications". "Particle size will be indicated when relevant for the functioning of the system in the case where substances are deliberately manufactured at a particular dimension to give them specific functional physical and chemical properties significantly different to those of larger particle sizes (substances in a nanometric form)" (translation by author).

59 In fact, in its latest version 5.2, the software IUCLID 5, which is the software used to collect data about chemical substances, specifically indicates the "nano" form of the chemical substance. The guide titles *IUCLID 5 Guidance and Support, Nanomaterials in IUCLID 5.2*, ECHA, June 2010 specifies that records of specific studies must be created for each form of a substance developed in the registration dossier. This implies that the industries are involved in a non-autonomous registration of their nanomaterials and assess the specific the "nano" form of the substance.

possible downgrading of registration thresholds for chemical substances in the form of nanomaterials. However, a similar trend would mean returning to the standard legislative process since these measures are not featured in the technical appendices of the REACH regulation[60].

Alongside the technical adaptation of REACH, the European authorities have become involved in a more assertive approach to adapting sector legislation, complying with the principles featured in European documents for planning in nanotechnology[61].

15.3.3. *Legal regulation consolidated by non-technical amendments to sector legislation*

The scheduled revision of sector legislation has given the European Commission the opportunity to take into account the particularities of nanomaterials. The regulation amendments, which go beyond simple technical adaptations, depend on principles of transparency and preliminary assessment of the risks posed by nanotechnologies. The reform concluded in December 2008 for the regulation of food additives contributed to part of the European strategy for controlling risks of nanotechnology. The new European regulation details that food additives featured in positive lists of appendixes III and IV must be deemed *new* in the case of changes in particle size by the use of nanotechnology[62]. This comparison of nanomaterials to different substances is continued in the regulation 10/2011/EU, concerning plastic materials and objects that come in contact food products, and indicates in article 9 that "nanosubstances can only be used if they are clearly authorized and mentioned in the specifications featured in appendix I"[63], as well as in the regulation 450/2009/ concerning active and intelligent materials and objects that will come in contact with food products, which states its exemption from the requirement to feature in the list of authorized substances, established in article 5 paragraph 2 c, of the regulation can

60 The conventional legislative process is a much more cumbersome than the delegation and execution processes that follow the comitology since the Lisbon Treaty came into force.

61 See the communication from the Commission to the Council, to the European Parliament and to the social and economic committee in "Nanosciences and nanotechnologies: an action plan for Europe 2005-2009", COM (2005), 243 final, 07.06.2005.

62 See article 12 of the regulation 1333/2008/EC of the European Parliament and Council of 16 December 2008 on food additives, OFEC, no. L 354/2008, pp. 16–33.

63 Regulation 10/2011/EU of the Commission concerning plastic materials and objects that will come in contact with food products, OJEU, no. L 12/2011, pp. 1–89. A unique substance in the nanometric form is currently featured in the "union list" on line 807. It involves titanium nitride nanoparticles. It indicates that they can be used in PET bottles (for Polyethylene terephthalate) "at a maximum of 20 mg/kg".

benefit from "substances deliberately engineered to a particular dimension with functional physical and chemical properties substantially different from those of larger particles"[64]. Protecting the consumer with regard to food is reinforced by the regulation 1169/2011/EU on consumer information, which requires that foodstuff containing ingredients in the form of manufactured nanomaterials is labeled[65]. Implementation of regulation 1169/2011/EU occurred under unusual circumstances; it was rejected by the European Parliament in 2011 in the Commission proposal on new foods[66]. Just like with food products, in the similar attempt to protect consumers, the regulation 1223/2009/EC on cosmetics[67] states that substances listed in appendices III to IV do not feature nanomaterials, unless otherwise indicated[68]. Further notification is also required in the case where nanomaterials are present in

---

64 Article 5 paragraph 2 c) II) of regulation 450/2009/EC of the Commission of 29 May 2009 concerning active and intelligent materials and objects that will come in contact with food products, OJEU, no. L 135/2009, pp. 3–11.

65 Regulation 1169/2011/EU of the European Parliament and Council of 25 October 2011 concerning consumer information about foodstuff, modifying the regulations (EC) no. 1924/2006 and (EC) no. 1925/2006 of the European Parliament and Council and repealing directive 87/250/EEC of the Commission, directive 90/496/EEC of the Council, directive 1 999/10/CE of the Commission, directive 2000/13/EC of the European Parliament and Council, directives 2002/67/EC and 2008/5/EC of the Commission and the regulation (EC) no. 608/2004 of the Commission, OJEU, no. L 304/2011, pp. 18–63. The presence must be indicated clearly in the list of ingredients. Article 18 of the regulation specifies that "the name of ingredients must be follows by 'nano' in parentheses".

66 This proposed to qualify "a food product containing manufactured nanomaterials, or consists of manufactured nanomaterials, unused for the production of foodstuff within the Community before 15 May 1997" as "new food". Article 3 of the proposition of the regulation concerning new food (COM(2007) 872 Final) modified by the legislative resolution of the European Parliament of 25 March 2009. Note that the measures that led the European Parliament to reject the proposition by the Commission were not to do with nanomaterials, but instead food products produced from cloned animals. A new proposal must be presented by the Commission, free of stances on animal cloning. Another proposal containing measures for nanomaterials must be communicated. This involves a proposed regulation by the European Parliament and Council concerning food destined for babies and young infants as well as food for special medical conditions, COM(2011) 353 Final. This proposal is strange in that it compares consumers to vulnerable groups of the population, a particularly high level of protection is proposed, which is expressed by the fact that nanomaterials will not be included in the Union list of authorized substances until further assessment by the Scientific Committee on Emerging and Newly Identified Health Risks (CSRSEN or SCENIHR).

67 Regulation 1223/2009/EC of the European Parliament and Council of 30 November 2009 with regard to cosmetics, OJEU, no. L 342/2009, pp. 59–209.

68 See "preface of appendices II to VI" on page 82 of the regulation. Appendix III addresses the list of substances that cosmetics cannot contain apart from set restrictions whereas appendix IV states a list of colorants that cosmetics can contain.

cosmetics[69]. Finally, the mandatory labeling of cosmetics containing nanomaterials (mentioning the name of the ingredient followed by the word "nano") is mentioned in Article 19 of the regulation. The year 2012 is marked by the approval of regulation on biocidal products[70], which naturally states that "the authorization of an active substance does not concern the nanotechnological form, except when it is indicated in an explicit manner"[71] requiring that biocidal products provide information about "possible nanomaterials present in the product and the specific risks associated, as well as the word 'nano' in parentheses every time the nanomaterial is mentioned"[72]. Finally, the directive 2011/65/EU must mention limiting the use of certain hazardous substances in electrical and electronic equipment (RoHS)[73]. This directive, which is an example of waste prevention, limits the use of certain substances in electronic products (these substances are listed in appendix II of the directive). If the directive does not include specific obligations for nanosubstances, recital 16 indicates that using the reexamination process of appendix II (set before 22 July 2014) is required to assess the restriction of other hazardous substances, particularly substances in the form of nanomaterials[74].

15.4. Conclusion

European measures featured in the product legislations that we have just shown, many of which can be applied within member states[75], are associated in France with the mandatory declaration of substances in the nanoparticule form. This is stated in article 42 of Grenelle I law[76], and its implementation is clarified in article 185 of the

69 The notification is addressed to the Commission before entering the market. This notification in particular includes information about the identity, quantity and toxicological profile of the nanomaterial. It must facilitate the creation of a catalogue of all nanomaterials used in cosmetics on the market. When the Commission has doubts about the safety of nanomaterials, it is possible to register banned substances (appendix II) or restricted substances (appendix III).

70 Regulation 528/2012/EU of the European Parliament and Council of 22 May 2012 concerning the availability on the market and the use of biocides. Text of interest to EEE, OJEU, no. L 167/2012, pp. 1–123.

71 Considering 66 and article 4 paragraph 4 of the previously cited regulation.

72 Article 69 paragraph 2b of the previously cited regulation.

73 Directive 2011/65/EU of the European Parliament and Council of 8 June 2011 regarding the restriction of use of certain hazardous substances in electronic and electrical equipment, OJEU, no. L 174/2011, pp. 88–110.

74 This involves stating that nanomaterials will be specifically examined in future revisions of the directive. Regarding the legislation implemented by the Commission on Environment, Public Health and Food Safety, it is proposed that particular attention will be given to nanosilver and long carbon, multiple-walled nanotubes.

75 This is the case for measures included in the regulations.

76 Law no. 2009-967 of 3 August 2009 of programming associated with the implementation of Grenelle environment (Grenelle I Law), published in JORF no. 179 of 5 August 2009.

Grenelle II law[77]. It is a unique example of a horizontal regulation which could form the starting point at the European level of a regulation specific to nanomaterials of the model regulation developed for GMOs.

15.5. Bibliography

[BIL 09] BILLET P., DUROUSSEAU M., MARTIN G., *et al.*, *Droit de l'environnement et protection de la santé*, L'Harmattan, Paris, 2009.

[CAR 04] CARLAC'H D., HEMERY Y., *Etude prospective sur les nanomatériaux*, DIGITIP, Paris, 2004.

[BAU 07] BAUA-VCI, Guidance for handling and use of nanomaterials at the workplace, available at: http://www.baua.de/en/Topics-from-A-to-Z/Hazardous-Substances/Nanotechnology/pdf/guidance.pdf;jsessionid=BD5714CB9F77F4BCFFCEB0F95E1BC89E.1_cid253?__blob=publicationFile&v=2, 2007.

[INR 12] INRS, Nanomatériaux: prévention des risques dans les laboratoires, available at: http://www.inrs.fr/accueil/produits/mediatheque/doc/publications.html?refINRS=ED%206115, 2012.

77 Law no. 2010-788 of 12 July 2010 with a national commitment to the environment (Grenelle II Law), published in JORF no. 160 on 13 July 2010. A system was put in place for manufacturers, importers and distributers of substances in the nanoparticle form, in the form of or contained in blends, associated or not, or materials that reject such substances under normal or reasonably foreseeable conditions of use. The declaration includes elements that facilitate the identification of the substances concerned, their usages and quantities on the market, as well as the identities of corporate users who have sold them whether in or not in return for payment. At the request of the administrative authority, available information regarding the dangers of these substances can be transmitted. A decree (Decree no. 2012-232 of 17 February 2012 regarding the annual declaration of substances in the nanoparticle form, published in JORF no. 43 of 19 February 2012) of the legislative system was implemented to define the expression "substances in the nanoparticle form" (the definition of "nanomaterial" provided in the 2011 recommendation of the European Commission previously cited) or even sanctions (producer, importer, distributer or user who fails to provide full disclosure is liable for an administrative fine of 3,000 euros which can be coupled with a daily fine of 300 euros).

16

Teaching Sustainable Development

16.1. Introduction

The term "sustainable development" combines a dynamic concept, that of development, with a stationary concept, that of sustainability.

This dynamic concept of development implies a change over time from an initial situation to a different final situation by its topological state as well as the point in time. However, any change implies the presence of energy.

The concept of energy is a primary notion, i.e. a concept with a clear meaning, which allows us to define other concepts whose definitions would be provided by phrases built around other primary concepts.

Most primary notions, which form the basis of science such as space, time and energy and the like, are not always as clear. However, what is certain is that energy links an initial state to a different final state, whether it is effective or potential, and establishes a link between time (initial-final) and space (state). Sustainable development, which involves the same variables, time and space, is therefore intimately linked to the concept of energy, which must play a leading role in teaching. However, it is essential to introduce the measurability criteria of a variable including one-dimensionality (often continuity and derivability), the existence of a standard and a unit of measurement. As these criteria are not part of the same concept of sustainable development (it is at least two-dimensional since it simultaneously implies temporal and spatial aspects), only a subjective approach capable of measuring one-dimensional aspects can be used. Such an approach can lead to objectively contradictory conclusions although totally just in the sub-space of their subjective observational paradigm.

Chapter written by Serge WALTER.

The concept of sustainability underpins the concept of implicit scale. An identity is assumed between an initial parameter, which acts as a reference, and that which it ultimately becomes. However, with regard to this concept there is high incompatibility between what we are aiming to do (i.e. a comparison, which is always possible) and the implicit method used (i.e. one measurement, which is only possible for one-dimensional parameters). If there is one thing that is certain, it is the multi-dimensional character of the concept of development, since it can involve completely orthogonal parameters, that is to say linearly independent (in the mathematical sense) such as quantity, quality, skill, technique and many others.

Finally, we must not forget that the same concept of sustainable development addresses the human appreciation of development as well as sustainability. It is therefore essentially subjective (the observer subject implicitly being humanity) which renders all conclusions regarding this subject dependent on cultural stance (also including spatio-temporal aspects) of the observer.

16.2. The foundations of teaching sustainable development

16.2.1. *The need for the rigorous quantification of energy*

It follows from the foreword that teaching sustainable development should be the complete opposite of that which applies to monothematic disciplines. Of course, it should include one-dimensional aspects such as the assessment of energy flows (i.e. power) essential for populations to achieve certain objectives (in particular the quality of life, a purely subjective notion, whose appreciation also depends on location, local culture as well as moment in human history) which requires the quantification (measurable) of energy demands as well as the consequences (often measurable, for some aspects at least) of producing and using this energy.

In this sense, teaching sustainable development may resemble many other technical or scientific teachings on the production of energy and its effects. It must cover the energy availability, but also our ability to use them, both in terms of control in the case of accidental contamination and the overall assessment of production, including the constant availability of sources (constant wind or solar energy is unrealistic and the implementation of these technologies is often contradictory to their environmental or ecological objectives based on the inability to store electrical energy), their predictable random fluctuations and the respective time constants of different production lines whose incompatibility may decrease the interest in combining certain renewable sources and conventional fossil fuels (including nuclear fission which is a fossil energy).

From this perspective, Earth can be considered as a spaceship with only two types of resources: resources on board (geothermal and nuclear energy) and renewable energy of solar and lunar origin: tidal, hydro, wind, thermal in all its forms. This spaceship must remain in thermal equilibrium by the constant balance between losses by radiation and radiation emitted back to earth with gains coming from the sun and the moon in the form of gravitational and radiant energy. These essentially depend on the albedo of the planet, i.e. its reflecting power on the whole solar spectrum.

In any case, sustainable development is only possible provided that there is rigorous quantification of all contributions, not excluding natural contributions beyond human control: note that volcanism represents more than 90% of the planet's CO_2 emissions; methane emissions are variable depending on the latitude and change in local temperature; terrestrial and oceanic water vapor plays a vital role in the greenhouse effect regulating the superficial temperature of the planet and that it is directly dependent on the average air temperature in the double interfacial layer of the ocean surface. Attention must also be drawn to the fact that natural water vapor is essentially heavier than air due to the formation of oligomers by hydrogen bonds. Finally, volcanism can play a dominant role in the change in albedo with explosive eruptions that send clouds of microparticles into the upper atmosphere. This influence may originate from the intrinsic reflective character, but also in the seeding and exogenous growth of fine ice crystals in the upper atmosphere supplied in water vapor by gaseous molecules of a monomer of water in equilibrium with oligomers.

Without a comprehensive quantified study of human energy production available from both renewable resources mentioned above, all solutions often proposed by so-called ecologists often only appear as decoys or illusions; the total energy balance including manufacture, mining, maintenance and recycling of systems often become negative if instead of considering the nominal power of these devices, the effective balance is seriously assessed under actual implementation and aging conditions.

16.2.2. *The problem of flow management*

The first conclusion to be drawn is that it is difficult to accurately decipher the states of the existing natural flows, whose knowledge is essential for assessing, even if approximately, the extremely serious short-, medium- and long-term policy decisions for the functioning of our societies. We can both reduce the production of CO_2 and do without nuclear energy. With regard to waste from nuclear energy, it is insufficient to consider the seriousness of local dangers without taking into account the small amount of matter involved at a global level. An equivalent summary would be to replace three large aircraft with many small ones, with the argument that it would be unacceptable to contemplate a crash of an Airbus A380 simply because an accident of such a plane could cause 500 deaths at once, and to replace it with 200 tourist airplanes with four seats each. Each accident would certainly cause fewer deaths at once, but it would create air traffic congestion which would result in accidents being much more frequent in terms of deaths; the larger air carrier is undeniably preferable. Similarly, replacing nuclear energy with coal or oil would cause more deaths by inhalation of powders or carcinogens, mutagens or reprotoxic agent (CMR) products in mines or processing sites of raw materials with facilities running on nuclear energy, whether in the environment or mining or reprocessing sites, whether under standard operating conditions or by accident. With the exploitation of nuclear resources and coal and oil spanning over 50 and 100 years, respectively, a realistic comparison of both approaches can be made, none of which is ideal. The media often estimate the deaths per year due to coal mining to be 18,000, whether directly on the mining sites or indirectly by illnesses caused by powders and other volatile compounds present in natural or manufactured coal. With regard to solar energy, we must not forget that a realistic assessment of the output of solar cells shows that it is difficult under manageable operating conditions to generate more energy than what would have been consumed to manufacture, install, maintain and recycle them, and therefore one must be very careful before stating their ecological cleanliness.

16.3. Conclusion

In conclusion, future energy will likely come from elsewhere, perhaps from the slow decay of matter (without high energy radiation), which is more efficient than heat exchange which, for thermodynamic and chemical reasons, loses a minimum of 60–70% of energy initially available. Courses on sustainable development must therefore make students aware of the complexity of the problem to avoid misleading remarks, often advocated by demagogic environments that are not based on any serious quantification of the advantages and disadvantages of the apparently attractive solutions proposed.

The process is difficult and subjected to technical improvements that aim to reduce losses, with the simplest being the most economic and most efficient as a whole, and were often discovered intuitively by our ancestors, whose main aim was not to reduce the amount of energy produced, but instead to reduce the physical effort required for its production or use. A useful method, but which would involve reducing the activities generating only financial benefit without any associated usage value, in particular advertising activities and long-distance transport of equivalent materials. Gross national product alone does not accurately represent wealth. It only represents money flows that may be perfectly associated with a zero profit usage value. Conversely, activities that do not generate any cash flow may prove to be of at most importance in the sustainable development of a society: some striking examples of this can be found in the human relation aspects of education, teaching and medicine.

To conclude, teaching sustainable development must therefore be a warning, an incentive for scientific caution rather than a transmission of established and duly verified knowledge, since it is still difficult to verify with our lack of knowledge about the complexity of our ecosystem. Complex systems are often locally very unstable and it is essential that students of sustainable development are aware of deterministic chaos introduced by Ecole de Prigogine as well as the existence of possible bifurcations in the changes to the systems when they have three or more control variables, associated with nonlinear transfer functions. The existence of such properties can be reduced to very few predictive horizons, rendering decision-making in sustainable development extremely difficult.

List of Authors

Luc AVÉROUS
University of Strasbourg
France

Sylvain CAILLOL
Institut Charles Gerhardt
Montpellier
France

Guy CÉSAR
SERPBIO
La Bastide Clairence
France

Rémi DETERRE
University of Nantes
France

Claude DUVAL
CNAM
Paris
France

Jean-François FELLER
University of South Brittany
Lorient
France

François GALGANI
Ifremer
Bastia
France

Audrey GOUJON
INRA
Massy
France

Thierry HAMAIDE
University Lyon 1
Villeurbanne
France

Eric JUET
University of Rennes
France

Stéphane LE POCHAT
EVEA
Nantes
France

Michel LOUBRY
Plastics Europe
Puteaux
France

Valérie MASSARDIER
INSA
Lyon
France

Laura MAXIM
ISCC
Paris
France

Jean-Charles MOUGENEL
ENSCM
Mulhouse
France

Gérard PIGNAULT
ESCPE
Villeurbanne
France

Arnaud ROQUESALANE
EVEA
Nantes
France

Bernard SILLON
ISA
Villeurbanne
France

Olivier TALON
Materia Nova
Mons
Belgique

Frédéric VIOT
Plastic Omnium
Sainte-Julie
Canada

Olivier VITRAC
INRA
Massy
France

Serge WALTER
ENSCM
Mulhouse
France

Index

D, E

F, G, H, I

L, M, N

O, P

R

S